STUDY GUIDE

TO ACCOMPANY
VANDER-SHERMAN-LUCIANO

HUMAN PHYSIOLOGY

FIFTH
EDITION

SHARON RUSSELL
University of California, Berkeley

McGRAW-HILL, INC.

New York St. Louis San Francisco Auckland Bogotá
Caracas Hamburg Lisbon London Madrid Mexico
Milan Montreal New Delhi Paris San Juan São Paulo
Singapore Sydney Tokyo Toronto

Study Guide to Accompany
Vander/Sherman/Luciano:
Human Physiology

1 2 3 4 5 6 7 8 9 0 MAL MAL 8 9 4 3 2 1 0 9

ISBN 0-07-066971-6

The editor was Kathi M. Prancan;
the production supervisor was Diane Renda.
Malloy Lithographing, Inc. was printer and binder.

TABLE OF CONTENTS

❒ About This Study Guide

When I was first approached about writing a Study Guide to accompany the Fifth Edition of Vander, Sherman, and Luciano's *Human Physiology*, I was honored at the prospect of being associated with such a highly regarded textbook but also somewhat daunted by the challenge that confronted me. The new edition of the text contains at the end of every chapter the staples of a standard study guide — a detailed summary outline of the chapter's contents, a list of key words to define, a series of review questions, and thought questions with answers. What could I add to facilitate a student's learning of the material offered in the text?

I decided to draw upon my experience with teaching physiology in a self-paced, tutorial format. For several years, human physiology was taught in this manner at the University of California, Berkeley, using the third edition of Vander, Sherman, and Luciano's *Human Physiology* as the text. Much of the teaching was in the form of asking questions of students, helping them to find the right answers, and occasionally supplementing text material with additional examples or explanations from another perspective. This tutorial approach to teaching physiology was successful and rewarding but prohibitively labor-intensive, and has been replaced with a more standard lecture format in recent years.

This Study Guide is my idea of a tutor-in-writing: Questions closely follow the text to allow the student to confirm his/her understanding of the material as he/she reads. All of the questions are answered and, where appropriate, clarifications and further explanations are offered. "Typical" wrong answers are anticipated based on my teaching experience. Study and memorization hints I have found helpful in the past are offered as well. At the end of each chapter there is a "practice test," which should indicate to the student how well she/he has mastered the material.

My purpose in writing this book is to help students to learn about — and even more important, perhaps, to *appreciate* — physiology. To me, an understanding of human body function is the most exciting knowledge one can have. I welcome this opportunity to share my enthusiasm for this subject, and to that end I also welcome any comments that might help improve future editions of this Guide.

❒ Acknowledgments

I wish to express my appreciation to several people who contributed significantly to this book, including Barbara Ingram, whose assistance was invaluable in preparing the final manuscript. I am also very grateful to the text's authors, Drs. Arthur J. Vander, James H. Sherman, and Dorothy S. Luciano, for their careful reading of the manuscript and helpful suggestions about its content. I offer my most profound gratitude to my editor, Irene Nunes, without whose skill, patience, and encouragement I could never have completed this project. Finally, I welcome this opportunity to thank my husband, Charles Nicoll, for his support in all my efforts, and for making them all worthwhile.

S.M.R.
Berkeley, California,
August, 1990

1: A FRAMEWORK FOR HUMAN PHYSIOLOGY

A science as complex and all-encompassing as physiology, the study of body function, cannot have begun with one person, and yet a nineteenth-century Frenchman named *Claude Bernard* is often referred to as "the father of physiology." Bernard, who is regarded as one of the greatest scientists who ever lived, performed experiments that demonstrated many physiological principles common to animals and humans and laid the foundation for much of what is known today about physiology and scientific medicine. Bernard argued eloquently that valid medical practice must be based upon *science* (an idea not universally held a century ago): "Knowledge of pathological or abnormal conditions cannot be gained without previous knowledge of normal states, just as the therapeutic action of abnormal agents, or medicines, on the organism cannot be scientifically understood without first studying the physiological action of the normal agents which maintain the phenomena of life."

Among the concepts Bernard is credited with introducing is the idea that many cells of the body must work together to preserve the constancy of the fluid bathing them — the internal environment. This concept is introduced in this chapter and explored in more depth in Chapter 7.

Instructions for answering questions in this Study Guide

1. True or false: Correct each false statement. Whenever you must correct a false statement, there will almost always be more than one way to do so. In the Guide, the answer requiring the least correction will generally be given, with the understanding that other correct answers are possible. It is usually not sufficient, in terms of demonstrating understanding, to simply insert a "no" or "not," however.

2. Multiple choice: Any, all, or none of the responses may be correct. Explain why each incorrect response is incorrect.

3. Fill in the blanks: Choose the best word or words to complete each statement.

4. Directions will be given for other types of questions as they appear.

❒ **Mechanism and Causality** (text pages 1-2)

> *'It must be recognized that in the organism there are things that cannot be accomplished outside mechanical or physicochemical laws. God could not make an animal which did not depend, for its movements, on the laws of mechanics. So one must always seek for chemical and physical laws in the mechanisms of the phenomena of life.'*
>
> Claude Bernard

Teleology means "the explanation of events in terms of purpose." In contrast, *causality* means "the explanation of events in terms of the actual cause, or a causally linked sequence of physicochemical events." Statements 1-6 below are examples of either teleology or causality. Label them accordingly.

1. Blood clots as a result of damage to blood vessels because clotting is necessary to keep a person from bleeding to death. ______________________

2. Blood clots as a result of damage to blood vessels because such damage causes chemical changes in the blood that lead to clot formation. ______________________

3. A person cannot hold her or his breath indefinitely because changes in blood-gas concentrations trigger reflexes that overcome voluntary breath-holding and force breathing. ______________________

4. A person cannot hold his or her breath indefinitely because the body requires frequent replenishment of oxygen. ______________________

5. Cigarette smoking causes coughing because smoke is harmful to the lungs. ______________________

6. Cigarette smoking causes coughing because smoke is irritating to the lungs. ______________________

7. True or false: The view of life that holds that all biological phenomena are explainable by the laws of physics and chemistry is termed "vitalism."

❒ A Society of Cells (text pages 2-4)

8. List four basic functions common to virtually all cells.

(a)____

(b)____

(c)____

(d)____

9. All multicellular organisms, including humans, begin life as a single cell. Name and describe the two fundamental processes necessary for development of the complex organism. ____________________________________

____

____

Match the cell category on the left with the correct functional description on the right.

10. _____ Muscle

11. _____ Connective tissue

12. _____ Nerve

13. _____ Epithelial

a. Specialized for movement

b. Specialized for secretion

c. Specialized for support

d. Specialized for synthesis of complex molecules

e. Specialized for conduction of electrical impulses

14. True or false: The number of distinct cell types in the human body is approximately 20.

15. True or false: One function of epithelial cells is to form selective barriers regulating exchange of materials across them.

16. Connective tissue cells

(a) include bone cells.

(b) include skeletal muscle cells.

(c) include blood cells.

(d) secrete an extracellular matrix consisting of fibers and ground substance.

17. The lungs are one organ in the organ system called the ______________________ system. Lungs are made up of ______________________, ______________________, ______________________, and ______________________ tissues. Other components of the same organ system are the ______________________, ______________________, ______________________, and ______________________.

18. True or false: An important function of organ systems is to regulate the external environment of the body.

❐ Internal Environment (text pages 5-6)

'All of the vital mechanisms, however varied they may be, have always but one goal, to maintain the uniformity of the conditions of life in the internal environment.'

Claude Bernard

19. The body's extracellular fluid is composed of two separate compartments, the ______________________ and the ______________________. The major differences between the two are that the ______________________ (former/latter) has a higher protein concentration than the ______________________ (former/latter) and is about four times ______________________ (smaller/larger).

20. True or false: Homeostasis refers to the relative constancy of the external environment.

21. The internal environment
    (a) refers to the intracellular fluid compartment of the body.
    (b) is regulated to remain relatively constant by the specialized activities of many of the body's cells.
    (c) is the medium for exchange of nutrients and wastes for the body's cells.

ANSWERS

Boldface type indicates the answers you should have given. Words in medium-face (ordinary) type explain or expand upon the answer.

1. **T**; 2. **C**; 3. **C**; 4. **T**; 5. **T**.

6. **Causality**. Note, however, that this explanation is not complete since it does not explain *how* lung irritation leads to coughing (irritation triggers reflexes you will learn about later). It nevertheless is an example of causality, since lung irritation is the stimulus that results in coughing. You may wonder why the answer to Question 5 is not also "causality" since something that is irritating is also likely to be harmful. While this is true, the term "harmful" is not specific enough to explain the phenomenon in a scientific way.

   The exercise of answering Questions 1 through 6 is very important for your understanding of what will be required of you in your study of physiology. While teleological explanations have utility in helping you remember *what* happens in various situations, they are of no use when you are asked to explain *how* and *why* something happens in the body. Be on guard against thinking in teleologic terms when you are asked to explain a physiological process.

7. **False.** The view of life that holds that all biological phenomena are ultimately explainable by the laws of physics and chemistry is termed ~~"vitalism"~~ "**mechanism**" (or "**the mechanistic view** ").

   Alternatively, the answer below is also correct:

   False. The view of life that holds that ~~all~~ biological phenomena ~~are ultimately explainable by~~ **require a force beyond** the laws of physics and chemistry is termed "vitalism."

8. Virtually all cells

   (a) exchange material (take up food, excrete waste products) with their immediate environment.

   (b) obtain energy from nutrients extracted from their environment.

   (c) synthesize complex molecules from simpler ones.

   (d) duplicate (reproduce) themselves. (But see answer to Question 9.)

9. **Cell division and cell differentiation. Following differentiation, cells become specialized to perform certain specific functions and become organized into tissues.** (Differentiated cells continue to perform the four basic functions listed in the answer to Question 8, except for certain highly specialized cells which have lost the ability to reproduce.)

10. **a** Muscle
11. **c** Connective tissue
12. **e** Nerve
13. **b** Epithelial

Note that option "d" does not describe any of the cell categories. It is a function common to all cells (see answer to Question 8) and thus is not a specialty of any of them.

14. **False.** The number of distinct cell types in the human body is approximately ~~20~~ **200**. (But recall that these 200 cell types can be grouped into four broad categories of function.)

15. **True.**

16. (a) **Correct.**
 (b) **Incorrect. Skeletal muscle cells are a type of muscle cell, as the name suggests.**
 (c) **Correct.** (Blood cells are grouped with the connective-tissue cells, even though they do not provide a supporting framework in the usual sense. The category "connective-tissue cells" includes all the cells that do not belong to the other groups.)
 (d) **Correct.** (This matrix is very important for keeping cells together as functional units — tissues and organs.)

17. The lungs are one organ in the organ system called the **respiratory** system. Lungs are made up of **connective**, **nervous**, **muscle**, and **epithelial** tissues. Other components of the same organ system are the **nose**, **pharynx**, **larynx**, **trachea**, and **bronchi**.

18. **False.** An important function of organ systems is to regulate the ~~external~~ **internal** environment of the body.

19. The body's extracellular fluid is composed of two separate compartments, the **interstitial fluid compartment** and the **blood plasma compartment**. The major differences between the two are that the **latter** has a higher protein concentration than the **former** and is about four times **smaller**.

20. **False.** Homeostasis refers to the relative constancy of the ~~external~~ **internal** environment.

21. (a) **Incorrect. Internal environment refers to the extracellular fluid compartment, which bathes the body's cells.**
    (b) **Correct.**
    (c) **Correct.**

PRACTICE

True or false (correct the false statements):

P1. The statement "Cigarette smoking can cause lung cancer" is an example of teleology.

P2. Differentiation is necessary before a cell can exchange material with its environment.

P3. Organs are generally composed of only one kind of tissue.

P4. The composition of the fluid bathing the cells of the body is essentially the same as that within the cells.

Fill in the blanks:

P5. The internal environment consists of the ______________________ fluid compartment of the body.

P6-9. The four general types of cells found in the human body are ____________, ________________, ______________, and ________________ cells.

P10. Preservation of the relative constancy of the internal environment is called ________________.

2: CHEMICAL COMPOSITION OF THE BODY

In 1962, the American James Watson and the Englishman Francis Crick were awarded the Nobel Prize for Physiology or Medicine for elucidating the structure of *deoxyribonucleic acid* — DNA. This molecule is the physical basis for heredity, and the discovery of its structure has led to a veritable explosion of knowledge about the fundamentals of life processes.

A knowledge of basic chemistry is necessary for understanding the complex structures and functions of the human body. In this chapter, therefore, you will learn the basics of general, organic, and biological chemistry, including the structure of DNA.

Instructions for answering questions in this Study Guide

1. True or false: Correct each false statement. Whenever you must correct a false statement, there will almost always be more than one way to do so. In the Guide, the answer requiring the least correction will generally be given, with the understanding that other correct answers are possible. It is usually not sufficient, in terms of demonstrating understanding, to simply insert a "no" or "not," however.

2. Multiple choice: Any, all, or none of the responses may be correct. Explain why each incorrect response is incorrect.

3. Fill in the blanks: Choose the best word or words to complete each statement.

4. Directions will be given for other types of questions as they appear.

❐ Atoms (text pages 11-13)

1. Can the terms "atom" and "element" be used interchangeably? Explain.

____

____

____

____

____

2. In the "miniature solar system" model of the atom, ____________________ orbit the nucleus, which is composed of particles called ____________________ and ____________________.

3. True or false: An atomic nucleus is electrically neutral.

4. True or false: Protons and neutrons have roughly the same charge.

5. True or false: The atomic number of an element refers to the number of particles in its atomic nucleus.

Use this table to answer Questions 6 - 11:

	Protons	Neutrons	Electrons
Hydrogen (H)	1	0	1
Carbon (C)	6	6	6
Oxygen (O)	8	8	8
Sodium (Na)	11	12	11
Potassium (K)	19	20	19
Calcium (Ca)	20	20	20

6. What is the atomic mass of H? ______

7. What is the atomic mass of Na? ______

8. What is the total (net) charge of an atom of K? ______

9. What is the gram atomic mass of C? ______

10. True or false: Forty grams of Ca contains the same number of atoms as one gram of H.

11. In an atom
    (a) the number of protons always equals the number of neutrons.
    (b) the number of protons always equals the number of electrons.

12. True or false: The four most common elements in the body are hydrogen, carbon, calcium, and oxygen.

13. True or false: The trace elements listed in text Table 2-1 are found in minute quantities in the body but do not serve any known function.

14. True or false: Any element found in the body can be presumed to have an important function for life.

❒ **Molecules** (text pages 13-15)

15. A molecule of water consists of two atoms of ____________________ and one atom of ____________________. The chemical formula for this molecule is ____________________.

16. A covalent bond between two atoms
    (a) is formed when each atom shares one of its inner-orbit electrons with the other atom.
    (b) is formed when each atom shares one of its outer-orbit electrons with the other atom.
    (c) is the strongest of the chemical bonds.

17. True or false: Nitrogen atoms can form a maximum of four covalent bonds with other atoms.

18. True or false: The shape of a molecule may change as atoms rotate about their covalent bonds.

❐ **Ions** (text pages 15-16)

19. Ions are
    (a) electrically neutral.
    (b) electrically charged.
    (c) formed by the gain or loss of protons from the nucleus.

20. Electrolytes
    (a) are ions.
    (b) conduct electricity when dissolved in water.
    (c) are found in pure water.

21. True or false: All of the physiologically important atoms of the body readily form ions.

22. Ions that have lost electrons, and thus are ______________________(positively/negatively) charged, are called ______________________. Ions that have gained electrons, and thus are ______________________(positively/negatively) charged, are called ____________________.

23. In the reaction R-COOH ↔ R-COO^- + H^+

    (a) COOH is the carboxyl group
    (b) COO^- is the ____________________
    (c) H^+ represents ____________________
    (d) R refers to ____________________
    (e) The symbol ↔ indicates that the process is ____________________________ (reversible/irreversible)

24. In the reaction R-NH_2 + H^+ ↔ R-NH_3^+
    (a) NH_2 is the ____________________
    (b) NH_3^+ is ____________________

25. What do you predict would be the result if you added R-NH_2 molecules to a solution of R-COOH molecules?

❒ Polar Molecules (text pages 16-17)

26. A polar chemical bond
    (a) is covalent.
    (b) is ionized.
    (c) has opposite electrical charge at each end.
    (d) has no net electrical charge.

27. A polar molecule
    (a) contains a significant proportion of polar bonds relative to nonpolar bonds.
    (b) may contain ionized groups.
    (c) can form hydrogen bonds with other polar molecules.

28. True or false: Water molecules can form covalent bonds with other water molecules.

29. Hydrolysis
    (a) involves removal of water molecules from larger molecules.
    (b) involves breaking of covalent bonds within water molecules and transfer of the resulting ions to other molecules.
    (c) results in the breakdown of large molecules in the body.

❒ Solutions (text pages 17-21)

30. A solution consists of substances called ______________ that are dissolved in a liquid called the ______________. The most important such liquid in living systems is ______________. The ability of a given molecule to dissolve in a given liquid is termed the molecule's ______________ in that liquid.

31. True or false: In general, polar molecules will dissolve in polar solvents, while nonpolar molecules cannot.

32. Consider the adage familiar to anyone who has observed oil spills in the ocean or has made a salad dressing: "Oil and water do not mix." Which one or more of the following help explain this observation?
    (a) Oil is hydrophobic.

(b) Oil is nonpolar.

(c) Oil is composed largely of carbon and hydrogen.

(d) Water is hydrophilic.

33. Molecules that have properties of both polar and nonpolar molecules are called

(a) hydrophobic.

(b) hydrophilic.

(c) amphipathic.

34. NaCl (sodium chloride; table salt) has a molecular mass of 58.5 (23 + 35.5).

(a) What is the mass in grams of one *mole* of NaCl?________

(b) What is the *molarity* (number of moles of solute per liter of solution) of the resulting solution when 9 g of NaCl is dissolved in enough water to give a final volume of one liter?________

35. Molecules that release H^+ ions in solution are called ______________. Those that accept H^+ ions are called ______________. Both kinds of molecules can be further subdivided into strong and weak classes, with strong indicating ______________ (complete/incomplete) ionization in water and weak indicating ______________ (complete/incomplete) ionization in water.

36. Based on the definitions of Question 35, the carboxyl group we encountered in Question 23 is a(n) ______________ and the amino group of Question 24 is a(n) ______________.

37. The pH of a solution

(a) is a measure of the concentration of H atoms in solution.

(b) is a measure of the concentration of bound H^+ ions in solution.

(c) is a measure of the concentration of free H^+ ions in solution.

(d) increases as the acidity of the solution increases.

38. A neutral (neither acidic nor alkaline) solution has a pH of ______________.

39. The pH of extracellular fluid is slightly ________________ (greater than/less than) that of a neutral solution and thus is slightly ________________ (acidic/alkaline).

❐ Classes of Organic Molecules (text pages 21-33)

40. Organic molecules
    (a) always contain O.
    (b) always contain C.
    (c) are always macromolecules.

41. Carbohydrates
    (a) are composed of equal parts of C atoms and water molecules.
    (b) are the major organic molecules of the body.
    (c) are nonpolar.

42. True or false: Sucrose is called "blood sugar" because it is the most abundant carbohydrate in the blood.

43. The general term for carbohydrate macromolecules is ________________. Examples are starch in plants and ________________ in animals. These macromolecules are broken down into constituent subunits, called ________________, by the process of ________________.

44. List the three major classes of lipids.
    (a) ________________
    (b) ________________
    (c) ________________

    (d) One class differs from the other two with regard to solubility in water. Explain.

    ____

    ____

    ____

45. We hear a great deal from dieticians about how it is better to eat unsaturated rather than saturated fats.

(a) What chemical quality distinguishes the two kinds of fats? ____________________

____

____

____

(b) Which kind of fat is more common in meat and dairy products than in vegetables? ____________________________

46. True or false: Cholesterol is a phospholipid.

47. One major chemical difference between proteins on the one hand and lipids and carbohydrates on the other is that all proteins contain ________________________ atoms.

48. The subunits that form proteins are called ________________________. There are ____________________how many?) different subunits, each of which contains a(n) ______________________ group, a(n) ______________________ group, and a specific ______________________________, which distinguishes one subunit from another.

49. Diagram the formation of a peptide bond between the two amino acids:

```
H   R1   O              H   R2   O
|   |    ||             |   |    ||
N - C  - C - O     +    N - C  - C - O
|   |        |          |   |        |
H   H        H          H   H        H
```

50. Differentiate among *polypeptide*, *peptide*, and *protein*.

____

____

____

____

51. True or false: Glycoproteins are protein molecules with molecules of glycogen attached to the amino acid side chains.

52. True or false: The sequence of amino acids in a protein is known as the secondary structure.

53. The three-dimensional structure of a protein is referred to as its ______________. Three kinds of noncovalent bonding forces contribute to this structure. The kind that occurs at regular intervals along the polypeptide chain and induces the formation of a helical structure is a ______________________ bond. Stronger bonds, called ____________________ bonds, are formed between ________________ (ionized/unionized) groups in the side chains. The weakest bonds, called ______________ forces, are formed between ________________ (polar/nonpolar) side chains and tend to stabilize the molecule. Special covalent bonds, termed __________________ bonds because of the sulfur atoms involved, link two __________________ amino acids together. Being covalent, these bonds are ________________ (stronger/weaker) than the others described above.

54. True or false: Protein molecules may consist of more than one polypeptide chain.

55. True or false: Substitution of one amino acid for another in a given protein will inevitably alter the conformation of that protein to a significant degree.

56. Name the two classes of nucleic acids.

____

____

57. The subunits of both classes of nucleic acids are termed __________________ and consist of a ______________________ group, a sugar, and one of five possible carbon-nitrogen rings called ______________________. The subunits for the two

classes of nucleic acids differ in the composition of the ______________________ molecule and also in one of the ______________________. The subunits are linked together by covalent bonds between the ________________ and ________________ groups of adjacent subunits.

58. The 3-dimensional structure of DNA is a ______________________. The two strands are linked together by ____________________ bonds between pairs of bases, a __________________(what class?) to a ________________(what class?). The bonds formed are specific: the __________________ base adenine always pairs with the __________________ base ________________; likewise, the ____________________ base cytosine always pairs with the __________________ base ____________________.

59. RNA molecules are ______________________(double/single) stranded and can form base pairs with DNA as above, except that RNA has the pyrimidine base __________________ instead of __________________, which nevertheless pairs with ______________________ on the DNA.

ANSWERS

Boldface type indicates the answers you should have given. Words in medium-face (ordinary) type explain or expand upon the answer.

1. **No. "Element" refers to matter that is composed of only one type of atom; it does not imply quantity. Thus, elemental gold may by one atom of gold or many grams of gold, so long as it is pure.**

2. In the "miniature solar system" model of the atom, **electrons** orbit the nucleus, which is composed of particles called **protons** and **neutrons**.

3. **False.** An atomic nucleus is electrically ~~neutral~~ **positive.** (It is composed of positively charged particles and particles with no charge, and so its net charge is positive.)

4. **False.** Protons and neutrons have roughly the same ~~charge~~ **mass.**

5. **False.** The atomic number of an element refers to the number of ~~particles~~ **protons** in its atomic nucleus.

6. **1** (1 proton + 0 neutrons = atomic mass of 1)

7. **23** (11 protons + 12 neutrons = a.m. 23)

8. **0** (all atoms have a total charge of zero, by definition)

9. **12 g**

10. **True.** (40 g Ca = 1 mole of Ca, 1 g H = 1 mole of H, and 1 mole of *any* substance contains 6 X 10^{23} atoms.)

11. (a) **Incorrect. The number of protons may be less than, equal to, or greater than the number of neutrons.**

    (b) **Correct.**

12. **False.** The four most common elements in the body are hydrogen, carbon, ~~calcium~~ **nitrogen**, and oxygen.

13. **False. Even though** the trace elements listed in text Table 2-1 are found in minute quantities in the body, ~~but do not serve any known function~~ **nevertheless they are known to serve important functions.** (Dietary studies show that every element listed in the table is necessary for good health in animals and humans.)

14. **False.** Any element found in the body can **not necessarily** be presumed to have an important function for life. (Some unnecessary or even harmful elements, such as mercury and lead, can be ingested.)

15. A molecule of water consists of two atoms of **hydrogen** and one atom of **oxygen.** The chemical formula for this molecule is **H_2O.**

16. (a) **Incorrect. A covalent bond is formed when an atom shares one of its outer-orbit electrons with another atom.**

    (b) **Correct.**

    (c) **Correct.**

17. **False**. Nitrogen atoms can form a maximum of ~~four~~ **three** covalent bonds with other atoms.

    Alternatively, the answer below is also correct:

    False. ~~Nitrogen~~ **Carbon** atoms can form a maximum of four covalent bonds with other atoms.

18. **True.**

19. (a) **Incorrect. Ions are electrically charged.**
    (b) **Correct.**
    (c) **Incorrect. Ions are formed by the gain or loss of electrons from the outermost orbit.**

20. (a) **Correct.**
    (b) **Correct.** (This is the definition of electrolyte.)
    (c) **Incorrect. Water is pure only when it contains no other molecules. Water molecules themselves are not ions, do not conduct electricity, and thus cannot be considered electrolytes.**

21. **False.** ~~All~~ **Many** of the physiologically important atoms of the body readily form ions, **but others** (including C, N, and O) **do not.**

22. Ions that have lost electrons, and thus are **positively** charged, are called **cations.** Ions that have gained electrons, and thus are **negatively** charged, are called **anions.** (Memorization hint: Calcium ions [Ca^{2+}] are positively charged, and thus are cations.)

23. (b) **carboxyl ion**
    (c) **hydrogen ion**
    (d) **the "remainder" of the molecule**
    (e) **reversible**

24. (a) **amino group**
    (b) **amino ion**

25. $\mathbf{R\text{-}COOH + R\text{-}NH_2 \leftrightarrow R\text{-}COO^- + R\text{-}NH_3^+}$

26. (a) **Correct.**
    (b) **Incorrect. A polar bond is not ionized.** (Ions are formed when one atom *captures* an electron from another. In polar bonds, electrons are still shared, but unequally.)
    (c) **Correct.**
    (d) **Correct.**

27. (a) **Correct.**
    (b) **Correct.**
    (c) **Correct.**

28. **False.** Water molecules can form ~~covalent~~ **hydrogen** bonds with other water molecules.

29. (a) **Incorrect.** (See note below.)

    (b) **Correct.**

    (c) **Correct.**

    Note: Combining answers (b) and (c) with the word "which" describes the process of hydrolysis.

30. A solution consists of substances called **solutes** that are dissolved in a liquid called the **solvent**. The most important such liquid in living systems is **water**. The ability of a given molecule to dissolve in a given liquid is termed the molecule's **solubility** in that liquid.

31. **True.**

32. (a) **Correct.** (Although oil cannot *literally* fear water, oil is repelled by water.)

    (b) **Correct.** (This explains why oil is hydrophobic.)

    (c) **Correct.** (This explains why oil is nonpolar.)

    (d) **Correct.** (Each water molecule has both regions of positive charge and regions of negative charge; the negative charges of one molecule attract the positive charges of another, and this water-water attraction effectively "squeezes out" nonpolar molecules.)

33. (a) **Incorrect.** (Although part of the molecule is hydrophobic.)

    (b) **Incorrect.** (Although part of the molecule is hydrophilic.)

    (c) **Correct.**

34. (a) **58.5 g**

    (b) **0.15 M/L.** (This concentration is determined by dividing 9 g by the mass in grams of one mole [58.5 g].)

35. Molecules that release H^+ ions in solution are called **acids**. Those that accept H^+ ions are called **bases**. Both kinds of molecules can be further subdivided into strong and weak classes, with strong indicating **complete** ionization in water and weak indicating **incomplete** ionization in water.

36. Based on the definitions of Question 35, the carboxyl group we encountered in Question 23 is an **acid** and the amino group of Question 24 is a **base**.

37. (a) **Incorrect. The pH of a solution is a measure of H^+ ions, not atoms.**

    (b) **Incorrect. The pH of a solution is a measure of free ions, not bound ones.**

    (c) **Correct.**

    (d) **Incorrect. The pH of a solution decreases as the acidity of the solution increases.**

38. **7.0**

39. The pH of extracellular fluid is slightly **greater than** that of a neutral solution and thus is slightly **alkaline**. (The term "basic" means the same thing as "alkaline.")

40. (a) **Incorrect.** (For example, methane [CH_4] is organic.)

    (b) **Correct.**

    (c) **Incorrect.** (See example "a.")

41. (a) **Correct.**

    (b) **Incorrect. Carbohydrates are less abundant than proteins and lipids in the body.**

    (c) **Incorrect. They are polar.**

42. **False.** ~~Sucrose~~ **Glucose** is called "blood sugar" because it is the most abundant carbohydrate in the blood. (Sucrose is called "table sugar." It is a disaccharide composed of glucose and fructose, and is hydrolized by the digestive system when it is eaten; sucrose is not found in the blood.)

43. The general term for carbohydrate macromolecules is **polysaccharide**. Examples are starch in plants and **glycogen** in animals. These macromolecules are broken down into constituent subunits, called **monosaccharides**, by the process of **hydrolysis**.

44. (a) **Triacylglycerols**

    (b) **Phospholipids**

    (c) **Steroids**

    (d) **Lipids are nonpolar and thus not soluble in water. However, phospholipids contain a polar side chain or group,and thus are amphipathic.** (This property is extremely important for the function of phospholipids in the formation of cell membranes, as we shall see in Chapter 3.)

45. (a) **Saturated fats are triacylglycerols in which all the carbon atoms in the constituent fatty acids are linked by single bonds; unsaturated fats have fatty acids with one or more double bonds joining the carbons.**

    (b) **Saturated**

46. **False.** Cholesterol is a ~~phospholipid~~ **steroid**.

47. One major chemical difference between proteins on the one hand and lipids and carbohydrates on the other is that all proteins contain **nitrogen** atoms.

48. The subunits that form proteins are called **amino acids.** There are **20** different subunits, each of which contains an **amino** (NH_2) group (except for *proline* — see Text Figure 2-13), a **carboxyl** (COOH) group, and a specific **amino acid side chain,** which distinguishes one subunit from another.

49.

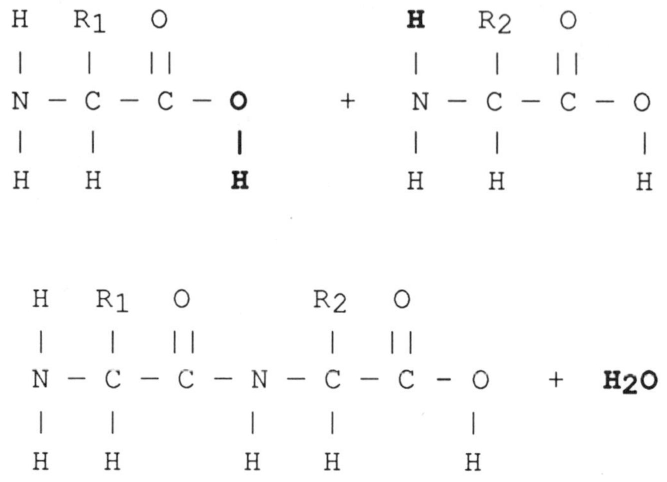

```
H   R1   O            H   R2   O
|   |    ||           |   |    ||
N - C  - C - O    +   N - C  - C - O
|   |        |        |   |        |
H   H        H        H   H        H


H   R1   O        R2   O
|   |    ||       |    ||
N - C  - C  - N - C  - C - O    +  H2O
|   |         |   |        |
H   H         H   H        H
```

50. **Two or more amino acids linked by peptide bonds form a polypeptide (a polymer of amino acids). Peptides are polypeptides containing fewer than about 50 amino acids, and proteins are polypeptides containing more than 50 amino acids.**

51. **False.** Glycoproteins are protein molecules with molecules of ~~glycogen~~ **monosaccharides** attached to the amino acid side chains.

52. **False.** The sequence of amino acids in a protein is known as the ~~secondary~~ **primary** structure.

53. The three-dimensional structure of a protein is referred to as its **conformation**. Three kinds of noncovalent bonding forces contribute to this structure. The kind that occurs at regular intervals along the polypeptide chain and induces the formation of a helical structure is a **hydrogen** bond. Stronger bonds, called **ionic** bonds, are formed between **ionized** groups in the side chains. The weakest bonds, called **Van der Waals** forces, are formed between **nonpolar** side chains and tend to stabilize the molecule. Special covalent bonds, termed **disulfide** bonds because of the sulfur atoms involved, link two **cysteine** amino acids together. Being covalent, these bonds are **stronger** than the others described above.

54. **True.**

55. **False.** Substitution of one amino acid for another in a given protein ~~will inevitably~~ **may or may not** alter the conformation of that protein to a significant degree, **depending on the chemical composition of the side chain of each amino acid.** (Substitution of one neutral amino acid for another — for example, alanine in place of leucine — may not have much effect.)

56. **Deoxyribonucleic acid (DNA)**
    Ribonucleic acid (RNA)

57. The subunits of both classes of nucleic acids are termed **nucleotides** and consist of a **phosphate** (PO_4^{3-}) group, a sugar, and one of five possible carbon-nitrogen rings called **bases**. The subunits for the two classes of nucleic acids differ in the composition of the **sugar** molecule and also in one of the **bases**. The subunits are linked together by covalent bonds between the **sugar** and **phosphate** groups of adjacent subunits.

58. The 3-dimensional structure of DNA is a **double helix.** The two strands are linked together by **hydrogen** bonds between pairs of bases, a **purine** to a **pyrimidine.** The bonds formed are specific: the **purine** base adenine always pairs with the **pyrimidine** base **thymine**; likewise, the **pyrimidine** base cytosine always pairs with the **purine** base **guanine**. (Do not confuse the special purine and pyrimidine bases with the generic term "base," which refers to any hydrogen-ion acceptor. The nucleotide bases are bases in that sense, but they are very special molecules. Pay attention to the context in which these terms are used.)

    Memorization hint: Remember that the purines (smaller word) are the larger (double-ringed) bases. Think of the expression "ta-ta" to remember that thymine (T) pairs with adenine (A), leaving C and G to pair with each other.

59. RNA molecules are **single** stranded, and can form base pairs with DNA as above, except that RNA has the pyrimidine base **uracil** instead of **thymine,** which nevertheless pairs with **adenine** on the DNA.

PRACTICE

True or false (correct the false statements):

P1. An atom is electrically neutral.

P2. The mass of an atom is the sum of its protons and electrons.

P3. The atomic number of an element is given by the number of electrons in the atom.

P4. Important mineral elements in the body include Na, Ca, and O.

P5. The number of covalent bonds that can be formed by a given atom depends upon the number of electrons present in the outermost orbit.

P6. In a molecule of water, an oxygen atom forms a double bond with each of two hydrogen atoms.

P7. The carboxyl ion is an anion.

P8. NaCl is a molecule formed by the covalent bonding of a sodium atom to a chlorine atom.

P9. All covalent bonds are polar.

P10. Water molecules can form ionic bonds with each other.

P11. During hydrolysis, hydrogen atoms and hydroxyl groups are formed.

P12. Solutes that do not dissolve in water are called hydrophilic.

P13. Molecules with both polar and nonpolar regions are called ambidextrous.

P14. The molarity of a solution is a measure of the concentration of the solvent.

P15. A solution with a pH of 8 is more acidic than one with a pH of 3.

P16. Organic chemistry is the study of oxygen-containing compounds.

P17. Polysaccharides are polymers of sugar molecules.

P18. Triacylglycerol is one subclass of lipid molecules.

P19. Saturated fats contain carbon atoms linked by double bonds.

P20. In DNA, thymine binds with adenine, and cytosine binds with uracil.

Multiple choice (correct each incorrect choice):

P21-24. Proteins are

P21. critically important for physiological processes.

P22. composed of fatty acids.

P23. composed of nucleic acids.

P24. macromolecules with subunits linked by polypeptide bonds.

P25-27. Protein conformation is

P25. independent of the sequence of subunits forming the protein.

P26. dependent upon a combination of covalent and noncovalent bonds.

P27. affected by interactions with water molecules.

P28-30. Nucleic acids are

P28. macromolecules.

P29. composed of nucleotides.

P30. distinguished from each other in part by the composition of the sugar they contain.

3: CELL STRUCTURE

Robert Hooke was a seventeenth-century Englishman who built one of the earliest compound microscopes (a device invented in 1609 by Galileo Galilei). Hooke used his microscope to study the microscopic structure of matter. When he magnified a piece of cork (the light outer bark of a species of oak tree), he noticed that it was composed of "little organic boxes or cells distinct from one another." Although he was seeing not actual cells but rather only the hollow walls that indicated where the cells had been in the plant when it was living, he was correct in supposing that higher organisms are composed of subunits, and the term "cell" has stuck. Today, we know that all living things consist of cells — from one to many trillions — and that the cell is the basic unit of life. Technological improvements in microscopy have allowed us to peer literally into the inner workings of cells and to (begin to) extract from them the secrets of life's basic processes.

Instructions for answering questions in this Study Guide

1. True or false: Correct each false statement. Whenever you must correct a false statement, there will almost always be more than one way to do so. In the Guide, the answer requiring the least correction will generally be given, with the understanding that other correct answers are possible. It is usually not sufficient, in terms of demonstrating understanding, to simply insert a "no" or "not," however.

2. Multiple choice: Any, all, or none of the responses may be correct. Explain why each incorrect response is incorrect.

3. Fill in the blanks: Choose the best word or words to complete each statement.

4. Directions will be given for other types of questions as they appear.

❐ Introduction (text page 37)

1. The cells of a human being
   (a) are generally larger than those of a mouse.
   (b) are greater in number than those of a mouse.
   (c) can generally be seen with the naked eye.
   (d) are the structural and functional units of the body.

❐ Microscopic Observations of Cells (text pages 37-38)

2. Light microscopes use ____________________ to form an image of a microscopic structure, whereas ________________ microscopes use ____________________. The wavelength of the radiation the latter kind of microscope uses is _______________ (longer/shorter) than that of the former; thus the resolution achieved by the latter kind is _____________ (lesser/greater).

3. True or false: A light microscope can be used to view very large proteins in a cell.

4. True or false: Living cells cannot be viewed under a microscope.

❐ Cell Compartments (text pages 38-45)

5. Most human cells consist of three major components: the ______________ membrane, the centrally located ___________________, and the __________________. The last-named component consists of a fluid called the ____________________ as well as structures known as ___________________, which perform specific functions within the cell.

6. In the plasma membrane, ______________________ molecules are organized into a bimolecular layer such that the ____________________ (polar/nonpolar) regions lie facing the membrane surfaces and the ________________ (polar/nonpolar) regions lie __________________________. This orientation allows the membrane to act as an effective barrier against ______________________ (polar/nonpolar) molecules, while permitting it to make contact with the aqueous extra- and intracellular fluids.

7. True or false: One striking feature of plasma membrane structure is its symmetry, with the extracellular and cytoplasmic surfaces virtual mirror images of each other.

8. Indicate whether the following are descriptive of integral membrane proteins (I), peripheral membrane proteins (P), or both kinds of membrane proteins (B):

_____ Are amphipathic

_____ May span entire membrane

_____ May be able to move laterally in the membrane

_____ Are located primarily on the inner membrane surface

_____ May form channels through the membrane

_____ May help confer cell shape and motility

9. Cholesterol

(a) is amphipathic.

(b) is a prominent lipid component of all cell membranes.

(c) confers rigidity to the plasma membrane.

10. The intracellular fluid in any given cell is not uniform in its chemical composition. Explain.

____

____

____

____

____

11. The diagrams below illustrate three ways in which cells are connected for various functions. Name each type of junction, briefly describe its nature and function, and name one type of tissue in which it is found.

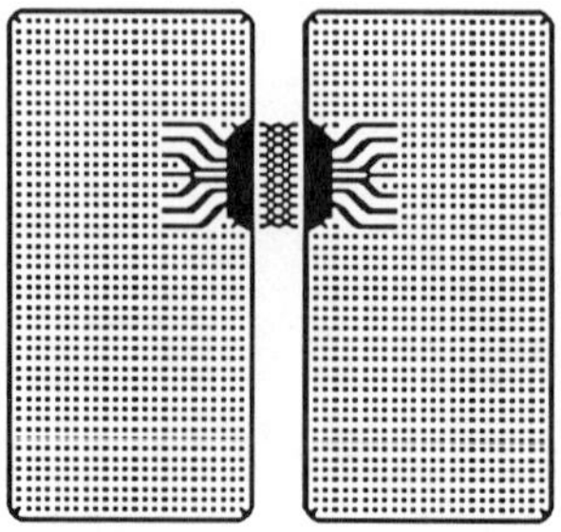

(A) Name:_______________________________

Description: ____

____

____

____

____

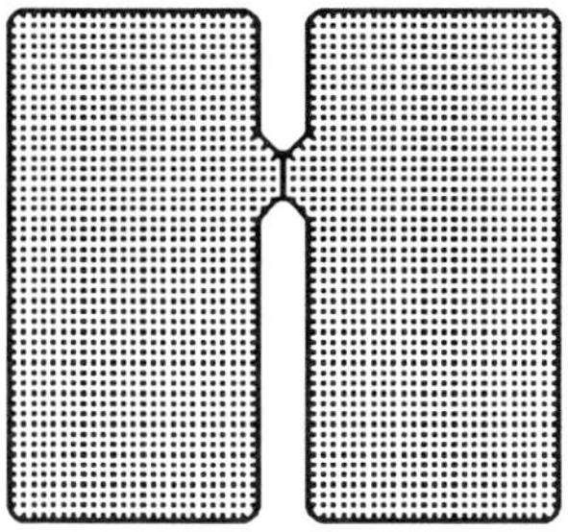

(B) Name: ______________________

Description: ____

____

____

____

____

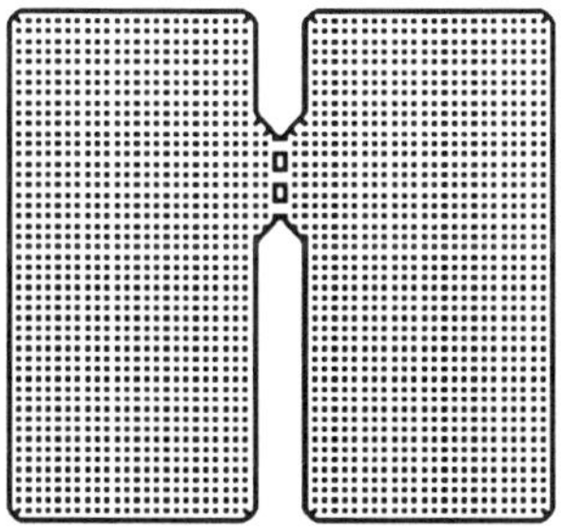

(C) Name: ______________________

Description: ____

____

____

____

____

❐ **Cell Organelles** (text pages 45-50)

12. Match the structure on the left with the best description on the right.

_____ Nucleolus	a.	Contain(s) chromatin
_____ Ribosomes	b.	Composed of DNA and protein
_____ Nuclear envelope	c.	Site(s) of protein synthesis
_____ Nucleus	d.	Double membrane
_____ Chromatin	e.	Site(s) of ribosome assembly
_____ Nuclear pores	f.	Allow(s) passage of large molecules between nucleus and cytoplasm

13. Some biology teachers find it useful to draw an analogy between the parts of a cell and the parts of a factory. Match the organelle on the left with its appropriate "factorial" analog:

_____ Golgi apparatus	a.	Business office
_____ Lysosomes and peroxisomes	b.	Production plant
_____ Nucleus	c.	Power plant
_____ Endoplasmic reticulum	d.	Packaging plant
_____ Mitochondria	e.	Waste management

14. True or false: Cells that require large amounts of energy for their activities contain more ribosomes than do less active cells.

15. Cell A is a gland cell that makes and secretes proteins. Cell B is another type of cell that is not specialized for secretion. If you were to look at electron micrographs of the two cells, what differences would you expect to see?_______________________________

____

____

____

16. The smallest in diameter of the cytoskeletal structures are the ____________________, which are composed of the protein ____________________. Larger structures, called ______________________________, are found only in muscle cells and are composed of ________________________. The largest of the cytoskeletal structures are the ________________________, which are composed of the protein ______________. Three examples of cell structures formed from this last cytoskeletal element are _______________, found on the surface of some cells, and the ____________________ and ____________________, which are important for cell division.

ANSWERS

Boldface type indicates the answers you should have given. Words in medium-face (ordinary) type explain or expand upon the answer.

1. (a) **Incorrect. Generally, cell size does not differ from one animal to another.** (There are some exceptions, of course. The eggs of birds, reptiles, and other nonmammalian animals, which are single cells, are much bigger than any human or other mammalian cell, even the largest one — the mammalian "egg," or ovum. With regard to mice and men, there are also differences in the length of some nerve cells, for example, which can reach several feet in humans but only a few inches in the mouse.)

   (b) **Correct.**

   (c) **Incorrect. Only the largest cells** (the ovum) **can be seen** (barely. The human ovum is about 120 μm in diameter. The limit of resolution of the human eye with 20/20 vision is 100 μm.)

   (d) **Correct.**

2. Light microscopes use **light rays** to form an image of a microscopic structure, whereas **electron** microscopes use **a beam of electrons**. The wavelength of the radiation the latter kind of microscope uses is **shorter** than that of the former; thus the resolution achieved by the latter kind is **greater**.

3. **False.** ~~A light~~ **An electron** microscope can be used to view very large proteins in a cell. (The lower limit of size of an object that can be seen through a light microscope is about 0.2 μm [200 nm]. Even very large proteins do not exceed 6 nm. This level of resolution is toward the lower limits of the electron microscope [see text Figure 3-2]. Another way of thinking about limits of resolution is to think of the human eye as having a resolving power of 1. A light microscope can magnify that power by 500. [100 μm is 500 times greater than 0.2 μm.] Similarly, an electron microscope increases our eyes' resolution 50,000 times.)

4. **False.** Living cells ~~cannot~~ **can** be viewed under a **light** microscope.

   Alternatively, the answer below is also correct:

   False. Living cells cannot be viewed under **an electron** microscope.

   Living cells can be viewed under a light microscope because light can penetrate the thickness of an intact cell or even several layers of intact cells. (One can easily see living single-celled organisms in a drop of pond water, for example, or living cells from a scraping of the lining of one's mouth on a microscope slide.) Cells from any tissue can be disassociated from other cells, allowed to grow in nutrient medium in plastic culture dishes, and observed. Frequently, however, tissues are chemically "fixed" (in a process analogous to fixing a photographic plate during the development of a picture) so that they can be preserved on a microscope slide. Fixed tissues can also be stained with dyes so that the different parts of the cells can be distinguished from one another. For electron microscopy, tissues must be fixed and embedded in plastic so that they can be sliced *very* thinly — to about 10 nm, one-twentieth of the diameter of the smallest cell. Electrons cannot penetrate thicker sections.

5. Most human cells consist of three major components: the **plasma** membrane, the centrally located **nucleus**, and the **cytoplasm**. The last-named component consists of a fluid called the **cytosol** as well as structures known as **organelles**, which perform specific functions within the cell. (The nucleus is an organelle.)

6. In the plasma membrane, **phospholipid** molecules are organized into a bimolecular layer such that the **polar** regions lie facing the membrane surfaces and the **nonpolar** regions lie **in the middle of the membrane**. This orientation allows the membrane to act as an effective barrier against **polar** molecules, while permitting it to make contact with the aqueous extra- and intracellular fluids.

7. **False.** One striking feature of plasma membrane structure is its ~~symmetry~~ **asymmetry**, with the extracellular and cytoplasmic surfaces ~~virtual mirror images of~~ **differing considerably from** each other. (Re-examine text Figure 3-7 if you answered incorrectly. Recall that, among other differences, the carbohydrates associated with integral proteins are found only on the extracellular membrane surface, whereas peripheral proteins are more abundant on the cytoplasmic side.)

8. **I** Are amphipathic
   I May span entire membrane
   I May be able to move laterally in the membrane
   P Are located primarily on the inner membrane surface
   I May form channels through the membrane
   P May help confer cell shape and motility

9. (a) **Correct.**
   (b) **Incorrect. Cholesterol is a prominent component only of plasma membranes, with relatively little found in any organelle membrane.**
   (c) **Incorrect. Cholesterol confers fluidity to the plasma membrane.** (By combining answers b and c, you should be able to infer that organellar membranes are more rigid than is the plasma membrane.)

10. **Intracellular fluid refers to all the fluid within the cell — the cytosol and the fluid contained within all the organelles. Its composition differs from one part of a cell to another because of the various membranes present in the cell.** (The membrane defining the Golgi apparatus, say, is not the same as the membrane surrounding the

nucleus.) **Differences in membrane structure mean that different membranes allow different molecules to pass through them.** (For example, suppose the Golgi apparatus membrane allows molecule A to pass from the cytosol into the organelle but not molecule B while the nuclear membrane allows passage of molecule B into the nucleus but not molecule A. Thus the intracellular fluid inside the Golgi apparatus is high in A and deficient in B, and that inside the nucleus is high in B and deficient in A. This same argument holds for all the other organelles in the cell, leading to the differences in intracellular fluid composition. Similarly, the plasma membrane differs from all of the organellar membranes, and so some molecules that pass into the cell across the plasma membrane cannot also pass through any organelle membrane, a condition that also contributes to differences in intracellular fluid composition in different parts of the cell.)

11. A. **Desmosome. These structures are like "spot welds" of protein filaments that hold adjacent cells tightly together. Desmosomes are not hollow and do not allow ready transcellular communication. Desmosomes are important where tissues are subjected to stretch, such as the skin and lining of the gastrointestinal tract. They also help to attach epithelial cells to underlying connective tissue.**

    B. **Tight junction. It is formed by fusion of the extracellular polar regions of the plasma membranes of two adjacent cells. Like desmosomes, tight junctions confer strength and stability to tissues, but they are also important for forming selective barriers to regulate the exchange of materials across them.** (Recall from Chapter 1 that formation of such barriers is a function of epithelial cells. It is not surprising, therefore, that) **tight junctions are common in epithelial tissues, including that lining the small intestine. Many cells with tight junctions also have desmosomes for added stability.**

    C. **Gap junction. Proteins from the plasma membranes of each of two adjacent cells interact to form the walls of narrow channels that penetrate the membranes, thus connecting the cytoplasms of the two cells and allowing ions and other small molecules to pass from the cytoplasm of one cell to the other. Gap junctions are particularly important in tissues in which cells must act as a unit, such as the heart.**

12. **e** Nucleolus
    c Ribosomes
    d Nuclear envelope
    a Nucleus
    b Chromatin
    f Nuclear pores

13. **d** Golgi apparatus
e Lysosomes and peroxisomes
a Nucleus
b Endoplasmic reticulum
c Mitochondria

14. **False.** Cells that require large amounts of energy for their activities contain more ~~ribosomes~~ **mitochondria** than do less active cells. (Ribosomes are important for protein synthesis; an active cell need not be one that synthesizes much more protein than other cells, although some very active cells do perform much protein synthesis. The hallmark of an active cell is the large number of mitochondria it contains, because these organelles supply most of the energy [in the form of ATP] that is necessary for cellular activity.)

15. **Cell A would have more granular endoplasmic reticulum than would cell B because this organelle** (or more specifically, the ribosomes on it) **is where proteins "for export" are synthesized. The Golgi apparatus may also be more extensive in A than in B because this organelle is required for packaging proteins for secretion. Another striking difference would be the obvious presence in A of secretion granules.**

16. The smallest in diameter of the cytoskeletal structures are the **microfilaments**, which are composed of the protein **actin**. Larger structures, called **muscle thick filaments**, are found only in muscle cells and are composed of **myosin**. The largest of the cytoskeletal structures are the **microtubules**, which are composed of the protein **tubulin**. Three examples of cell structures formed from this last cytoskeletal element are **cilia**, which are found on the surface of some cells, and the **mitotic spindles** and **centrioles**, which are important for cell division.

PRACTICE:

True or false (correct the false statements):

P1. In general, the larger an animal is, the larger are its individual cells.

P2. Small structures and large molecules in cells can be viewed with an electron microscope.

P3. Intracellular fluid is the fluid in the cytoplasm.

P4. The major lipids in cellular membranes are phospholipids.

P5. The special functions of plasma and organelle membranes depend primarily on the specific composition of the phospholipids of those membranes.

P6. Desmosomes are structures that permit direct communication between cells by allowing the cells to exchange small molecules in their cytoplasms.

P7. Chromosomes are condensed forms of chromatin.

P8. Free ribosomes differ from membrane-bound ribosomes in that free ribosomes specialize in synthesizing proteins for export (secretion) from the cell.

P9. Lysosomes are organelles specialized for breaking down intracellular debris or malfunctioning parts of cells.

Fill in the blanks:

P10 -13. Rough-surfaced endoplasmic reticulum, which is also called ________________ endoplasmic reticulum, is important for the synthesis and packaging of ___________________ molecules. The rough appearance stems from the association of ____________________________ with the endoplasmic reticulum membranes. Smooth endoplasmic reticulum is important for the synthesis of ________________ molecules.

Multiple choice (correct each incorrect choice):

P14-16. Cell membranes

P14. are components of cytosol.

P15. are passive barriers against the passage of molecules from one side to the other.

P16. consist primarily of protein and carbohydrate.

P17-20. The cytoskeleton

P17. refers to the cellular components of bone.

P18. refers to a network of cytoplasmic filaments.

P19. is important for cellular movement.

P20. helps to determine a cell's shape.

4: MOLECULAR CONTROL MECHANISMS: DNA AND PROTEIN

The discovery of the mechanisms by which genes direct protein synthesis has given molecular biologists an immensely powerful tool for probing the fundamental secrets of life. The techniques of recombinant DNA and genetic engineering have profound implications for improving our health and well-being. Plants have been treated with "frost-ban" — a "designer" bacterium that keeps the plant from being damaged by cold weather. Other recombinant bacteria are being developed to disperse oil spills. Human genes have been given to mice to make the mice susceptible to certain human diseases in order to study these diseases and to search for cures. Research is under way to learn how to correct damaged genes, and thus hereditary defects, in animals and humans.

In this chapter, then, you will learn the basics of some of the most exciting research in the history of science.

Instructions for answering questions in this Study Guide

1. True or false: Correct each false statement. Whenever you must correct a false statement, there will almost always be more than one way to do so. In the Guide, the answer requiring the least correction will generally be given, with the understanding that other correct answers are possible. It is usually not sufficient, in terms of demonstrating understanding, to simply insert a "no" or "not," however.

2. Multiple choice: Any, all, or none of the responses may be correct. Explain why each incorrect response is incorrect.

3. Fill in the blanks: Choose the best word or words to complete each statement.

4. Directions will be given for other types of questions as they appear.

❒ **Introduction** (text pages 53-54)

1. The fundamental differences between a man and a mouse can be ascribed to differences in
   (a) proteins.
   (b) DNA.
   (c) hereditary information.

SECTION A: PROTEIN BINDING SITES

❒ **Binding-Site Characteristics** (text pages 54-57)

2. A ligand is a molecule that binds to
   (a) a protein by covalent bonds.
   (b) a protein by either ionic or hydrogen bonds.
   (c) any other molecule by means of noncovalent bonds.

3. A binding site on a protein is
   (a) an area of the protein with a shape complementary to that of a ligand.
   (b) determined by the amino acid sequence of the protein.
   (c) formed by a region of amino acids adjacent to each other on a polypeptide chain.

4. True or false: Any given protein contains binding sites for only one kind of ligand.

5. True or false: The ability of a binding site on a protein to "recognize" and bind with only certain ligands is called the affinity of the binding site.

6. The affinity a binding site on a protein has for a ligand
   (a) can be influenced by the shape of the binding site.
   (b) can be influenced by the presence of charged groups on the ligand and the binding site.
   (c) is a measure of how readily a bound ligand can be released from the protein.

7. Rank the three ligand/binding site combinations according to decreasing affinity:

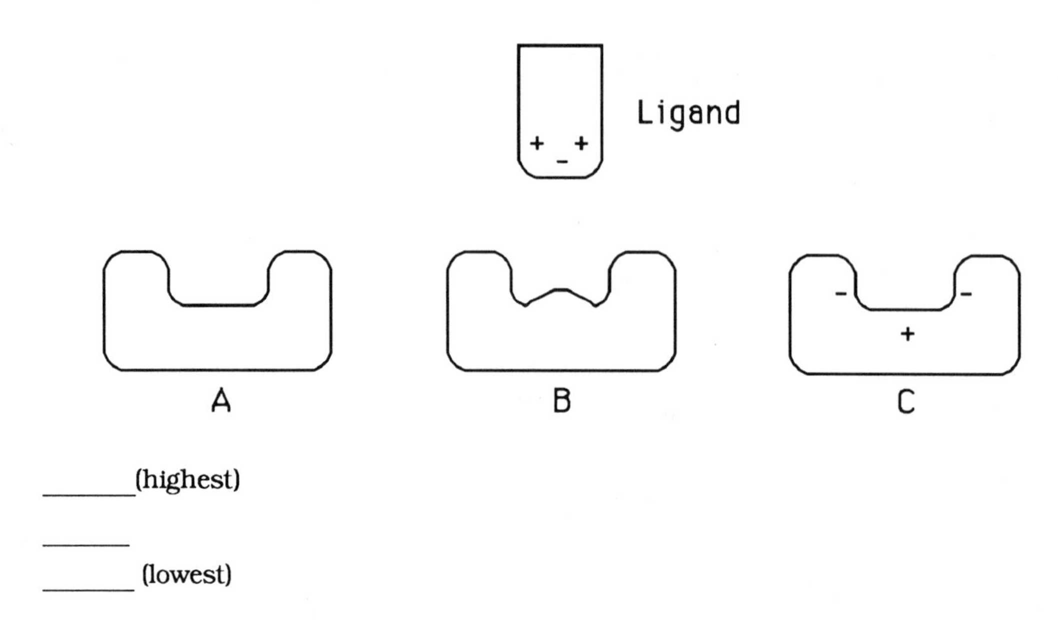

_______(highest)

____

_______ (lowest)

8. The 50%-saturation point of a solution of ligands and proteins
   (a) refers to the amount of protein required to bind 50% of the ligand.
   (b) refers to the amount of ligand required to bind 50% of the binding sites.
   (c) refers to the amount of ligand required to bind one binding site 50% of the time.
   (d) can be used to measure the affinity of a binding site for a ligand.

9. Two ligands, A and B, bind to a site on protein X. The two binding curves are shown below. When ligand A binds, it produces a biological effect. Ligand B produces no effect when it binds. Using this information, tell whether each statement below is correct:

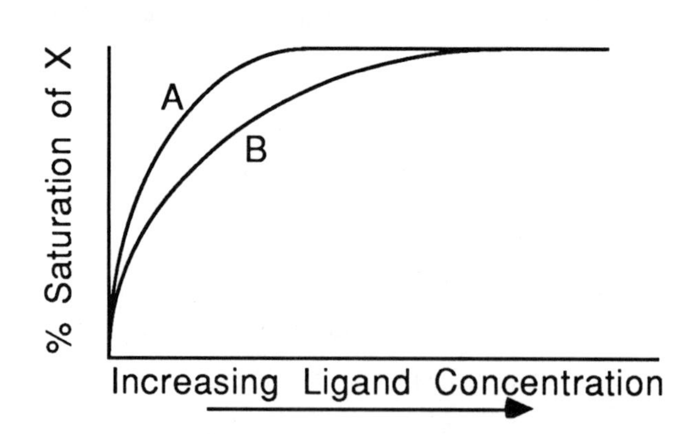

   (a) Increasing the concentration of A will result in an increase of the biological effect.
   (b) The binding site on X has a higher affinity for A than for B.
   (c) Increasing the concentration of B in the presence of A will enhance the biological effect.

❑ Regulation of Binding-Site Characteristics (text pages 57-59)

10. The two general ways of controlling the activity of a protein are to change its ________________________ and to regulate its ___________________________.

11. Allosteric proteins
    (a) contain more than one kind of binding site.
    (b) undergo a change of shape when a ligand binds to the regulatory binding site.
    (c) are always activated when a modulator molecule binds to the regulatory site.
    (d) constitute the majority of proteins.

12. True or false: Allosteric modulation of a protein involves covalent binding of a ligand to a regulatory binding site.

13. The most common kind of covalent protein modulation involves the addition of a ________________________ group, which has a net ________________________ (positive/negative) charge; the process is called ____________________________.

14. Fill in the missing terms:

    protein kinase
    Protein + (a) ____________________ → (b) ___________________ + ADP

    In general terms, the protein and (a) are the ______________________ of the reaction, (b) and ADP are the ___________________, and protein kinase is a(n) ___________________.

15. Indicate which of the following are descriptive of protein kinases (k), phosphoprotein phosphatases (p), both (b), or neither (n):
    (a) _______use ATP as substrate
    (b) _______use phosphorylated protein as substrate
    (c) _______act on many kinds of protein
    (d) _______are very specific
    (e) _______are continuously active
    (f) _______are allosteric proteins

16. The process of covalent modulation is itself regulated by allosteric mechanisms. Explain. ____________________

____

____

SECTION B: GENETIC INFORMATION AND PROTEIN SYNTHESIS

❒ Genetic Information (text pages 59-60)

17. A gene
    (a) contains information necessary for the synthesis of carbohydrates.
    (b) is a sequence of nucleotides in DNA.
    (c) encodes the amino acid sequence for a protein molecule.
    (d) is composed of many molecules of DNA.
    (e) is a unit of hereditary information.

18. Most genes are located in the ____________________ of cells, whereas protein synthesis occurs in the ____________________.

19. True or false: The gene for brown eyes encodes the structure of brown pigment.

20. What is the central dogma of molecular biology? ____________________

____

21. Name the "letters" of the genetic language.
    (a) ____________________ (c) ____________________
    (b) ____________________ (d) ____________________

22. With respect to the genetic language:
    (a) Of how many letters are the words composed? ____________________
    (b) How many words are there? ____________________
    (c) What do the words mean or stand for? ____________________

    ____

    ____

    (d) Does the language differ from one organism to another? ____________________

(e) What significance does the answer to (d) have with regard to theories of the origin of life on earth? ____________________

____

____

23. Assume the base sequence T-A-C-C-C-A-A-A-A-C-A-T is the beginning of a gene. How many code words are there in the sequence? __________

❒ Protein Synthesis (text pages 61- 67)

24. The information in a gene that codes for the synthesis of a polypeptide is transferred in the nucleus to a molecule of ____________________, a process called ____________________ of the genetic message. This molecule then moves into the cytoplasm, where it directs the polypeptide synthesis, a process called ____________________ of the message.

25. Match the terms/processes on the right with the appropriate descriptions on the left.

______early step in transcription of a DNA message	(a) guanine
______base in mRNA that pairs with adenine	(b) RNA polymerase
______enzyme that catalyzes the splitting of H bonds in DNA	(c) splitting of H bonds in DNA
______base in mRNA that pairs with cytosine	(d) uracil
______modulator molecule for RNA polymerase	(e) DNA
______region of DNA to which RNA polymerase initially binds	(f) thymine
	(g) promoter
	(h) formation of peptide bonds

26. Describe the mechanism that keeps RNA subunits from linking up in random order.

____

____

____

27. True or false: The promoter sequence of nucleotides in a gene is present on both strands of the DNA molecule, allowing transcription of both strands.

28. A codon is a
 (a) triplet of deoxyribonucleotides.
 (b) triplet of ribonucleotides.
 (c) sequence of ribonucleotides complementary to a triplet of bases in DNA.

29. The codon that corresponds to the DNA sequence G-T-A is
 (a) G-T-A.
 (b) A-T-G.
 (c) C-A-T.
 (d) C-A-U.

30. True or false:
 (a) All the nucleotide sequences in a gene are transcribed into mRNA.
 (b) All the transcribed message of a gene is translated into amino acid sequences.
 (c) The mRNA that enters the cytosol is considerably longer than that synthesized during transcription.

31. The site of protein synthesis is organelles called ____________. These organelles can be either free in the cytosol or bound to ____________________. Proteins synthesized in the free organelles are released into the cytosol, and those synthesized in the bound organelles are ______________________________.

32. True or false: Unlike mRNA, the structure of ribosomal RNA is not coded for by DNA.

33. True or false: Ribosomal subunits are synthesized in the cytosol.

34. A functional ribosome is made up of a 40S component bound to ____________ and ________________.

35. A third kind of RNA (besides mRNA and rRNA) is called ________________. It can bind both to ____________________ in mRNA and to a specific ________________. There are at least __________ different forms of this RNA molecule present in the cytosol of all cells.

36. True or false: The nucleotide triplet in tRNA that base-pairs with a complementary triplet in mRNA is called the codon.

37. During protein synthesis,
    (a) rRNA and ribosomal proteins are required to identify the beginning of the mRNA coding sequence.
    (b) ribosomal enzymes catalyze the formation of hydrogen bonds between amino acids.
    (c) several ribosomes may move along the mRNA molecules at once.
    (d) mRNA is used up during the formation of peptide bonds.

38. True or false: Once protein synthesis is completed, the protein that was synthesized may undergo further changes prior to its secretion or use within the cell.

39. The rate of protein synthesis can be regulated at four levels:
    (a)______________________________
    (b)______________________________
    (c)______________________________
    (d)______________________________

❒ Replication and Expression of Genetic Information (text pages 67-73)

40. Each time a cell divides, it must duplicate (replicate) all of its DNA. DNA replication is similar to RNA transcription except that in DNA replication the enzyme catalyzing the process is ____________________ rather than ____________________, and the replication involves ____________________ (one/both) strand(s) of DNA rather than ____________________ (one/both) strand(s). DNA replication takes place during the period known as the ____________________ of the cell-division cycle.

41. Match the description on the left with the correct term on the right:

______	condensed, rod-shaped chromatids	(a) centromere
______	replicated strand of DNA	(b) chromosome
______	point where two replicated DNA strands are joined	(c) chromatin
______	association of DNA with protein dispersed throughout nucleus	(d) chromatid
		(e) centriole

42. Cell division consists of two separate processes: ______________________, which refers to the division of the contents of the nucleus, and ______________________, which refers to the division of the cytoplasm. The resultant cells are called ______________________.

43. The mitotic apparatus, which is necessary for the physical separation of the nuclear contents, consists of two ______________________ and their associated ______________________, which are composed of microtubules.

44. True or false: During cell differentiation, certain genetic information is lost, which accounts for the difference between, for example, a liver cell and a lung cell.

45. Any alteration in DNA structure that occurs during replication is called a ______________________. Environmental factors that increase the rate of this process are known as ______________________, and include ______________________ and ______________________.

46. True or false: The consequences of mutation are invariably harmful.

47. The inborn error of metabolism phenylketonuria (PKU) results from a hereditary defect in the gene coding for the enzyme that converts the amino acid phenylalanine to tyrosine. The normal enzymatic pathway for phenylalanine metabolism is shown below. Modify the diagram to illustrate what happens to phenylalanine in PKU, and use your diagram to explain why people who have this condition generally have light colored hair and blue eyes.

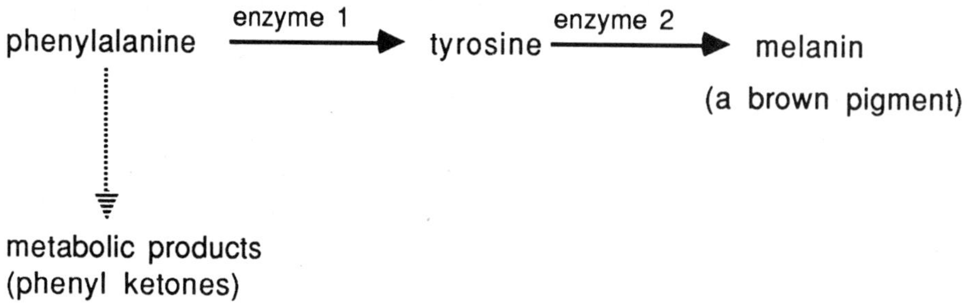

phenylalanine —enzyme 1→ tyrosine —enzyme 2→ melanin (a brown pigment)

phenylalanine ⇣ metabolic products (phenyl ketones)

____

____

____

____

48. Describe how tumors are formed, distinguishing between *benign* and *malignant* tumors. What is *metastasis*?

____

____

____

____

49. True or false: Genetic mutations can result in cancer.

50. An oncogene may
(a) be inherited.
(b) be environmentally induced.
(c) cause cancer.
(d) interfere with a cell's growth-regulatory mechanisms.

51. Construct your own gene! A restriction nuclease recognizes the nucleotide sequence GAATTC in DNA. This nuclease cleaves DNA between the G and the first A, as indicated in the diagram below. After cleavage, you are left with three pieces of DNA, two small ones and one large one. The large piece contains the gene for human insulin.

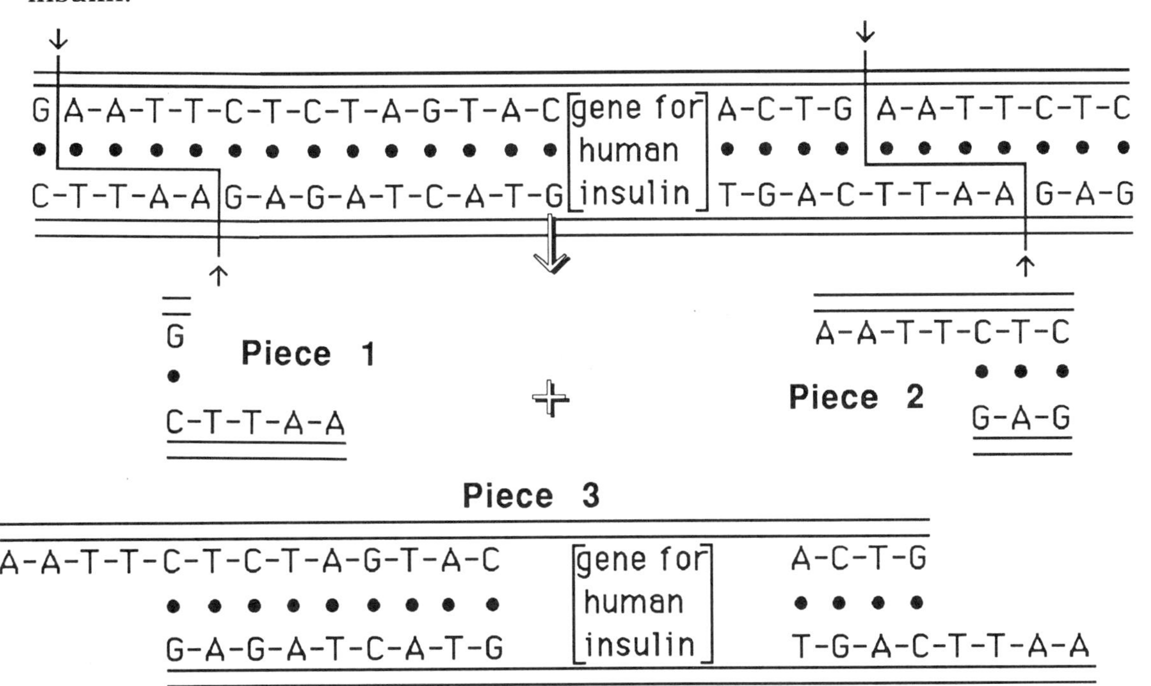

Notice that the restriction nuclease "reads" the first strand of DNA in one direction and reads the second strand in the opposite direction.

A hypothetical sequence of bacterial DNA is shown below. Indicate where the same nuclease will cleave this molecule.

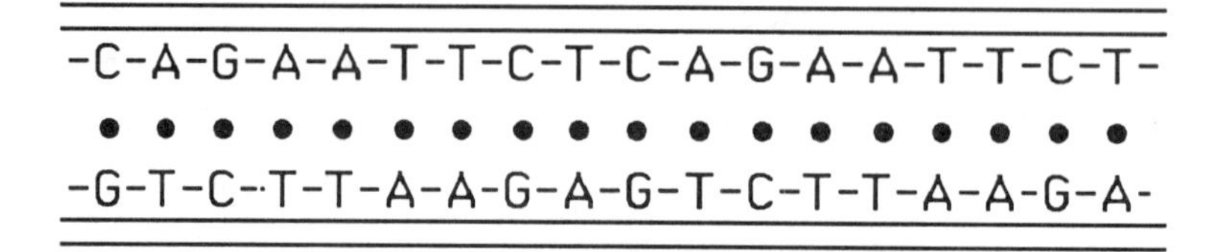

Combine Piece 3 of the human DNA with two of the bacterial fragments so that the complementary bases can pair. These complementary bases are called "sticky ends." The sugar-phosphate backbones of the two kinds of DNA are bonded together by enzymes called ________________.

52. In order to <u>express</u> (transcribe and translate) the human insulin gene, a genetic "engineer" must be sure that there is a ________________ region "upstream" of the beginning of the gene.

ANSWERS

Boldface type indicates the answers you should have given. Words in medium-face (ordinary) type explain or expand upon the answer.

1. (a) **Correct.**
   (b) **Correct.** (Since it is DNA that directs protein synthesis.)
   (c) **Correct.** (DNA *is* the hereditary information for all cells.)

2. (a) **Incorrect. The bonds are noncovalent.**
   (b) **Correct.** (Binding may also result from Van der Waals forces.)
   (c) **Incorrect. Ligands bind to proteins only.** (The importance of a ligand lies in its ability to interact with a protein — the "premier" molecules with respect to body function.)

3. (a) **Correct.**
   (b) **Correct.**

(c) **Incorrect. The amino acids that form the binding site must be juxtaposed** (close to one another in the folded-up protein), **but they do not necessarily have to be adjacent to each other on the polypeptide chain.**

4. **False.** Any given protein **may** contain binding sites for **several** ~~only one~~ kinds of ligands. (Moreover, a given binding site may bind to more than one kind of ligand.)

5. **False.** The ability of a binding site on a protein to "recognize" and bind with only certain ligands is called the ~~affinity~~ **chemical specificity** of the binding site.

6. (a) **Correct.**
   (b) **Correct.**
   (c) **Correct.**

7. **C** (highest)
   A
   B (lowest)

8. (a) **Incorrect. The degree of saturation refers to the amount of ligand, as in (b) and (c).**
   (b) **Correct.**
   (c) **Correct.**
   (d) **Correct.**

9. (a) **Incorrect. Increasing the concentration of A beyond the point at which X is 100% saturated will result in no further increase in biological effect**
   (b) **Correct.**
   (c) **Incorrect. It is A that elicits the biological effect, not B. Thus increasing the concentration of B will cause the effect to *decrease* as more B molecules are available to interfere with the binding of A.** (There is an adage to keep in mind – "A ligand binds where it acts, but it does not necessarily act where it binds." In this case, B is a *competitive antagonist* of A. Knowing about such antagonisms is very important when scientists are developing drugs for treating diseases in which, for example, the body may produce too much of a given biologically active molecule.)

10. The two general ways of controlling the activity of a protein are to change its **shape** and to regulate its **rate of synthesis**.

11. (a) **Correct.** (Allosteric proteins must, by definition, contain one regulatory site, which influences their shape, and one functional site, which when bound produces a biological effect.)

(b) **Correct.**

(c) **Incorrect. Some allosteric proteins are activated by binding of a modulator molecule, but some are inactivated.**

(d) **Incorrect. Most proteins are not allosteric.** (The minority that are allosteric are very important, however.)

12. **False.** Allosteric modulation of a protein involves ~~covalent~~ **noncovalent** binding of a ligand to a regulatory binding site. (Remember that a ligand is defined as a molecule that binds to protein by forces other than covalent bonds.)

13. The most common kind of covalent protein modulation involves the addition of a **phosphate** group, which has a net **negative** charge; the process is called **phosphorylation.**

14.

$$\text{Protein} + \text{(a) } \textbf{ATP} \xrightarrow{\text{protein kinase}} \text{(b) } \textbf{protein-PO}_4^{2-} + \text{ADP}$$

In general terms, the protein and (a) are the **substrates** of the reaction, (b) and ADP are the **products**, and protein kinase is an **enzyme**.

15. (a) **k**
(b) **p**
(c) **b** (Both the very specific kinases and the much less specific phosphatases act on many different proteins.

Whether you have many kinases:

kinase a → protein a
kinase b → protein b
kinase c → protein c
kinase d → protein d
kinase e → protein e
kinase f → protein f

Or a few phosphatases:

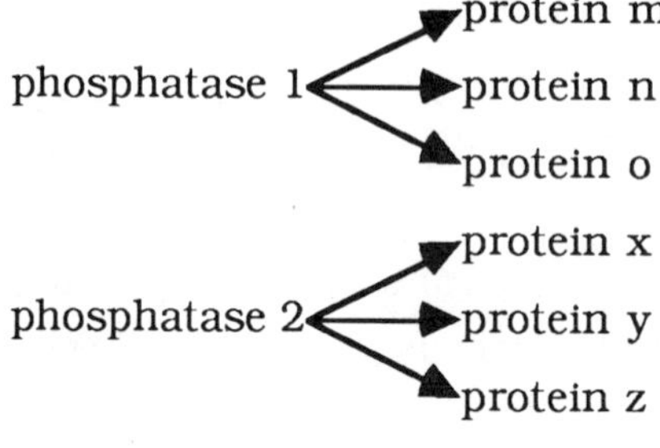

the result is the same.)

(d) **k**
(e) **p**

(f) **k**

16. (As indicated in the answer to 15-f) **protein kinases, the enzymes that catalyze protein phosphorylation** (the most common form of covalent modulation), **are allosteric. Thus, their activity is regulated by allosteric mechanisms.**

    Note: The significance of this statement is that it underscores how important the regulation of protein phosphorylation is. As we shall see in Chapter 7, protein kinase activity is vital for the functioning of a host of intercellular and intracellular messengers.

17. (a) **Correct.** (Genes are responsible for carbohydrate synthesis *indirectly*, by coding for enzymes [proteins] that mediate the synthesis. The purpose of this question is to emphasize the importance of proteins for virtually every cell function.)

    (b) **Correct.**

    (c) **Partially correct.** One gene *may* code for a protein molecule, but recall that some protein molecules are composed of more than one polypeptide chain, and thus would require more than one gene for their synthesis. (Remember: One gene equals one polypeptide chain.)

    (d) **Incorrect. Each strand of DNA** (humans have 46 strands) **contains many, many genes.**

    (e) **Correct.**

18. Most genes are located in the **nucleus** of cells, whereas protein synthesis occurs in the **cytoplasm**.

19. **False.** The gene for brown eyes encodes the structure of ~~brown pigment~~ **enzymes that mediate reactions leading to pigment synthesis.** (Pigment is not a protein. Note the difference in wording between this question and Question 17-a. Genes *contain the information necessary* for synthesis of carbohydrates, but they *encode* the structure only of proteins, including enzymes, that in turn can promote synthesis of nonprotein molecules.)

20. **Genetic information flows from DNA to RNA to protein.**

21. (a) **adenine** (c) **thymine**
    (b) **guanine** (d) **cytosine** (in any order)

22. (a) **3**
    (b) **64 (4 X 4 X 4)**

(c) **61 stand for specific amino acids.** (All but two amino acids have more than one code word; three of them have six.) **Three sequences are "stop signs," or termination sequences.**

(d) **no**

(e) **All organisms share the same genetic code. This is seen as evidence that all life forms originated from one common ancestor.**

23. **4** This DNA base sequence codes for the amino acid sequence: methionine - glycine - phenylalanine - valine (see text Figure 4-10). Note that all amino acid sequences start with "met," because the code for methionine is also the code for "start here."

24. The information in a gene that codes for the synthesis of a polypeptide is transferred in the nucleus to a molecule of **messenger (m)RNA**, a process called **transcription** of the genetic message. This molecule then moves into the cytoplasm, where it directs the polypeptide synthesis, a process called **translation** of the message.

25. **c** early step in transcription of a DNA message*
d base in mRNA that pairs with adenine
b enzyme that catalyzes the splitting of H bonds in DNA
a base in mRNA that pairs with cytosine
e modulator molecule for RNA polymerase
g region of DNA to which RNA polymerase initially binds

*The *first* step in the transcription of a DNA message is binding of RNA polymerase to the DNA promoter sequence. Only then, after the enzyme has been allosterically modulated, can it catalyze the splitting of H bonds and then the linkage of the ribonucleotides. RNA polymerase is required for both of these processes.

26. **RNA polymerase, the enzyme that catalyzes the linkage of ribonucleotides (RNA subunits), does so only when bound to DNA. This arrangement allows for base-pairing of the ribonucleotides to the complementary deoxyribonucleotides, preserving the sequence of the DNA message.**

27. **False.** The promoter sequence of nucleotides in a gene is present on ~~both strands~~ **only one strand** of the DNA molecule, allowing transcription of ~~both strands~~ **only that strand.**

28. (a) **Incorrect. By definition, a codon is the sequence of a triplet of nucleotides on RNA.**

(b) **Correct.**

(c) **Correct.** (This is a more complete answer than [b]).

29. (a) **Incorrect. The codon is complementary to the DNA sequence, not a copy of it.** "Complementary" means that base A (on RNA) base-pairs with T; U with A; C with G; G with C.
    (b) **Incorrect. Complementary does not mean "reading backwards."**
    (c) **Incorrect. RNA does not have T (thymine). Instead, U (uracil) base-pairs with A.**
    (d) **Correct.**

30. (a) **True.**
    (b) **False.** ~~All~~ **Only** the transcribed message **present in the exon regions** of a gene is translated into amino acid sequences.
    (c) **False.** The mRNA that enters the cytosol is considerably ~~longer~~ **shorter** than that synthesized during transcription (since the mRNA has undergone processing to remove the noncoding intron segments).

31. The site of protein synthesis is organelles called **ribosomes**. These organelles can be either free in the cytosol or bound to **endoplasmic reticulum**. Proteins synthesized in the free organelles are released into the cytosol, while those synthesized in the bound organelles are **secreted from the cell or transferred to other organelles**.

32. **False.** ~~Unlike~~ **Like** mRNA, the structure of ribosomal RNA is ~~not~~ coded for by DNA.

33. **False.** Ribosomal subunits are synthesized in the ~~cytosol~~ **nucleolus**.

34. A functional ribosome is made up of a 40S component bound to **mRNA** and **the 60S subunit**.

    More about ribosomes: The designations "40S" and "60S" refer to "sedimentation units," which are measures of particle mass — the larger the S number, the greater the mass. The 40S subunit is composed of one molecule of RNA and 33 protein molecules. The 60S subunit contains 3 RNA molecules and more than 40 proteins, including enzymes that mediate the peptide bonding of amino acids.

35. A third kind of RNA (besides mRNA and rRNA) is called **transfer (t)RNA**. It can bind both to **a specific codon** in mRNA and also to a specific **amino acid**. There are at least **61** different forms of this RNA molecule present in the cytosol of all cells.

    You may have answered "20" to the last question, because that is the number of kinds of amino acids present in cytosol. However, there must be at least as many different tRNA molecules as there are codons, and there are as many codons as there are triplets of DNA bases that code for amino acids — that is, 61. For example, there are four tRNAs for the amino acid glycine, because there are four codons for it: GGU, GGC, GGA, and GGG. (Verify these codons by checking the DNA triplets for glycine given on text page 60.)

Note: You may be wondering where the cytosolic amino acids that are used in protein synthesis come from. As you will learn in Chapter 5, our cells (specifically liver cells) can manufacture only some of the twenty amino acids needed for protein synthesis. We must obtain the others from protein we eat and then digest into component amino acids. These amino acids from ingested protein food are absorbed into the blood and, along with amino acids secreted by the liver, are taken up into cells by mechanisms described in Chapter 6.

36. **False.** The nucleotide triplet in tRNA that base-pairs with a complementary triplet in mRNA is called the ~~codon~~ **anticodon**.

37. (a) **Correct.** (This is the rate-limiting first step in protein synthesis.)
    (b) **Incorrect. Ribosomal enzymes catalyze the formation of peptide bonds.**
    (c) **Correct.**
    (d) **Incorrect. mRNA molecules are eventually broken down by enzymes, but during protein synthesis they can be used over and over again by multiple ribosomes.**

38. **True.** (See text Figure 4-17.)

    Note: The first amino acid of a newly synthesized protein is always methionine, because the DNA triplet that codes for methionine also serves as the signal for the beginning of transcription. Post-translational processing usually removes this amino acid, however. Thus, most functional proteins do not begin with a methionine residue.

39. (a) **Regulation of rate of transcription into mRNA**
    (b) **Regulation of post-transcriptional mRNA processing**
    (c) **Regulation of mRNA stability**
    (d) **Regulation of rate of mRNA translation**

    The importance of regulating protein synthesis should be obvious, since the only difference between, say, liver cells and lung cells is a difference in the proteins each synthesizes. Having four levels of regulation gives cells good control of synthetic processes.

    Recall, too, that the only difference between a human and a hippopotamus is a difference in genetic information, and consequently in proteins produced. And it is not merely a matter of the more evolved life forms containing greater amounts of the same genetic material. Although humans have 20 times as much DNA as fruit flies, for example, salamanders and other lower forms have many times more DNA per cell than do mammals. (Why there are these differences in amounts of DNA is one of life's mysteries — for now.)

40. Each time a cell divides, it must duplicate (replicate) all of its DNA. DNA replication is similar to RNA transcription except that in DNA replication the enzyme catalyzing the process is **DNA polymerase** rather than **RNA polymerase,** and the replication involves **both** strands of DNA rather than **one** strand. DNA replication takes place during the period known as the **interphase** of the cell-division cycle.

41. **b** condensed, rod-shaped chromatids
    d replicated strand of DNA
    a point where two replicated DNA strands are joined
    c association of DNA with protein dispersed throughout nucleus

42. Cell division consists of two separate processes: **mitosis,** which refers to the division of the contents of the nucleus, and **cytokinesis,** which refers to the division of the cytoplasm. The resultant cells are called **daughter cells**.

    Note: By convention, divided cells are referred to as "daughters" no matter which sex the organism is. Similarly, replicated DNA strands are always called "sister" chromatids.

43. The mitotic apparatus, which is necessary for the physical separation of the nuclear contents, consists of two **centrioles** and their associated **spindle fibers,** which are composed of microtubules.

44. **False.** During cell differentiation, ~~certain~~ **no** genetic information is lost, **but many genes are inactivated ("turned off") while others are activated ("turned on"). This differential activation** ~~which~~ accounts for the difference between, for example, a liver cell and a lung cell.

45. Any alteration in DNA structure that occurs during replication is called a **mutation**. Environmental factors that increase the rate of this process are known as **mutagens**, and include **chemicals** and **radiation.**

46. **False.** The consequences of mutation **may be** ~~are invariably~~ harmful, **neutral, or even beneficial.** (If no mutations had ever occurred, life on earth would consist only of the virus-like particle that was our common ancestor.)

    One can study molecular evolution by analyzing changes in the sequences of proteins along the evolutionary "tree." Such analysis shows that some proteins have changed very little in structure over the course of evolution, while others have changed a great deal. For example, proteins called *histones*, which are very important for the structure of a cell's nucleus, have not changed very much. Other proteins, however, such as the venoms of poisonous snakes, have changed markedly since they first appeared. This sort of comparison tells us that there are differences in the amount of variation that can be allowed in the composition of a protein. Histones apparently can tolerate very little variation. Since no animals exist that have significantly different histones, we can assume that changes in the histone amino acid sequence are incompatible with life. Snake venoms, on the other hand, apparently can function well even with many amino acid substitutions.

47.

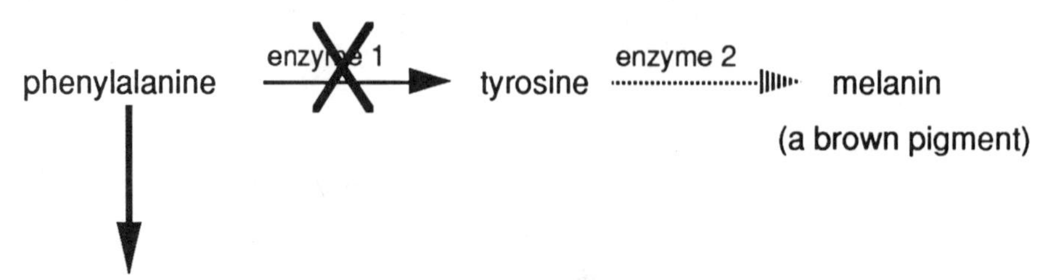

The inability to convert phenylalanine to tyrosine means that there is less than the normal amount of substrate for the conversion of tyrosine to melanin. Melanin is necessary for brown eye and hair color. Therefore, even if a PKU patient has the genes for brown eyes and hair (i.e., the gene for the enzyme that converts tyrosine to melanin), **she or he will have light-colored hair and blue eyes.** (This fact can be helpful in making the diagnosis of PKU in a child expected to have brown eyes and hair.)

Note: Metabolic products of phenylalanine, including phenyl ketones, accumulate because of the blockage of the conversion to tyrosine. These compounds interfere (in ways that are not yet understood) with maturation of the brain and are responsible for the mental retardation that results from PKU if dietary phenylalanine is not restricted.

Second note: Fortunately for victims of PKU, tyrosine is available in food. It is important not only as an amino acid building block for protein, but also for the formation of hormones and other chemical messengers. It is not ingested in sufficient quantities to be used for much conversion to melanin, however.

48. **A tumor in a tissue results from an alteration in the balance between old cells dying and new cells being produced by cell division. Because of this imbalance, the number of cells in the tissue increases. A benign** (relatively harmless) **tumor does not spread to other tissues. A malignant tumor does spread and causes cells in other tissues to become cancerous. The spread of cancer cells is called metastasis.**

49. **True.** (As the text states, many mutagenic agents are also carcinogenic [cancer-causing].)

50. (a) **Correct.** (Certain cancers tend to run in families, indicating that cancers can be inherited.)
    (b) **Correct.** (By mutagens that are also carcinogens.)
    (c) **Correct.** (This is indicated by the name "oncogene.")
    (d) **Correct.** (This is the underlying defect that results in cancer formation.)

51. You end up with three bacterial fragments:

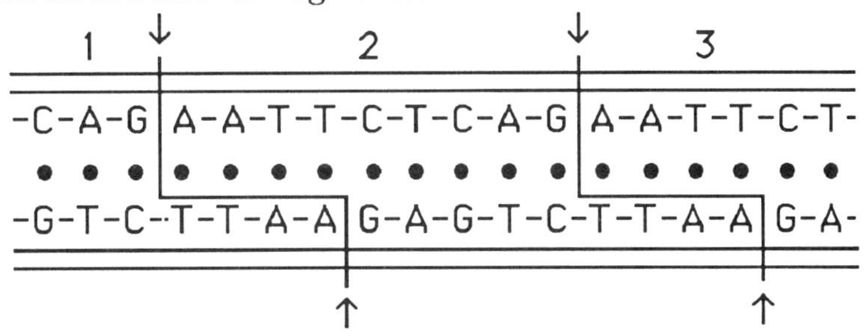

The human-bacterial combination is:

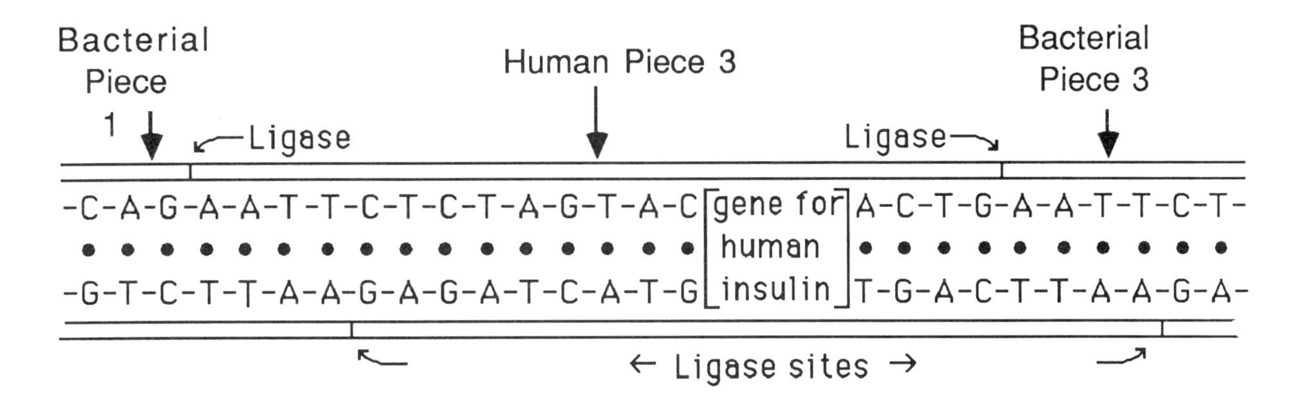

The sugar-phosphate backbones of the two kinds of DNA are bonded together by enzymes called **ligases**.

Note: The "sticky ends" of the DNA fragments are sticky only in the sense that there is a strong attraction for one sequence of complementary bases to another.

52. In order to *express* (transcribe and translate) the human insulin gene, a genetic "engineer" must be sure that there is a **promoter** region "upstream" of the beginning of the gene. (Recall that the promoter region is the region of the DNA to which RNA polymerase binds. It would do no good to splice a gene from one DNA molecule into another DNA molecule if there were no way to transcribe the gene's message.)

PRACTICE

True or false (correct the false statements):

P1. In general, the larger the number of different ligands that can bind to a particular binding site on a protein, the greater the specificity of that binding site.

P2. The shape of a ligand and the shape of its binding site influence both the specificity and the affinity of binding.

P3. The greater the ligand concentration needed to achieve 50% saturation of a particular binding site, the higher the affinity of the binding site for that ligand.

P4. The role of modulator molecules is to enhance the binding affinity of the functional site of an allosteric protein.

P5. Proteins that mediate, or catalyze, reactions in which the covalent bonding of the reactant molecules changes are called substrates.

P6. The site of protein synthesis is the endoplasmic reticulum.

P7. A ribosome is composed of one molecule of RNA and several proteins.

P8. The most likely time for a genetic mutation to occur is during mitosis.

P9. A tumor is a result of malignant, unregulated growth in a tissue.

P10. An inborn error of metabolism results from hereditary defect(s) in genes that code for enzymes.

P11. "Recombinant DNA" refers to genes from different organisms that are spliced together by hydrogen and covalent bonding.

Multiple choice (correct each incorrect choice):

P12-15. Protein kinases

P12. are enzymes.

P13. are proteins.

P14. remove phosphate groups from proteins.

P15. are allosteric.

P16-19. The genetic code

P16. consists of one code word for each amino acid.

P17. contains stop and start messages.

P18. is the same for all cells.

P19. is the same for all organisms.

P20-23. Messenger RNA synthesis from a DNA template

P20. is called translation of the message.

P21. requires DNA polymerase.

P22. is regulated.

P23. occurs in the nucleolus.

Fill in the blanks:

P24-25. The DNA sequence that codes for the codon AUG is ______________. The anticodon on tRNA that binds to the codon AUG is ______________.

P26-28. During RNA processing, nucleotide sequences corresponding to "nonsense" sequences of DNA, called ____________________, are split from the coding regions, known as __________________________ The latter are then spliced together to form ______________________.

P29-30. Mutated genes that allow cell division to escape normal regulatory mechanisms are called _______________________. Invasion of normal tissues by these unregulated cells is called ________________________.

5: ENERGY AND CELLULAR METABOLISM

A longstanding adage made popular in the 1960s by "health-food" enthusiasts states that "You are what you eat." This adage is literally true — that is, all of the molecules in our bodies (which we learned about in Chapter 2) come from the food we eat. These molecules are broken down and built up again (i.e., *metabolized*) over and over during the course of our lives, which is one reason that we must keep eating to stay alive. Another important reason to consume food is that we use food molecules for *fuel* — as sources of energy for the activities our cells perform, including synthesis of macromolecules. This chapter describes metabolism and the mechanisms our bodies use to obtain energy from the fuel we consume — food.

Instructions for answering questions in this Study Guide

1. True or false: Correct each false statement. Whenever you must correct a false statement, there will almost always be more than one way to do so. In the Guide, the answer requiring the least correction will generally be given, with the understanding that other correct answers are possible. It is usually not sufficient, in terms of demonstrating understanding, to simply insert a "no" or "not," however.

2. Multiple choice: Any, all, or none of the responses may be correct. Explain why each incorrect response is incorrect.

3. Fill in the blanks: Choose the best word or words to complete each statement.

4. Directions will be given for other types of questions as they appear.

❐ Introduction (text pages 77-78)

1. Metabolism
   (a) refers to all the chemical reactions that occur in the body.
   (b) includes the synthesis of complex molecules from simpler molecules.
   (c) includes the breakdown of complex molecules into simpler molecules.
   (d) includes anabolism and catabolism.

2. You may waken in the morning and think, "I am a new person today!" — and be right. Explain. ______________________________

   ____

   ____

❐ Chemical Reactions (text pages 78-80)

3. Consider the reaction: $H_2CO_3 \leftrightarrow CO_2 + H_2O + 4$ kcal/mol

   Which of the following statements is/are true?
   (a) The reaction is anabolic.
   (b) The reaction is catabolic.
   (c) CO_2 and H_2O are the reactants.
   (d) CO_2 and H_2O are the products.
   (e) The energy content of the reactants is greater than the energy content of the products.

4. True or false: One kilocalorie is the amount of heat energy required to raise the temperature of one gram of water one centigrade degree.

5. With regard to chemical reaction rates, which of the following statements is/are true? Explain each answer.

   (a) Increasing the concentration of the reactants will increase the reaction rate.

   ____

   ____

   ____

(b) Increasing the activation energy will increase the reaction rate.

____

____

____

(c) Increasing the temperature will increase the reaction rate.

____

____

____

6. A catalyst for a chemical reaction
   (a) lowers the activation energy of the reaction.
   (b) is modified by the reaction.
   (c) may be used to convert many molecules of reactants to products.
   (d) changes the energy content of the reactants.

7. When a chemical reaction is at equilibrium
   (a) the rate of the reaction in one direction is equal to the rate in the reverse direction.
   (b) the concentration of the reactants is equal to the concentration of the products.
   (c) the ratio of reactants to products varies according to the amount of energy released in one direction.
   (d) there may be almost total conversion of reactants to products.

8. Consider the reaction: $M + N \leftrightarrow Y + Z$

   Which of the following is/are correct? Explain.

   (a) The reaction is reversible. ______________________________________

   ____

   (b) At chemical equilibrium, increasing the concentration of M will drive the reaction to the left. ____

   ____

   (c) At chemical equilibrium, decreasing the concentration of M will drive the reaction to the left. ____

   ____

❒ **Enzymes** (text pages 80-81)

9. Enzymes
   (a) are catalysts in chemical reactions.
   (b) can be carbohydrate molecules.
   (c) are broken down during chemical reactions they catalyze.
   (d) have names generally ending in the suffix "-ace."

10. Quick review: List four enzymes you encountered in Chapter 4.
   (a) ____________________ (c) ____________________
   (b) ____________________ (d) ____________________

11. True or false: The enzyme carbonic anhydrase catalyzes the catabolism of H_2CO_3 to H_2O and CO_2.

12. True or false The active site of an enzyme is a binding site.

13. A cofactor
   (a) may alter the conformation of an enzyme.
   (b) may be a metal such as iron.
   (c) must be present in high concentrations to affect the rate of chemical reactions.
   (d) may be a substrate in a catalyzed reaction.

14. A coenzyme
   (a) is an organic cofactor.
   (b) may be a metal such as iron.
   (c) is a substrate for a catalyzed reaction.
   (d) must be present in high concentrations to affect the rate of chemical reactions.

15. True or false: An important function of coenzymes is to act as carriers to transport hydrogen atoms.

16. Vitamins
   (a) may be precursors of coenzymes.
   (b) may be synthesized by the body from precursor subunits.
   (c) are required in large amounts for metabolism.

17. True or false: Niacin is a coenzyme.

❒ **Regulation of Enzyme-Mediated Reactions** (text pages 81-83)

18. The graph below describes the rates of an enzymatic reaction as a function of increasing substrate concentration under two different conditions.

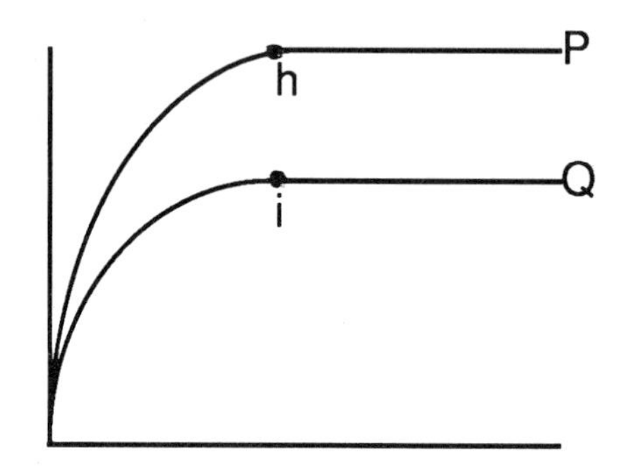

(a) Label the axes.

(b) The two curves P and Q give the reaction rate at two different ____________________ ____________________.

(c) The curve P represents ____________________ ____________________.

(d) Points h and i mark the substrate concentration at which the reactions are said to be ____________________ and will not proceed any faster.

19. The diagram below also describes the rates of a reaction catalyzed by an enzyme.

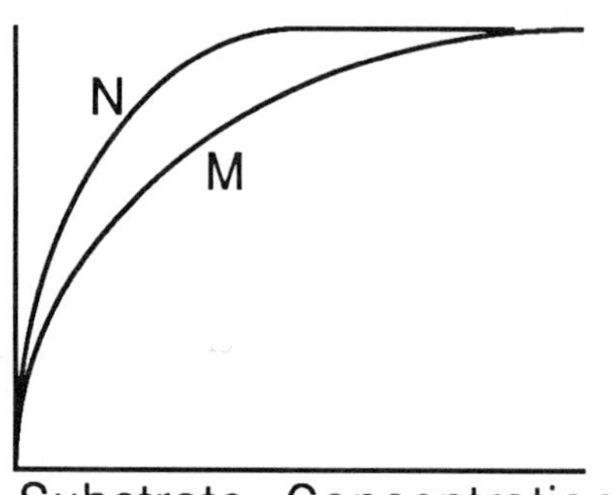

The two curves M and N depict the rates of reaction where N depicts a

(a) higher substrate concentration than M.

(b) greater enzyme concentration than M.

(c) reaction catalyzed by an enzyme that has a higher affinity for the substrate than in reaction M.

❒ **Multienzyme Metabolic Pathways** (text pages 83-84)

20. True or false: In a metabolic pathway, the same enzyme ordinarily catalyzes several different reactions.

21. True or false: The rate-limiting reaction in a metabolic pathway is one that occurs at a faster rate than the other steps in the pathway.

22. Consider the following metabolic pathway:

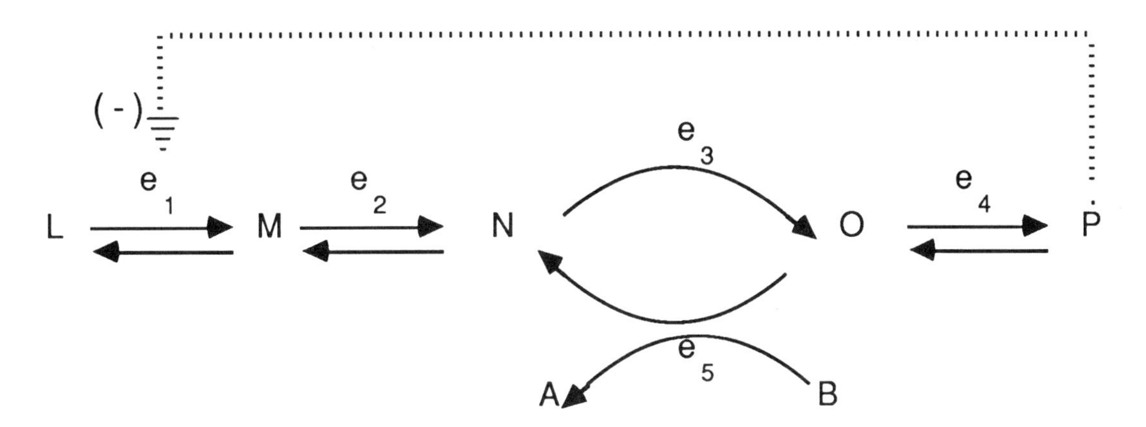

Which of the following statements is/are correct? Explain.

(a) Product P exerts end-product inhibition on enzyme e_1. ____________________

____

(b) Enzyme e_1 catalyzes a rate-limiting step in the pathway. ____________________

____

(c) The energy content of metabolite O is probably considerably greater than that of metabolite N. ____________________

____

(d) The energy content of metabolite B is probably considerably greater than that of metabolite A. ____________________

____

(e) The overall rate of this metabolic pathway, from L to P, is probably controlled at more than one step. ______________________________

____

❐ ATP and Cellular Energy Transfer (text pages 84-90)

23. ATP
    (a) is formed during the hydrolysis of ADP.
    (b) is used by cells for the storage of energy.
    (c) represents the energy "currency" of all cells.

24. List three major functions of ATP in cells.
    (a)______________________________
    (b)______________________________
    (c)______________________________

25. (a) Complete the following equation:

    ATP + H_2O $\leftrightarrow$ ADP + ____________ + ____________(energy).

    (b) If all the energy (686 kcal/mol) liberated by the catabolism of one mole of glucose to CO_2 and H_2O were transferred to ATP, how many moles of ATP would be formed?____________

    (c) In reality, only about 40% of the energy derived from the metabolism of fuel molecules can be transferred to ATP. The remainder is released in the form of ____________. This energy also is important to the body. Explain.

    ____

    ____

26. Transfer of energy to ATP occurs with the formation of a high-energy ____________ bond, a process called ____________. This process can be further described as ____________ or ____________, depending upon whether the phosphate group comes from an organic molecule or from inorganic phosphate, respectively. A further difference is that the latter process requires ____________, whereas the former does not.

27. Match each item on the left with its most appropriate description from the list on the right:

____ coenzyme	a. donor of electrons to O_2
____ 2 e^-	b. reactants in an "uphill" (energy-requiring) reaction
____ ATP	c. acceptor of hydrogen atoms during breakdown of fuel molecules
____ 1/2 O_2	
____ cytochrome	d. formed during passage of electrons down cytochrome chain
____ ADP + P_i	e. donor of hydrogen atoms during breakdown of fuel molecules
____ coenzyme-2H	f. final acceptor of electron from cytochrome chain
____ 2 H^+	g. two protons
	h. donated to O_2
	i. reactants in a reaction that releases energy

28. Cyanide poisoning causes death in a manner similar to suffocation. Explain.

____

____

____

____

____

____

29. True or false: Glycolysis refers to the breakdown of glucose to two 3-carbon molecules.

30. True or false: Substrate phosphorylation takes place in the ribosomes of cells, whereas oxidative phosphorylation takes place in mitochondria.

31. Glycolysis
   (a) does not occur in the absence of O_2.
   (b) does not occur in the presence of O_2.
   (c) may result in the formation of two molecules of lactic acid for each molecule of glucose.
   (d) results in the formation of two molecules of pyruvic acid for each molecule of glucose.

32. True or false: During anaerobic glycolysis, a net of two molecules of ATP are generated for each molecule of glucose broken down.

33. True or false: More ATP is generated from glycolytic reactions under aerobic than under anaerobic conditions.

34. The reactions of the Krebs cycle
    (a) take place in the cytosol of cells.
    (b) generate ATP directly by substrate phosphorylation.
    (c) are important for the metabolism of carbohydrates but not other fuel molecules.
    (d) require molecular oxygen.

35. True or false: In the Krebs cycle, an acetyl fragment is broken down to CO_2 and H atoms.

36. True or false: The major waste product of metabolism is H_2O.

❐ **Carbohydrate, Fat, and Protein Metabolism** (text pages 90-101)

37. Write the equation for the catabolism of one molecule of glucose to carbon dioxide and water.

38. Describe the relationship between glycolysis and the Krebs cycle.

____
____
____
____
____

39. True or false: Under aerobic conditions, metabolism of one molecule of glucose will generate 20 molecules of ATP.

40. Glucose
    (a) is the only carbohydrate utilized as an energy source.
    (b) must be phosphorylated before it can be metabolized.
    (c) can be synthesized from other types of molecules, including amino acids and triacylglycerol.

41. True or false: Glucose phosphorylation is a reversible reaction in all tissues.

42. True or false: Most of the energy stored in the body is in the form of glycogen.

43. Adipocytes are cells specialized for the synthesis and storage of molecules of ______________________________. During periods of fasting, these molecules are broken down in the adipocytes to ____________________ and ____________________, which are secreted into interstitial fluid.

44. Fatty acid catabolism
    (a) takes place in the cytosol of cells.
    (b) is initiated by the binding of a molecule of coenzyme A.
    (c) requires the breakdown of ATP.
    (d) proceeds by a process called beta-oxidation.
    (e) generates one molecule of acetyl coenzyme A for every carbon atom in the fatty acid.

45. Calculate the number of ATP molecules formed from the catabolism of one molecule of a six-carbon fatty acid. How does this compare with the amount generated by one molecule of (six-carbon) glucose?

46. Describe the significance of the fact that animals store most of their fuel in the form of fat. ____
____
____

47. Fatty acid synthesis
    (a) takes place in the cytosol of cells.
    (b) is initiated by the linking of two molecules of coenzyme A.
    (c) results in molecules with an even number of carbon atoms only.
    (d) requires more energy than is produced by the catabolism of the same fatty acid.

48. Sugar can be stored as fat, but fat cannot be stored (completely) as sugar. Explain.

    ____
    ____
    ____
    ____
    ____

49. True or false: Proteins are broken down to amino acids by enzymes called transaminases.

50. True or false: Removal of the amino group from an amino acid is necessary before the amino acid can be metabolized for energy.

51. Two methods for removing the amino group from amino acids are ______________________, which involves substituting an oxygen molecule derived from water for the amino group, and ______________________, which involves exchanging the amino group for the oxygen molecule of a keto acid. The ____________ (former/latter) reaction requires the presence of a coenzyme, whereas the other does not.

52. True or false: Pyruvic acid is a keto acid.

53. Ammonia is
    (a) a waste product of fatty acid metabolism.
    (b) excreted by the kidneys into urine.
    (c) processed by the kidneys to form a less toxic compound.

54. True or false: Human beings can synthesize all twenty amino acids the body must have.

❒ Essential Nutrients (text pages 101-103)

55. True or false: Glucose is not considered an essential nutrient because it is not essential for good health.

56. Why is the protein found in animal products considered higher quality (or more "complete") in terms of food for human consumption than the protein found in plants?

____

____

____

____

57. True or false: Fat-soluble vitamins are required for health because they act as coenzymes.

<u>ANSWERS</u>

Boldface type indicates the answers you should have given. Words in medium-face (ordinary) type explain or expand upon the answer.

1. (a) **Correct.**
   (b) **Correct.**
   (c) **Correct.**
   (d) **Correct.** (Anabolism is defined by [b] and catabolism by [c]. In general, anabolism [or anabolic reactions] requires energy — energy must be added. During catabolic reactions, energy is released and can be used for anabolism [and other energy-requiring functions], and so the two kinds of metabolic reactions are linked.)

   Unfortunately, you may be aware of the term "anabolic" in a negative context — from hearing about athletes who take anabolic steroids to increase the size and strength of their muscles. (We will come back to this topic in later chapters.)

2. As stated in the text, **"The body's composition is in a state of dynamic equilibrium." Our various macromolecules are constantly being torn down and rebuilt.**
   In adults, anabolism and catabolism normally proceed at roughly the same rate. In children, however, anabolism outpaces catabolism, and the result is body growth.

3. (a) **Incorrect. A more complex molecule is being broken down into simpler ones, and energy is released, both indicating a catabolic reaction.**

   (b) **Correct.**

   (c) **Incorrect. They are the products. The reactants are H_2CO_3.** (It may seem odd to speak of one molecule as the "reactants" of a reaction, but this phenomenon is common in catabolism.)

   (d) **Correct.**

   (e) **Correct.** (Precisely 4 kcal/mol greater.)

4. **False.** One kilocalorie is the amount of heat energy required to raise the temperature of one ~~gram~~ **kilogram** of water one centigrade degree

5. (a) **Correct. Increasing the concentration of reactants increases the chances that they will collide (with each other or with other molecules) and thus speeds up the reaction.**

   (b) **Incorrect. Activation energy is energy that must be added in order for a reaction to occur. Increasing it will decrease the likelihood that the reactants can acquire the necessary energy, and thus slow the rate of reaction.**

   One common misunderstanding about chemical reactions is confusing activation energy with either the energy required for anabolic reactions to proceed or the energy that is released during catabolism. These latter energies, which are often expressed as part of chemical equations, are consequences of the relative energy contents of the products and reactants. The following example may help to clarify the meaning of the different kind of energies:

   A boulder on the top of a hill possesses *potential* energy directly related to the height of the hill. If the boulder is pushed over the edge, it will roll down the hill and the potential energy will be released as *kinetic* energy. The higher the hill, the more kinetic energy will be released as the boulder rolls. At the bottom, the boulder will no longer possess as much potential energy as it did at the top of the hill. Thus, its energy content will be less than it was before (analogous to the products in a catabolic reaction). The "push" required to start the boulder rolling is the activation energy, which is unrelated to the height of the hill and thus unrelated to the amount of kinetic energy dissipated during the roll. Getting the boulder back up the hill would require external energy (for example, from the gasoline burned by a helicopter that could hoist the boulder). Some of this energy expended would then be stored in the boulder as potential energy, a situation analogous to an anabolic reaction. In this hoisting example, ignition of the helicopter fuel by a spark from a battery would be the activation energy.

   (c) **Correct. Increasing the temperature increases the reaction rate because it makes molecules move faster and thus enhances both the force of**

collisions between molecules (which impart energy to them) and the probability that two molecules will collide.

Heating is one way of giving chemical reactants sufficient activation energy to react. Ordinarily, for humans and most other warmblooded animals (mammals and birds), body temperature is kept almost constant (as we shall see in Chapter 7), and so changing temperature to increase or decrease the likelihood of reactions occurring is not an option. However, we are all aware that some animals hibernate in the winter. They lower their body temperature to such an extent that many metabolic reactions can occur only very slowly. Thus, these animals are able to exist for months without "refueling" (eating food — our source of energy for anabolism).

Humans do not hibernate, of course, but recent research has shown that artificially lowering body temperature during some surgical procedures allows the surgeon more time to perform delicate operations without harming any sensitive body tissues that require a constant source of fuel.

6. (a) **Correct.**

   (b) **Incorrect. The catalyst is not chemically altered by the reaction** (although it may be changed transiently as it interacts with the reactants).

   (c) **Correct.**

   (d) **Incorrect. The energy content of the reactants and products depends only on their chemical composition, which is not changed by the catalyst.** (Keep in mind the difference between activation energy and the energy content of molecules.)

7. (a) **Correct.**

   (b) **Incorrect. The concentrations will be equal only if there is no difference in the energy content of reactants and products** (an unusual occurrence).

   (c) **Correct.** (The greater the energy released in one [the "forward"] direction, the less chance there is of the "reverse" reaction occurring, and the greater will be the ratio of products to reactants.)

   (d) **Correct.** (This occurs when a large amount of energy is released by the reaction — i.e., in an irreversible reaction.)

8. (a) **Correct. The double arrow signifies that the reaction is reversible.**

   (b) **Incorrect. By the law of mass action, increasing the concentration of a reactant will drive the reaction to the right, increasing formation of product until a new equilibrium is reached.**

   (c) **Correct. Also by the law of mass action, decreasing the concentration of a reactant will drive the reverse reaction, just as if one had increased the concentration of a product.**

9. (a) **Correct.**

   (b) **Incorrect. Enzymes are protein catalysts, by definition.** (And no carbohydrates acts as a catalyst in a chemical reaction.)

   (c) **Incorrect. They are not broken down during the reactions they catalyze and so can catalyze the conversion of many reactant molecules to products.** The text mentions that one molecule of carbonic anhydrase can cause the breakdown of 100,000 molecules of carbonic acid each second. (Of course, enzyme molecules are eventually metabolized themselves, as occurs with all molecules.)

   (d) **Incorrect. The suffix is "-ase"** (although it is usually pronounced "ace").

10. (a) **Protein kinase** (c) **RNA polymerase**

    (b) **Phosphoprotein phosphatase** (d) **DNA polymerase**

    Also: **Restriction nuclease; ligase; aminoacyl-tRNA transferase**

11. **True.** (It also catalyzes the reverse reaction, which is generally true for any enzyme-mediated reversible reaction.)

12. **True.**

13. (a) **Correct.**

    (b) **Correct.** (Iron is one of the several metals needed in trace amounts in our diets because they function as cofactors. Iron also has other important functions involving the transport of oxygen and carbon dioxide in the blood, as we shall see in Chapter 14).

    (c) **Incorrect. As indicated in (b), cofactors are needed in small amounts.**

    (d) **Correct.** (In this case, they are called "coenzymes.")

14. (a) **Correct.**

    (b) **Incorrect. The mineral cofactors are not coenzymes.**

    (c) **Correct.** (Note the difference in wording in Questions 13d and 14c.)

    (e) **Incorrect. Because they are "recycled," coenzymes need not be present in high concentrations even though they function as substrates.**

15. **True.**

16. (a) **Correct.** All coenzymes are derived from vitamins, but not all vitamins form coenzymes. A general definition for a vitamin is "an organic molecule that (1) is necessary for life and health, (2) must be consumed in the diet, and (3) does not function by supplying energy" (as does most of what we think of as "food").

(b) **Incorrect. Part of the definition of "vitamins" is that they must be supplied in the diet: the body cannot make them.**

(c) **Incorrect. As in answers to question 13 and 14, coenzymes (and the vitamins from which they are derived) need not be present in large quantities.**

The Food and Nutrition Board of the National Research Council has published vitamin tables listing minimum daily amounts needed to maintain health; the recommendations range from a few micrograms to a few milligrams, depending on the vitamin. As we shall see later in this chapter, ingesting large quantities of certain vitamins is harmful, not helpful, to health.

17. **False.** Niacin is **a vitamin that is the precursor of** a coenzyme.

18.

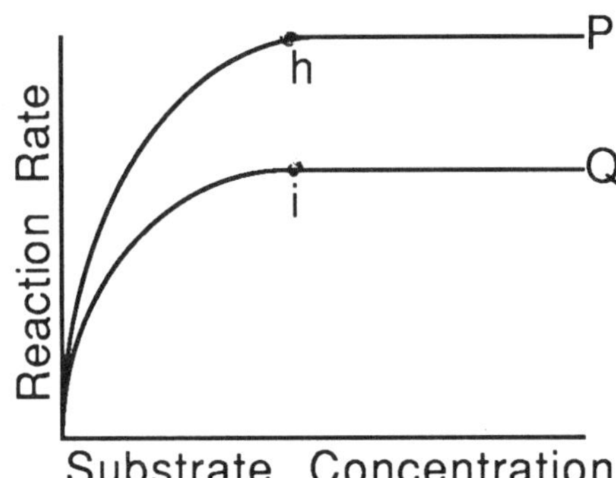

(a) **See graph**. Proper labeling of the axes of a graph is simple if you keep in mind that the *dependent variable* (that is, the property that is the function of, or dependent upon, something else, in this case the rate of reaction) is plotted on the vertical (Y) axis. The *independent variable* (here, the concentration of substrate) is plotted on the horizontal (X) axis.

(b) **enzyme concentrations**

(c) **the reaction rate at twice the enzyme concentration as in Q**

(d) **saturated**

19. (a) **Incorrect. Each curve depicts increasing reaction rate with increasing substrate concentration.**

(b) **Incorrect. This situation is pictured in the graph for question 18, which shows that increasing the enzyme concentration increases the maximal reaction rate (not the case here).**

(c) **Correct.** (Recall that affinity of an enzyme can be modified either allosterically or covalently.)

20. **False.** In a metabolic pathway, ~~the same enzyme~~ **different enzymes** ordinarily catalyze~~s several~~ different reactions. (Enzymes are generally quite specific for substrate.)

21. **False.** The rate-limiting reaction in a metabolic pathway is one that occurs at a ~~faster~~ **slower** rate than the others in the pathway.

22. (a) **Correct.** **Inhibition of e_1 by P is indicated by the dashed line in the diagram.**
    (b) **Correct.** **This can be deduced from the fact that product P inhibits e_1.**
    (c) **Incorrect.** **The reverse is true. According to the diagram, O can be converted to N only if considerable energy is supplied, in this case by the conversion of B to A.**
    (d) **Correct.** (See explanation to "c".)
    (e) **Correct.** **The diagram shows that the conversion of L to M is controlled. Furthermore, one can infer that enzymes e_3 and e_5 are also regulated, since such linked reactions, involving relatively large amounts of energy, are common sites of regulation.**

23. (a) **Incorrect.** **The opposite is correct: ADP is formed during the hydrolysis of ATP.**
    (b) **Incorrect.** **ATP is used by cells to transfer chemical energy either (1) from one molecule to another or (2) to an energy-requiring process.** (Storage of energy is the function of "fuel" molecules — primarily carbohydrates and fat.)
    (c) **Correct.** All cells "spend" energy in the form of ATP, and all fuel molecules are sources of energy for generating ATP. (Remembering the text's analogy of ATP to quarters needed to feed the energy-requiring "meters" of cells is useful.)

24. (a) **ATP provides energy for the synthesis of large molecules.**
    (Keep in mind that this synthesis is done in small steps, just as large molecules are broken down in small steps.)
    (b) **ATP provides energy for membrane active transport.**
    (The transport of molecules across membranes against the concentration gradient of the molecules is vital for cell functions and requires energy in the form of ATP, as you shall see in Chapter 6.)
    (c) **ATP provides energy for cell movement.**
    (Muscle cells require a great deal of energy for contraction, as you shall learn in Chapter 11.)

25. (a) ATP + H_2O $\leftrightarrow$ ADP + $\mathbf{P_i}$ + **7 kcal/mol**

(b) 686 kcal ÷ 7 kcal/mol = **98 mols** (This answer depends upon understanding that the breakdown of ATP is, as indicated by the double arrow, reversible and that the same amount of energy must be supplied to form ATP as is generated by its hydrolysis.)

(c) **heat. Metabolism provides the heat necessary to keep our body temperature stable and optimal** (and generally higher than ambient temperature).

26. Transfer of energy to ATP occurs with the formation of a high-energy **phosphate** bond, a process called **phosphorylation.** This process can be further described as **substrate phosphorylation** or **oxidative phosphorylation**, depending upon whether the phosphate group comes from an organic molecule or from inorganic phosphate, respectively. A further difference is that latter process requires **oxygen**, whereas the former does not.

27. **c** coenzyme
h 2 e^-
d ATP
f $1/2\ O_2$
a cytochrome
b ADP + P_i
e coenzyme-2H
g 2 H^+ (Recall from Chapter 2 that hydrogen ions are protons.)

28. **Suffocation causes death from lack of oxygen, which is necessary for the generation of ATP by oxidative phosphorylation. Without ATP from this source, (most) cells cannot continue to function. Cyanide interferes with the transfer of electrons from the cytochromes to oxygen and thus also blocks ATP production.**

29. **True.**

30. **False.** Substrate phosphorylation takes place in the ~~ribosomes~~ **cytosol** of cells, whereas oxidative phosphorylation takes place in mitochondria.

31. (a) **Incorrect. Anaerobic glycolysis takes place in the absence of air (O_2).**

(b) **Incorrect. Aerobic glycolysis takes place in the presence of air (O_2).**

(c) **Correct.** (In the absence of O_2.)

(d) **Correct.** Pyruvic acid is an inevitable product of glycolysis. Under anaerobic conditions, pyruvic acid accepts two hydrogen atoms from coenzyme-2H (which regenerates the coenzyme), forming lactic acid. (You may be interested to know that in yeast and certain other non-animal cells, pyruvic acid forms ethanol instead of lactic acid under anaerobic conditions, resulting in the formation of wine and beer.) Under aerobic conditions, pyruvic acid is the endpoint of the reactions referred to as glycolysis.

32. **True.** (Four molecules of ATP are generated, but two of those are "spent," for a net gain of two. Note in text Figure 5-12 that the first ATP is used to phosphorylate glucose to glucose 6-phosphate. This reaction occurs as soon as glucose enters a cell.)

33. **True.** (Even though the same net amount of ATP is formed via substrate phosphorylation under both conditions, H atoms accepted by the coenzyme at step 6 [see text Figure 5-12] can be donated to O_2 by way of the cytochrome chain when O_2 is present, forming three more molecules of ATP per molecule of glucose. You should understand that "glycolysis" is not synonymous with "substrate phosphorylation.")

34. (a) **Incorrect. They take place in the intramitochondrial fluid.** (Recall that the cytochromes are embedded in the inner mitochondrial membrane; this arrangement allows for convenient "processing" of the hydrogen derived from Krebs-cycle reactions.)

    (b) **Correct.** (One molecule per turn of the cycle.)

    (c) **Incorrect. Proteins and fats are also broken down to acetyl fragments that are carried to the Krebs cycle by coenzyme A.** The Krebs cycle represents a "final common pathway" for obtaining useful energy from all fuel sources for most cells.

    (d) **Correct.** (Although O_2 does not participate directly in Krebs-cycle reactions, it is the final acceptor for the H atoms that are generated in the cycle and transported to the cytochromes by coenzymes. If O_2 is not present, the coenzymes carrying 2H cannot be regenerated and the cycle will stop.)

35. **True.**

36. **False.** The major waste product of metabolism is ~~H_2O~~ **CO_2.** (While one could say that water is a major *end* product of metabolism, it is certainly not a waste product. Water is as necessary for life as ATP and is more immediately crucial for survival than any food source. In order to have enough, most animals must drink water as well as generate it during metabolism, but some desert-dwelling animals [such as the kangaroo rat] have evolved water-conservation mechanisms that allow them to make do with end-metabolic water only. You will be introduced to these mechanisms in Chapter 15.)

37. **$C_6H_{12}O_6 + 6\ O_2 \rightarrow 6\ H_2O + 6\ CO_2 + 686$ kcal/mol**

38. **Under aerobic conditions (i.e., in the presence of oxygen) the end product of glycolysis is pyruvic acid. This molecule enters mitochondria from the cytosol and reacts with coenzyme A, forming acetyl coenzyme A. This molecule then enters the Krebs cycle, combining with oxaloacetic acid to form citric acid (and liberate coenzyme A).**

39. **False.** Under aerobic conditions, metabolism of one molecule of glucose will generate ~~20~~ **38** (or 36; see note on text page 93) molecules of ATP.

40. (a) **Incorrect. Other sugars** (such as galactose and fructose) **can be broken down by glycolytic enzymes after being converted to glucose or other molecules in the glycolysis pathway.** (Note that fructose 6-phosphate is one intermediate in the pathway — see text Figure 5-12.)

    (b) **Correct.** (Note that this statement applies both to glucose breakdown and to glucose storage as glycogen. Recall that metabolism covers both anabolic and catabolic reactions.)

    (c) **Correct.** Several amino acids can be converted to Krebs-cycle intermediates and can enter the glucose synthetic pathway (the glycolytic pathway in reverse) via oxaloacetic acid. The glycerol from triacylglycerol can also be converted to glucose (although the fatty acid parts of the molecule cannot be). Why are these interconversions important? Some tissues, most notably nervous tissues, require glucose specifically for fuel — nothing else will do. For this reason, the body carefully regulates levels of glucose in the blood and stimulates synthesis of glucose by the liver when levels fall. This regulation is the subject of Chapter 17. (Note that there is no pathway from carbon dioxide [and water — not shown] back to glucose in Figure 5-19. Only plants, with the aid of the sun, can synthesize glucose from the "raw materials" — a process called *photosynthesis.*)

41. **False.** Glucose phosphorylation is a reversible reaction in ~~all tissues~~ **the liver only.** (Although all tissues can phosphorylate glucose, in order to metabolize it, only the liver has the enzyme that removes the phosphate group. The significance of this fact is that only free [that is, unphosphorylated] glucose can leave a cell. Thus, the liver is the only organ that can secrete glucose from its cells.

42. **False.** Most of the energy stored in the body is in the form of ~~glycogen~~ **fat** (triacylglycerol). (Only about 1% is stored as carbohydrate — see Table 5-5.)

43. Adipocytes are cells specialized for the synthesis and storage of molecules of **triacylglycerol**. During periods of fasting, these molecules are broken down in the adipocytes to **fatty acids** and **glycerol**, which are secreted into the interstitial fluid.

44. (a) **Incorrect. It takes place in the mitochondria.**

    (b) **Correct.**

    (c) **Correct.** (The initial binding of coenzyme A requires energy. Recall that, in this case, ATP is degraded to AMP rather than ADP.)

(d) **Correct.** ("Oxidation" refers either to the addition of oxygen or to the removal of hydrogen from a molecule.)

(e) **Incorrect. One molecule of acetyl CoA is generated for every pair of carbon atoms.** (An acetyl fragment has two carbons.)

45. **One acetyl CoA per pair C : 6 C ÷ 2 = 3 pairs C = 3 acetyl CoA;**

12 ATP per acetyl CoA: 3 acetyl CoA x 12 ATP per acetyl CoA = 36 ATP

2 coenzyme—2H per (pairs C - 1) : 2 coenzyme—2 H per (3 pairs - 1) = 2 *pairs* coenzyme—2H

5 ATP per pair coenzyme—2H : 5 ATP x 2 pairs coenzyme —2H = 10 ATP

Subtract 2 ATP equivalents. Total = 36 + 10 - 2 = *44 ATP*

(compared to 38 ATP maximum for glucose).

Note that the molecular mass of a 6-carbon fatty acid is considerably less than that of glucose because there are four fewer O molecules in the fatty acid:

$CH_3(CH_2)_4COOH$:

6 x 12 =	72
12 x 1 =	12
2 x 16 =	32
Molecular mass =	116

Therefore, the number of ATP molecules generated per unit molecular mass in this example is 44/116 = 0.38, compared to 38/180 = 0.21 for glucose. (The number of ATP molecules generated per unit molecular mass of fatty acid catabolized varies with the size of the fatty acid — it increases as chain length increases. The example in the text of an 18-C fatty acid would give 146/284 = 0.51 ATP molecules per unit mass.)

46. (As Question 45 shows) **fat can store more energy per unit weight than carbohydrates. Thus, fat is more economical as a fuel-storage molecule than is glycogen.** (For the same reason, fat in the diet gives more calories per unit weight than carbohydrate, a fact that is all too familiar to dieters.)

47. (a) **Correct.**
(b) **Correct.**
(c) **Correct.**
(d) **Correct.**

48. **The breakdown of glucose (sugar) results in the formation of acetyl coenzyme A molecules, the precursors of fatty acids. Glycerol, which is also necessary for triacylglycerol synthesis, is formed from glucose intermediates during glycolysis. Thus, glucose molecules can be converted to fat molecules. The reverse is not true because fatty acids cannot enter the (reverse) glycolytic pathway; the step converting**

pyruvic acid to acetyl CoA is not reversible. Glycerol from triacylglycerol catabolism can be used to synthesize glucose, however.

49. **False.** Proteins are broken down to amino acids by enzymes called ~~transaminases~~ **proteases**.

50. **True.**

51. Two methods for removing the amino group from amino acids are **oxidative deamination**, which involves substituting an oxygen molecule derived from water for the amino group, and **transamination**, which involves exchanging the amino group for the oxygen molecule of a keto acid. The **former** reaction requires the presence of a coenzyme, whereas the other does not. (The coenzyme is necessary to bind the hydrogen atoms that are released from the water molecule.)

52. **True.** (As such, it can undergo transamination to form the amino acid alanine.)

53. (a) **Incorrect. It is a waste product of amino acid metabolism.**
    (b) **Incorrect. Most ammonia is converted to urea before it is excreted.**
    (c) **Incorrect. This processing (to urea) occurs in the liver.**

54. **False.** Human beings can synthesize ~~all twenty~~ **about half of the** amino acids the body must have. (Nine are termed "essential" because they must be eaten in the diet.)

55. **False.** Glucose is not considered as essential nutrient ~~because~~ **even though** it is ~~not~~ essential for good health **because it can be synthesized in the body in adequate amounts.** (The same is true for the amino acids the body can synthesize — they are "nonessential" only because they don't have to be consumed.)

56. **Animal protein contains all the essential amino acids in the proper ratios for humans to utilize in synthesizing our own proteins. Few plants contain such nicely balanced proteins.**

    Many plant proteins lack one or more of the essential amino acids. In order to have a healthful diet that is strictly vegetarian, one must know which vegetables to combine in order to obtain all the essential nutrients. For example, legumes, such as beans and peas, lack tryptophan and methionine, whereas most cereals, such as wheat and rice, supply those amino acids but lack lysine and isoleucine. Therefore a combination (say, of beans and rice, popular in Mexican food) is necessary.

    Humans cannot have a strictly vegetarian diet (that is, one utilizing no animal products at all, including eggs and dairy products) and remain healthy unless they supplement that diet with synthetic cyanacobalamin (vitamin B_{12}). This vitamin is not found in plants, and lack of it leads to the disease pernicious anemia, which is fatal if untreated (and which you will learn more about in Chapter 13).

57. **False.** ~~Fat~~ **Water**-soluble vitamins are required for health because they act as coenzymes.

PRACTICE

True or false (correct the false statements):

P1. The activation energy of a reaction is often given in the chemical equation for the reaction.

P2. Coenzymes are organic cofactors.

P3. Allosteric inhibition of rate-limiting enzymes is a common means of regulating anabolic pathways.

P4. During oxidative phosphorylation, hydrogen atoms are passed serially from a coenzyme down a chain of molecules called cytochromes.

P5. In the absence of oxygen, cells can derive energy to form ATP from the metabolism of fatty acids.

P6. The first step in glycogen synthesis is the phosphorylation of glucose.

P7. Glucose cannot be synthesized from fatty acids because the reaction converting pyruvic acid to acetyl coenzyme A is not reversible.

P8. Although amino acids can be metabolized to form glucose, glucose cannot be metabolized to form amino acids.

P9. An essential nutrient is one necessary for good health that may or may not be synthesized by the body.

Multiple choice (correct each incorrect choice):

P10 -13. The rate of a chemical reaction may be influenced by

P10. the concentration of substrate

P11. the rate of synthesis of the enzyme catalyzing the reaction.

P12. covalent modulation of the enzyme catalyzing the reaction.

P13. the concentration of products.

P14 -17. Oxidative phosphorylation

P14. occurs in the cytoplasm of cells.

P15. requires the presence of unoccupied coenzymes.

P16. requires phosphorylated organic molecules.

P17. produces the great majority of useable energy for cell functions.

Fill in the blanks:

P18. Increasing the concentration of a substrate for a reaction will increase the rate of that reaction until the point of ________________________ is reached.

P19. The slowest reaction in a metabolic pathway is called the ________________________ reaction.

P20-21. ATP is formed from the ________________________ of ADP. This reaction requires ________________________ kcal/mol of energy.

P22. The series of reactions that result in ATP formation in the absence of oxygen is called ________________________.

P23-26. In the Krebs cycle, a molecule of ________________________ reacts with a molecule of oxaloacetic acid to form ________________________. At later stages in the cycle, ________________________ atoms are transported to coenzymes to react ultimately with oxygen, while molecules of ________________________ are generated as waste products.

P27-28. Most of the body's energy is stored in the form of ________________________ in ________________________ cells.

P29-30. The nitrogen from metabolized amino acids is excreted from the body as ________________________ in ________________________.

6: MOVEMENT OF MOLECULES ACROSS CELL MEMBRANES

In cystic fibrosis, an inherited and as yet incurable disease, the fluid secreted by the lungs and digestive tract is abnormally thick. Until the development of antibiotics, most victims of this disease died in infancy because their weakened lungs were easy targets for the microorganisms that cause pneumonia. Those patients who escaped that fate failed to thrive because they were unable to digest their food. Today, enzyme tablets that replace faulty digestive secretions allow the affected children to grow, many to adulthood, until they finally succumb to lung deterioration. Recently, the cause of the disease was discovered to be an abnormality in the protein that normally forms a channel that allows chloride ions to enter and leave cells. Thus, cystic fibrosis is caused by a genetic defect — a mutation in the gene that codes for the protein that forms chloride channels.

In this chapter you will learn how proteins and other components of cell membranes allow ions and other molecules to move back and forth across the membrane. The tragedy of cystic fibrosis is just one testament to the importance of understanding the details of membrane functions.

Instructions for answering questions in this Study Guide

1. True or false: Correct each false statement. Whenever you must correct a false statement, there will almost always be more than one way to do so. In the Guide, the answer requiring the least correction will generally be given, with the understanding that other correct answers are possible. It is usually not sufficient, in terms of demonstrating understanding, to simply insert a "no" or "not," however.

2. Multiple choice: Any, all, or none of the responses may be correct. Explain why each incorrect response is incorrect.

3. Fill in the blanks: Choose the best word or words to complete each statement.

4. Directions will be given for other types of questions as they appear.

❐ Introduction (text pages 107-108)

1. Plasma membranes are
   (a) responsible for the differences in composition between the extracellular fluid and the cytosol.
   (b) impermeable barriers.
   (c) permeable to some molecules and impermeable to others.

2. Quick review: Cell membranes are composed of a bimolecular layer of ____________________ molecules oriented such that their ________________ (polar/nonpolar) regions face the membrane surfaces and their ________________ (polar/nonpolar) regions lie in the middle. Associated with these molecules are ____________________________ that can form channels in the membrane.

❐ Diffusion (text pages 108-113)

3. Diffusion
   (a) depends upon the random motion of molecules.
   (b) results in net movement of molecules from regions of low concentration to regions of high concentration.
   (c) is important for moving molecules over large distances in the body.

4. Diffusion across any surface can be quantified by determining the flux, which is defined as ____ ___________________________. The _____________________ is the difference between two one-way fluxes. When this difference becomes zero, the system is said to be in ____________________________________.

5. True or false: The greater the concentration difference of a substance between two volumes separated by a permeable surface, the lesser the magnitude of the net flux of the substance.

6. True or false: At any given concentration difference, the speed of diffusion across a permeable surface will be greater for small molecules than for larger ones.

7. Consider the equation describing the net flux (F) of a substance (Z) across a plasma membrane:

$$F_Z = k_p A (C_o - C_i)$$

According to the equation, F_Z is proportional to A, which stands for (a) ______________________; to $(C_o - C_i)$, which refers to (b) ______________________; and to k_p, the (c) ______________________. What does a positive value for F indicate? (d) ______________________. A negative value? (e) ______________________. What might a value of zero indicate? (f) ______________________.

8. True or false: In general, polar molecules diffuse more rapidly across cell membranes than nonpolar molecules.

9. True or false: The component of the plasma membrane that acts as the selective barrier for diffusion of polar and nonpolar molecules is the integral proteins.

10. The permeability of the plasma membrane to mineral ions
    (a) is dependent upon channels formed by proteins.
    (b) varies from one cell to another.
    (c) is affected by differences in electrical charge on the two sides of the membrane.
    (d) can be regulated by the cell.

11. List the three factors that can open Na^+ channels in a plasma membrane.
    (a) ______________________
    (b) ______________________
    (c) ______________________

❐ **Mediated-Transport Systems** (text pages 113-121)

12. The mediated transport of a substance across a plasma membrane
    (a) depends upon binding of that substance to a specific site on the membrane protein.
    (b) depends upon movement of proteins from one side of the membrane to the other.
    (c) increases in direct proportion to the increasing concentration of the substance on one side of the membrane.

13. True or false: Mediated transport is required in order for glucose, amino acids, and fatty acids to pass into cells because these substances cannot diffuse through plasma membranes.

14. Graphs A and B depict two ways of determining how molecules (in this case, molecules X, Y, and Z) can gain entry into a cell: (A) Increase the extracellular concentration (C_o) of the molecule and measure flux and (B) keep C_o constant and measure the intracellular concentration (C_i) of the molecule over time until equilibrium is reached.

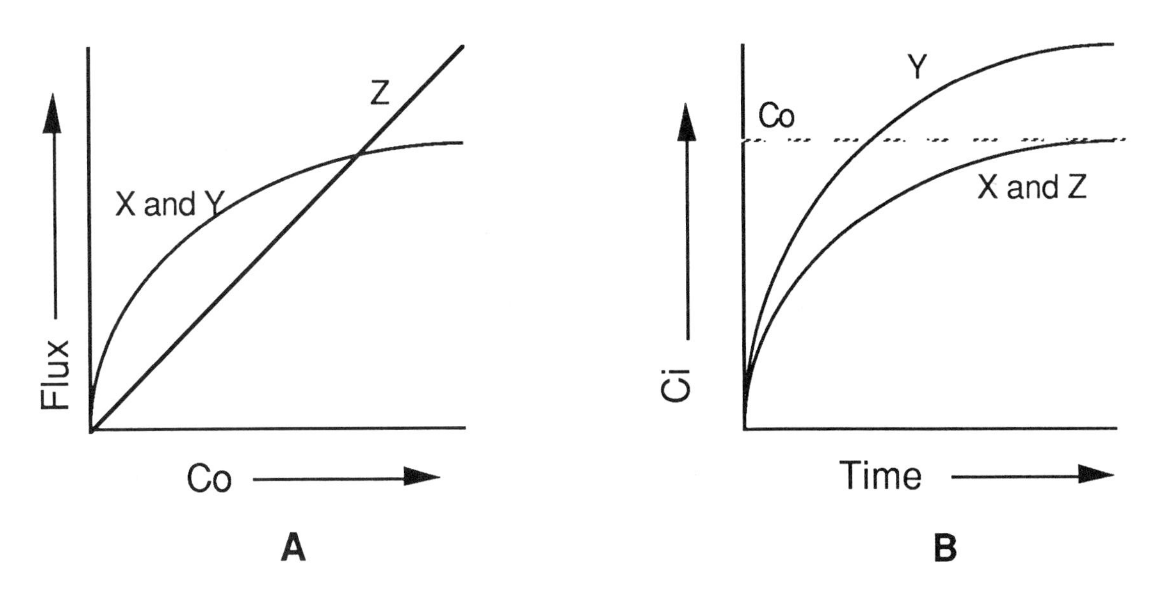

Which molecule enters by simple diffusion? (a) ____; by facilitated diffusion? (b) ____; by active transport? (c) _______

15. Glucose
    (a) enters most cells by simple diffusion.
    (b) enters most cells by active transport.
    (c) concentrations inside most cells are generally the same as those in the extracellular fluid.

16. True or false: In active transport, the affinity any given binding site has for the molecule to be transported changes as the site goes from facing one side of the membrane to facing the other side.

17. Primary active transport differs from secondary active transport in that the former uses
    (a) energy derived from ATP whereas the latter does not.
    (b) a carrier that is an ATPase whereas the latter does not.
    (c) a carrier that is phosphorylated whereas the latter does not.
    (d) a carrier whose binding affinity for the solute is modified by covalent modulation whereas the latter does not.

18. True or false: Because of the active transport of Na^+ and K^+, the intracellular concentration of Na^+ is lower than the extracellular concentration, whereas the reverse is true for K^+.

19. True or false: The Na,K-ATPase carrier transports sodium ions out of cells and potassium ions into cells on a one-to-one basis.

20. In secondary active-transport systems in which Na ions are bound to carriers, the
    (a) Na ions stay on one side of the membrane.
    (b) actively transported solute always follow Na ions across the membrane.
    (c) actively transported solute always moves across the membrane in the direction opposite of Na ion movement.

21. Calcium ions move across cell membranes by
    (a) simple diffusion.
    (b) primary active transport.
    (c) secondary active transport.

❐ **Osmosis** (text pages 121-127)

22. Water molecules
    (a) diffuse across the lipid regions of cell membranes.
    (b) diffuse through ion channels of cell membranes.
    (c) are actively transported across cell membranes.

23. Define "osmosis."

____

____

____

24. True or false: Adding one mole of NaCl to one liter of water will lower the water concentration twice as much as adding one mole of glucose.

25. True or false: Adding one gram of NaCl to one liter of water will lower the water concentration twice as much as adding one gram of glucose. (Hint: The molecular mass of NaCl is 58.5; that of glucose is 180.)

26. True or false: The higher the osmolarity of a solution, the higher the concentration of water in it.

27. The bottom of a tube with rigid walls is divided by a membrane that is permeable to water but not to solute:

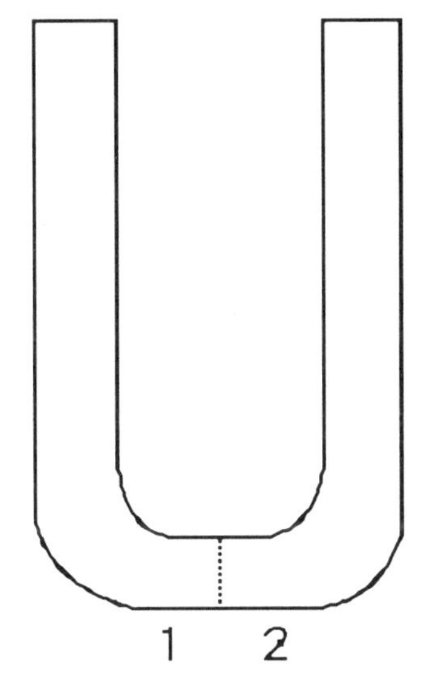

A liter of a 1 *M* NaCl solution is poured into arm 1 while at the same time a liter of a 1 *M* glucose solution is poured into arm 2. At equilibrium, what will be the osmolarity of the solutions in each arm?

(a) ________________ (arm 1)

(b) ________________ (arm 2)

What will be the volume in each arm? (Do not consider the effects of gravity on the tube.)

(c)________________ (arm 1)

(d)________________ (arm 2)

(e) Are Na^+ and Cl^- penetrating or nonpenetrating solutes? ____________________

(f) At equilibrium, is there a difference in osmotic pressure in the two arms? ____

(g) Is there a difference in osmotic pressure before equilibrium is reached? _______

Explain. ____

____

____

____

28. In the body,

(a) Na ions behave as if they are nonpenetrating solutes because they are actively transported out of cells.

(b) K ions behave as if they are penetrating solutes because they are actively transported into cells.

29. If a cell is placed in a hypotonic solution, it

(a) will swell.

(b) will shrink.

(c) may swell, shrink, or stay the same size, depending upon the concentration of penetrating and nonpenetrating solutes in the solution.

30. If a cell is placed in a hyperosmotic solution, it

(a) will swell.

(b) will shrink.

(c) may swell, shrink, or stay the same size, depending upon the concentration of penetrating and nonpenetrating solutes in the solution.

❐ Endocytosis and Exocytosis (text pages 127-131)

31. There are three types of endocytosis, all of which move substances in the extracellular fluid into cells: ______________________________ and ___________________________ (which collectively are referred to as ___________________________, or "cell drinking,") and ___________________________________, or "cell eating." Most cells in the body perform the ____________________ (former/latter), while only a few specialized cells perform the __________________ (former/latter).

32. True or false: The fate of all endocytotic vesicles is digestion of their contents by lysosomal enzymes.

33. Some cells in the body are called "phagocytes." Explain what their function might be.

    ____

    ____

    ____

34. True or false: A portion of a cell's plasma membrane is removed during endocytosis.

35. True or false: All protein synthesis in a cell begins on free ribosomes.

36. The signal sequence of a protein
    (a) identifies the protein as one that will be used within the cell.
    (b) is made up of the last 15 or 20 amino acids of the protein.
    (c) binds to proteins on the surface of the granular endoplasmic reticulum.
    (d) is present in the mature protein molecule.

37. True or false: Proteins that are to be secreted are packaged in vesicles formed from membranes of the Golgi apparatus.

38. True or false: The lipid portion of cell membranes is synthesized in the agranular endoplasmic reticulum.

❐ **Epithelial Transport** (text pages 131-135)

39. The diagram below represents a layer of epithelial cells lining a tube in the body. Label
    (a) the luminal side of one cell.
    (b) the blood side of the same cell.
    (c) the tubular lumen (see text page 45 if you have forgotten the meaning of "lumen").

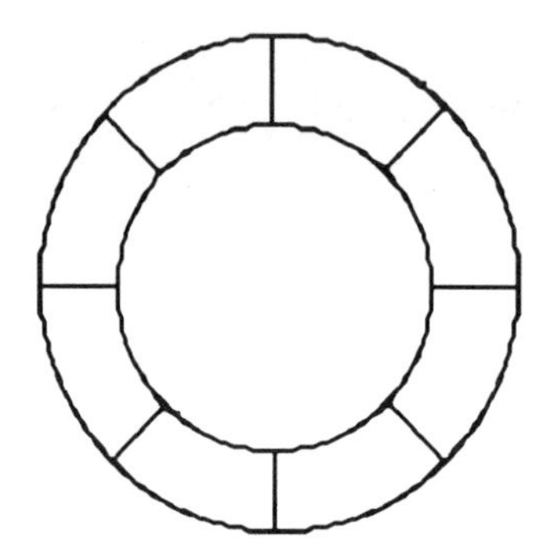

40. True or false: The properties of the plasma membranes of epithelial cells lining hollow organs or tubes are similar, regardless of the orientation of the cells.

41. True or false: Most organic solutes cross epithelial membranes by simple diffusion on the lumen side followed by active transport across the blood side of the membrane.

42. True or false: Net movement of solute across an epithelium that is permeable to water is always accompanied by movement of water in the opposite direction.

43. Exocrine glands differ from endocrine glands in that the former
    (a) are formed from infoldings of embryonic epithelium whereas the latter are not.
    (b) are specialized for secretion whereas the latter are not.
    (c) secrete their their products into ducts whereas the latter do not.
    (d) secrete hormones whereas the latter do not.

44. True or false: Ductless glands secrete only hormones.

ANSWERS

Boldface type indicates the answers you should have given. Words in medium-face (ordinary) type explain or expand upon the answer.

1. (a) **Correct.** Note in particular the differences in intracellular and extracellular concentrations of minerals (text Table 6-1).
   (b) **Incorrect. They are *selectively permeable* barriers.**
   (c) **Correct.**

2. Cell membranes are composed of a bimolecular layer of **phospholipid** molecules oriented such that their **polar** regions face the membrane surfaces and their **nonpolar** regions lie in the middle. Associated with these molecules are **integral proteins** that can form channels or pores in the membrane.

3. (a) **Correct.**
   (b) **Incorrect. The movement is from regions of *high* concentration to regions of *low* concentration.**
   (c) **Incorrect. It is important for moving molecules over *short* distances.** (Molecules cannot travel very far in a straight line because of collisions with other molecules. If we had to rely entirely on diffusion to bring nutrients to our cells, we could be only a few cells thick. Systems that provide for the *bulk flow* of nutrients — circulatory systems — have evolved so that large animals such as humans can exist.)

4. Diffusion across any surface can be quantified by determining the flux, which is defined as **the amount of material crossing a surface per unit of time**. The **net flux** (F) is the difference between two one-way fluxes. When F becomes zero, the system is said to be in **diffusion equilibrium**.

   The term "flux" is not part of most people's everyday vocabularies, but it need not be intimidating; it is simply a useful term describing the flow of material from one place to another, taking into account the time required for the flow to occur. For example, if three molecules of substance X cross a surface in one second, then the flux of X is 3/second. If three molecules of substance Y cross the same surface in 10 seconds, the flux of Y is 0.3/second. In this example, the flux of X is ten times greater than that of Y. Text Figures 6-3 and 6-4 are helpful for visualizing fluxes.

5. **False.** The greater the concentration difference of a substance between two volumes separated by a permeable surface, the ~~lesser~~ **greater** the magnitude of the net flux of the substance.

6. **True.**

7. (a) **the surface area of the membrane**

   (b) **the concentration difference of Z across the membrane** (the concentration outside the cell minus the concentration inside the cell)

   (c) **membrane permeability constant** (for substance Z)

   (d) **net movement of Z into the cell** (because the extracellular concentration is greater than the intracellular concentration)

   (e) **net movement of Z out of the cell** (because the intracellular concentration is greater than the extracellular concentration)

   (f) **Either $C_o = C_i$, or $k_p = 0$** (If the latter is the case, then the cell is impermeable to Z.)

8. **False.** In general, polar molecules diffuse ~~more rapidly~~ **more slowly** across cell membranes than nonpolar molecules (because polar molecules cannot readily diffuse through the lipid portion of the plasma membrane).

9. **False.** The component of the plasma membrane that acts as the selective barrier for diffusion of polar and nonpolar molecules is the ~~integral proteins~~ **lipid bilayer.** (Integral proteins are important for allowing selective diffusion of certain ions, however.)

10. (a) **Correct.** A single integral protein in the membrane may be cylindrical with a hollow, water-filled center that minerals (generally one kind) can dissolve in. The side chains of the amino acids making up the channel would of course be hydrophilic (polar) because they come into direct contact with water molecules. When several proteins constitute a channel, they are aligned so that their hydrophilic regions face each other, forming a channel. (This arrangement of several protein molecules forming a channel is also used by cells for intercellular communication via gap junctions, as we learned in Chapter 3.)

    (b) **Correct.** (The composition of membrane proteins varies from cell to cell.)

    (c) **Correct.** Even if a membrane has many Cl^- channels, for example, Cl^- would not diffuse readily from outside a cell to the inside if the interior of the cell was negatively charged. Like charges repel, remember. Conversely, positively charged ions would diffuse quickly through such a membrane if there were channels open for them and if the concentration difference was favorable.

    (d) **Correct.** As is implied in the remarks after answer (c), the opening and closing of membrane channels is regulated. See the answer to Question 11.

11. (a) **binding of a ligand to the Na^+ channel protein(s)** (receptor-operated channels)
    (b) **changing the electrical difference across the membrane** (membrane potential; voltage-sensitive channels)
    (c) **stretching the membrane** (stretch-activated channels)

12. (a) **Correct.** (This is one characteristic that distinguishes mediated transport from simple diffusion.
    (b) **Incorrect. The membrane carrier proteins change conformation, but they do not move through the membrane.** (Even though the proteins involved are referred to as "carriers," do not think of them as "ferry boats." Examine text Figure 6-9 closely to see how a change in conformation of a carrier protein allows a carrier binding site to be exposed first on one side of the membrane and then on the other. The "scissoring" effect in the figure is an artist's depiction of what such a conformational change might look like.)
    (c) **Incorrect. The transport (flux) increases with increasing substance concentration only until the membrane carriers are saturated.** (See text Figure 6-10. This property of carrier saturation is another distinguishing characteristic of mediated transport.)

13. **False.** Mediated transport is required in order for glucose **and** amino acids ~~and~~, **but not** fatty acids to pass into cells, because ~~these~~ **the glucose and amino acids** ~~substances~~ cannot diffuse through plasma membranes. (Fatty acids, being lipids, easily permeate the lipid bilayer whereas glucose and amino acids are polar molecules that cannot do so.)

14. (a) **Z**
    (b) **X**
    (c) **Y**

    Note that neither kind of measurement alone can conclusively demonstrate which method all of the molecules use to cross the membrane; both kinds of measurements are needed for complete differentiation. Looking only at Graph A, for example, we can determine that molecule Z crosses by simple diffusion because it has no maximal flux and that molecules X and Y cross by mediated transport. Graph B tells us that molecule Y crosses by active transport, and both graphs A and B are needed to clarify that molecule X crosses by facilitated diffusion.

15. (a) **Incorrect. Glucose enters most cells by facilitated diffusion.**
    (b) **Incorrect.** (see above)
    (c) **Incorrect. Glucose concentrations inside cells are generally lower.** After glucose enters a cell, it is quickly metabolized, either by anabolic pathways

(and then stored as glycogen) or by catabolic pathways (and then utilized for energy). This quick metabolism provides for a glucose concentration difference that favors entry of the molecule into most cells.

16. **True.** This difference in affinity as the binding site faces first one side of the membrane and then the other accounts for the ability of active-transport systems to transport molecules *against* a concentration difference in an "uphill" direction.

17. (a) **Incorrect. ATP supplies the energy for both transport systems.** (In secondary active transport, the energy comes *indirectly* from ATP).

    It must be emphasized that blocking ATP synthesis will inhibit both kinds of active transport because the secondary type depends upon primary active transport for establishment of the concentration difference (usually of sodium ions) that is the direct "driver" of secondary transport.

    (b) **Correct.** (Recall that there are three primary active transport carriers — for Na,K; Ca; and H ions — and each is an ATPase.)

    (c) **Correct.** (Recall that the carrier ATPases are self-phosphorylating enzymes.)

    (d) **Correct.** (Recall that phosphorylation is an example of covalent modulation of a protein. Secondary active transport relies on the allosteric modulation of ionic bonds.)

18. **True.**

19. **False.** The Na,K-ATPase carrier transports sodium ions out of cells and potassium ions into cells on a ~~one-to-one~~ **three-to-two** basis (three Na^+ out for every two K^+ in).

20. (a) **Incorrect. The Na ions move across the membrane along their concentration difference** (from the outside to the inside of the cell).

    (b) **Incorrect. The actively transported solute may follow the Na ions into the cell** (*cotransport*) **or it may move in the opposite direction** (*counter-transport*). (Thus, secondary active transport can move molecules either into cells or out of cells, but always from a region of low concentration to a region of high concentration.)

    (c) **Incorrect.** (See answer b.)

21. (a) **Correct.** (There are specific channels for Ca ions, just as there are for Na, K, and Cl ions.)
    (b) **Correct.** (A Ca-ATPase moves Ca^{2+} out of the cytosol, either across the plasma membrane into the extracellular fluid or across organellar membranes into cell organelles.)
    (c) **Correct.** (Ca^{2+} is *counter*transported with Na^{+} out of cells by secondary active transport. Note that this mechanism does not pump Ca^{2+} into organelles.)

22. (a) **Correct.**
    (b) **Correct.** (Recall that ion channels are water-filled; thus ions diffuse in *solution* — that is, dissolved in water.)
    (c) **Incorrect. Water molecules are not carried by any mediated transport system.**

23. **Osmosis is the movement** (diffusion) **of water molecules from a region of high *water* concentration** (dilute solute) **to a region of lower *water* concentration** (higher concentration of solute) **through a barrier that inhibits the ready movement of solute.**

24. **True.** (Recall from Chapter 2 that one *mole* of any substance is the number of grams of the substance equal to its molecular mass and that one mole of any substance contains the same number of molecules — 6 x 10^{23} — as one mole of any other substance. Thus one mole of dissolved glucose yields 6 x 10^{23} molecules but one mole of dissolved NaCl yields 6 x 10^{23} Na^{+} + 6 x 10^{23} Cl^{-} = 12 x 10^{23} ions.

25. **False.** Adding one gram of NaCl to one liter of water will dilute the water concentration ~~twice~~ **about six times** as much as adding one gram of glucose.

    Calculation: One mole is molecular mass in grams:

    1 mole NaCl = 58.5 grams   1 mole glucose = 180 grams

    $$\frac{1\text{ g NaCl}}{58.5\text{ g/mole}} = 0.017\text{ mole NaCl} \qquad \frac{1\text{ g glucose}}{180\text{ g/mole}} = 0.0056\text{ mole glucose}$$

    Now the problem is like Question 24 with different values for the moles:

    0.017 mole NaCl = 0.017 mole Na^{+} + 0.017 mole Cl^{-} = *0.034 osmole NaCl.*
    0.0056 mole glucose = *0.0056 osmole glucose*

    $$\frac{0.034\text{ osmole}}{0.0056\text{ osmole}} = 6$$

    There are six times as many particles in the NaCl solution as in the glucose solution.

26. **False.** The higher the osmolarity of a solution, the ~~higher~~ **lower** is the concentration of water in it.

Osmolarity is nothing more than a measure of the number of particles in solution taking into account that some molecules such as NaCl dissociate when dissolved in water, as we saw in Questions 24 and 25. Let's work through a few examples to help you become familiar with the concept of osmolarity. Remember that you need to have concentrations in moles to compare osmolarities, and that we are concerned with the final volume of the solution, not the volume of the solvent.

<u>Example 1</u>: 1 mole of glucose dissolved in enough water to make 1 L of solution:

Since glucose stays as a single particle when it dissolves,

1 mole glucose = 1 osmole glucose

$$\frac{1 \text{ osmole}}{\text{L}} = \textit{1 osmolar}$$

<u>Example 2:</u> 1 mole glucose dissolved in enough water to make 2 L of solution:

$$\frac{1 \text{ osmole}}{2 \text{ L}} = \textit{0.5 osmolar}$$

<u>Example 3:</u> 1 mole KCl dissolved in enough water to make 1 L of solution:

KCl dissociates completely to K^+ and Cl^-, and so we have 1 osmole K^+ + 1 osmole Cl^- =

$$\frac{2 \text{ osmoles}}{1 \text{ L}} = \textit{2 osmolar}$$

<u>Example 4:</u> 1 mole KCl + 1 mole glucose dissolved in enough water to make 1 L of solution:

1 osmole K^+ + 1 osmole Cl^- + 1 osmole glucose =

$$\frac{3 \text{ osmoles}}{1 \text{ L}} = \textit{3 osmolar}$$

<u>Example 5:</u> 1 mole $MgCl_2$ dissolved in enough water to make 1 L of solution:

Similarly, $MgCl_2$ dissociates completely in solution, yielding :

1 osmole Mg^{2+} + 2 osmole Cl^- =

$$\frac{3 \text{ osmoles}}{1 \text{ L}} = \textit{3 osmolar}$$

The importance of understanding osmolarity lies in its ability to predict *which way water will move* through a semipermeable membrane separating solutions of different osmolarities. Even though all of our calculations are based on concentrations of *solutes*, it is the *water* concentration that we are really interested in. Knowing how to calculate osmolarity is a tool to use in figuring out where water will move in a given situation. As you will see in later chapters, movement of water through membranes is a fundamental phenomenon of physiology, and understanding *why* it moves where it does is central for understanding many body functions.

27. (a) **1.5 Osm** (osmolar)

(b) **1.5 Osm**

As the text states on page 123 (and shows in Figure 6-21), when two solutions of different initial osmolarities separated by a permeable-to-water-only membrane reach equilibrium, the osmolarity on one side of the membrane is equal to that on the other side. Let us first calculate the initial osmolarities:

arm 1: 1 *M* NaCl means 1 mole NaCl per liter. Since we have 1 L of solution, we have 1 mole of NaCl and since NaCl dissociates completely in solution,

1 mole Na^+ + 1 mole Cl^- = 2 osmoles of solute in 1 L = 2 osmolar solution (initially).

arm 2: Similarly, 1 *M* glucose means 1 mole glucose per liter = 1 osmolar solution initially (see example # 1 after Question 26).

Initially we have an imbalance, with the solution in arm 1 (NaCl) having twice the osmolarity (number of solute particles per liter) as the solution in arm 2 (glucose). Since the membrane is permeable only to water, water molecules will diffuse from the region of higher water concentration (lower osmolarity — arm 2) to the region of lower water concentration (higher osmolarity — arm 1) until the water concentration is the same on the two sides of the membrane. In effect, water movement will dilute the solution in arm 1 and concentrate the solution in arm 2 until the osmolarities are the same. We calculate the osmolarity in the two arms at diffusion equilibrium by taking the average of the initial osmolarities:

$$\frac{2+1}{2} = 1.5 \text{ osmolar}$$

(This explanation may seem excessive when you realize that all we did was take the mean of the two initial osmolarities, but it is important for you to realize that, although it is water molecules that do the moving, we can measure and calculate only the solute concentrations.)

(c) **1.33 L**

(d) **0.67 L**

Remember that no NaCl can leave arm 1 and no glucose can leave arm 2 because the membrane is permeable to H_2O only. Therefore, the changes in osmolarity that we see are a result of changes in the amount of water in each arm.

Arm 1 always contains 2 osmoles of solute. The definition of osmolarity is "number of osmoles per volume of solution. " Thus, at equilibrium in arm 1,

$$\frac{\text{number of molecules}}{\text{volume of solution}} = 1.5 \text{ osmolar}$$

$$\frac{\text{2 osmoles}}{\text{volume}} = 1.5 \text{ osmolar} \quad (= 1.5 \text{ osmoles/liter})$$

Solving for volume, we have

$$\text{volume} = \frac{\text{2 osmoles}}{\text{1.5 osmoles/liter}} = \textit{1.33 liters}$$

In arm 2, which always contains 1 osmole of glucose,

$$\text{volume} = \frac{\text{1 osmole}}{\text{1.5 osmoles/liter}} = \textit{0.67 liter}$$

(e) **nonpenetrating**

(f) **no**

(g) **Yes. Before equilibrium is reached the solution in arm 1, which has the higher osmolarity, has the higher osmotic pressure. At equilibrium, the osmolarities of the solutions in the two arms are equal, and so osmotic pressure is equal.** As the text mentions, osmotic pressure is the pressure required to prevent the movement of water from areas of high water concentration (low osmolarity) to areas of low water concentration (high osmolarity) — in other words, the pressure required to physically prevent the diffusion of water molecules. If you combine this definition with another sentence in the text, namely: "The greater the osmolarity of a solution, the greater its osmotic pressure," then you can see that osmotic pressure is a force that counters the flow of water molecules *into* a solution of relatively high osmolarity.

The instruction to ignore the effect of gravity when calculating volume changes in this question is not trivial: In real life, gravity would push down on a column of water in a tube, and would affect the water movement across the membrane. If the force of gravity just equaled the osmotic pressure of the solution initially in arm 1, then there would be no net water movement. We shall return to the very important concept of osmotic pressure in Chapter 13, when we study the circulation of blood.

28. (a) **Correct.** Although Na^+ diffuses into cells both because of a concentration difference and because of an electrical-charge difference (the interior of cells is generally negatively charged with respect to the extracellular fluid), the Na,K pump makes Na ions effectively nonpenetrating.

(b) **Incorrect. K ions behave as if they are <u>nonpenetrating</u> solutes because they cannot effectively leave the cell** (because of the Na,K pump). Thus, since Cl ions also behave as if they are nonpenetrating solutes, and most of the solutes inside cells besides K ions are truly nonpenetrating, most of the solute particles inside and outside of cells behave as if they are nonpenetrating.

29. (a) **Correct.** (By definition, such a solution contains less than the cell's 300 mOsmol/L of nonpenetrating solutes; because the solute concentration is less in the solution than in the cell, the water concentration is *greater* in the solution than in the cell. Thus water would move from an area of high water concentration [outside the cell] to an area of low water concentration [inside the cell] and cause the cell to swell.)

(b) **Incorrect.** (See answer a.)

(c) **Incorrect.** (See answer a.)

30. (a) **Incorrect.** (See answer c below.)

(b) **Incorrect.** (See answer c below.)

(c) **Correct.** You cannot predict what will happen to cell size from information about osmolarity alone. Any solution that is greater than 300 milliosmolar is hyperosmotic relative to the osmolarity of cells. However, whether or not cell size changes depends upon the *tonicity* of the solution. A solution is hyper*tonic*, and will make a cell shrink (because of diffusion of water out of the cell), only if the solution contains more than 300 milliosmoles/L of *nonpenetrating* solutes. If the solution contains less than 300 milliosmoles/L of nonpenetrating solutes, then the cell will swell regardless of the concentration of penetrating solutes because the solution is hypotonic. Similarly, a solution containing 300 milliosmoles/L of nonpenetrating solutes is isotonic and will not cause a change in cell size again regardless of the concentration of the penetrating solutes in the solution. The reason we need consider only the concentration of nonpenetrating solutes when determining water movement back and forth across plasma membranes is that penetrating solutes follow the water: wherever water molecules go (and create a concentration difference across a membrane for a penetrating solute) the solute molecules will follow.

Let us look at one example: A cell is placed in a solution that is 400 milliosmolar (hyperosmotic) but all of the solutes are penetrating. The solutes would diffuse into the cell until their concentrations were equal inside and outside the cell. Meanwhile, water molecules, which initially began diffusing out of the cell in response to the hyperosmotic solution (lesser concentration of water outside the cell) would go back into the cell in response to diffusion of the solutes into the cell (causing a lesser concentration of water there). Solute would follow the water, attracting still more water molecules, and the cell would swell and ultimately burst. In other words, as far as a cell is concerned, a hyperosmotic solution of penetrating solutes is the same as pure water.

The best way to keep osmolarity and tonicity straight in your mind is to remember the definition of osmolarity: the number of moles of particles per liter solution, regardless of the properties (that is, whether they are penetrating or nonpenetrating) of the solute.

The text mentions the importance of ensuring that any liquid injected into the body is isotonic, so that the body's cells will not shrink or swell. "Physiological (isotonic) saline" is often the vehicle for injections of drugs such as antibiotics, vaccines, and so forth. On page 13 of this Study Guide, you were asked to determine the molarity of one liter of an aqueous solution containing 9 g of NaCl. The answer was 0.15 *M*, which is 300 mOsm (0.3 osmolar). Thus, physiological saline is a solution of 9 g NaCl/L.

31. Their are three types of endocytosis, all of which move substances in the extracellular fluid into cells: **fluid endocytosis** and **adsorptive endocytosis** (which collectively are referred to as **pinocytosis**, or "cell drinking,") and **phagocytosis**, or "cell eating." Most cells in the body perform the **former**, while only a few specialized cells perform the **latter**.

32. **False.** The fate of ~~all~~ **most** endocytotic vesicles **in most cells** is digestion of their contents by lysosomal enzymes. (Some epithelial cells transport proteins and other molecules intact in their endocytotic vesicles to the opposite side of the cell, where they exit by exocytosis.)

33. **A phagocyte is a cell** ("cyte" denotes cell) **that specializes in phagocytosis** ("cell eating"). Phagocytes are the body's "scavengers" or "garbage collectors." They eliminate damaged cells and harmful invaders such as bacteria by engulfing them and then breaking them down with lysosomal enzymes. You will meet them again in Chapter 13, in the discussion of blood cell types, and in Chapter 19, when we study the body's defense mechanisms.

34. **True.** (But the membrane is restored during exocytosis; that cell size does not fluctuate much suggests that the two processes occur to roughly the same extent.)

35. **True.**

36. (a) **Incorrect. It identifies the protein as one that will be secreted.**
    (b) **Incorrect. It is made up of the first 15 or 20 amino acids.**
    (c) **Correct.** (It forms a ligand that carries the ribosome into the reticulum.)
    (d) **Incorrect. It is cleaved from the protein by enzymes in the endoplasmic reticulum.**

37. **True.** (See text Figure 6-25.)

38. **True.**

39.

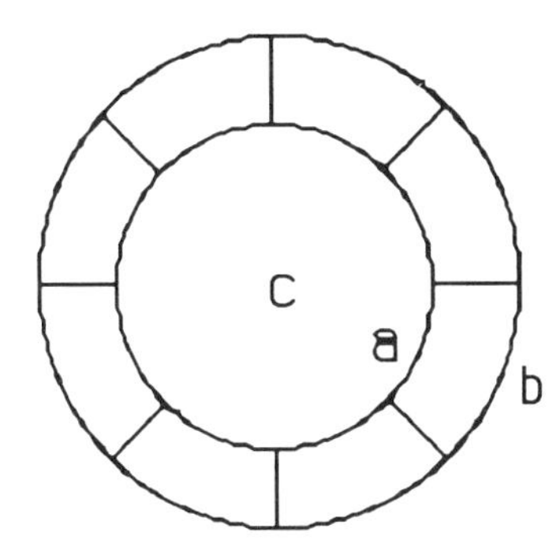

Recall that *lumen* is the cavity of a hollow organ or tube.

40. **False.** The properties of the plasma membranes of epithelial cells lining hollow organs or tubes are ~~similar, regardless of~~ **specialized according to** the orientation of the cells. (The membranes on the luminal and blood sides of an epithelium will also differ from the portions of the cells that touch on other cells in the epithelium [which may form tight junctions — see Chapter 3]. In text Figure 6-26, we see that Na^+ enters the luminal side of some epithelial cells by diffusion through channels but is actively transported out of cells on the blood side. Na^+ cannot be actively transported out of the luminal side because the membranes there do not have the Na,K-ATPase.)

41. **False.** Most organic solutes cross epithelial membranes by ~~simple diffusion~~ **secondary active transport** on the lumen side followed by ~~active transport~~ **facilitated diffusion** across the blood side of the membrane.

42. **False.** Net movement of solute across an epithelium that is permeable to water is always accompanied by movement of water in the ~~opposite~~ **same** direction.

43. (a) **Incorrect. Both kinds of glands are formed this way.**
    (b) **Incorrect. Both kinds of glands are specialized for secretion.**
    (c) **Correct.**

(d) **Incorrect. Exocrine glands do not secrete hormones; only endocrine glands do.** Hormones by definition are blood-borne messengers that are secreted not into ducts but rather into the interstitial fluid, from which they diffuse into the blood vessels.

The terms "endocrine" and "exocrine" are derived from the Greek roots "endon" (inside, within), "exo" (outside, external), and "krino" (to separate or sift). Thus an endocrine gland is one that "separates from itself" (secretes) a substance that stays within the interior of the body (the internal environment); an exocrine gland, however, secretes material into cavities that are continuous with the external environment — such as saliva from salivary glands into the mouth, tears from tear duct glands over the eye, and so on.

44. **False. Some** ductless glands secrete ~~only hormones~~ **nonhormonal substances.** Even though, as the text points out, endocrine glands have come to be considered synonymous with "hormone-secreting" glands, many substances secreted by endocrine glands into the interstitial fluid do not function as chemical messengers. A good example of this is the secretion of glucose by the liver, a liver function discovered by Claude Bernard.

PRACTICE

True or false (correct the false statements):

P1. Molecules increase their rate of diffusion as temperature increases.

P2. Although permeability to mineral ions does not vary much from one cell to another, different cells vary considerably in their permeability to nonpolar molecules.

P3. Movement of lipid-soluble molecules into cells is mediated by specific proteins in the membranes.

P4. Integral membrane proteins can form channels through which ions such as Na^+ and K^+ can diffuse.

P5. The final equilibrium state reached by a molecule that enters a cell by facilitated diffusion is the same as that for a molecule that enters the cell by diffusion.

P6. The Na,K pump is an enzyme that phosphorylates itself.

P7. In most of the cells in the body there is an electrical difference such that the inside of cells is positive with respect to the outside.

P8. The concentration of calcium in the cytosol of cells is very much lower than the concentration of extracellular calcium.

P9. The intracellular concentration of water in the cells of the body is the same as the extracellular concentration of water.

P10. Active transport, facilitated diffusion, and osmosis all require the expenditure of metabolic energy.

P11. If a cell were placed in a solution of 0.15 *M* NaCl, it would shrink.

P12. Pinocytosis is a method by which molecules can leave cells whose membranes are impermeable to the molecules.

P13. The signal sequence of a protein is recognized by proteins in the membrane of granular endoplasmic reticulum.

P14. Exocrine glands secrete into ducts, whereas endocrine glands are ductless.

Multiple choice (correct each incorrect choice):

P15-17. Curves A, B, C, and D below represent the entry of various molecules into a cell. Therefore,

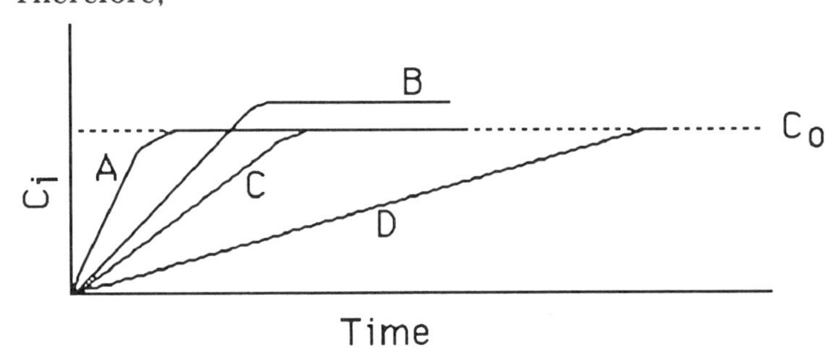

P15. if molecules C and D enter the cell by diffusion through the lipid portion of the cell membrane, then C is probably less polar than D.

P16. molecule B probably enters the cell by facilitated diffusion.

P17. if molecule A crosses the membrane by facilitated diffusion, then curve C may represent the entry of A in the presence of a similar molecule.

P18-20. Diffusion of molecules across a cell membrane

P18. may involve the formation of chemical bonds with proteins in the membrane.

P19. occurs at a rate determined by the permeability constant of the membrane and the magnitude of the concentration difference across it.

P20. is directly dependent upon energy derived from ATP.

P21-24. The concentration of molecule A is greater inside cells than in the extracellular fluid. This concentration difference can be accounted for by

P21. diffusion of A across the cell membrane.

P22. synthesis of A by the cell, whose plasma membrane is impermeable to the molecule.

P23. facilitated diffusion.

P24. active transport mechanisms.

P25-27. The movement of

P25. glucose across cell membranes involves binding with a carrier protein in the membranc.

P26. water across a cell membrane is from a region of high solute concentration to a region of low solute concentration.

P27. molecules by diffusion over a distance of 25 centimeters would take only a few seconds to reach equilibrium.

P28-30. If a cell is placed in a hypertonic solution

P28. the cell will swell.

P29. the concentration of water outside the cell is initially lower than the concentration inside the cell.

P30. the concentration of nonpenetrating solute inside the cell will decrease.

7: HOMEOSTATIC MECHANISMS AND CELLULAR COMMUNICATION

It is not surprising that humans and other animals have evolved with "biological clocks" governing cyclic functions that have a periodicity of about one day. People who attempt to keep work schedules that differ very much from a 24-h work/rest cycle have problems functioning efficiently. Medical interns, for example, are required to work very long hours, with a schedule of two days on call and one day off being common. Opportunities to sleep when on duty are few and irregular. Thus, these physicians are physically exhausted and stressed because their normal sleep, body temperature, and hormonal patterns are disrupted.

These symptoms are similar to those produced when several time zones are crossed ("jet lag"). Travellers simply need time to adjust, but interns and others with work/rest cycles spanning much more than 24 h cannot adjust completely because of the 24-h nature of our inherent biological rhythms. The physiology of biological rhythms and of their entrainment by environmental cues is one of the subjects you will understand after completing this chapter.

Instructions for answering questions in this Study Guide

1. True or false: Correct each false statement. Whenever you must correct a false statement, there will almost always be more than one way to do so. In the Guide, the answer requiring the least correction will generally be given, with the understanding that other correct answers are possible. It is usually not sufficient, in terms of demonstrating understanding, to simply insert a "no" or "not," however.

2. Multiple choice: Any, all, or none of the responses may be correct. Explain why each incorrect response is incorrect.

3. Fill in the blanks: Choose the best word or words to complete each statement.

4. Directions will be given for other types of questions as they appear.

❒ General Characteristics of Homeostatic Control Systems (text pages 141-146)

1. Define "homeostasis" and "homeostatic control system."

____
____
____
____
____
____

2. Having read the text's opening remarks on homeostasis, you should be beginning to get the idea of the importance of homeostasis and of how homeostatic mechanisms control virtually everything about the body. List all the body variables that might be controlled by homeostatic systems. Doing so will test your physiological intuition and help you preview and focus on what you will be learning in the rest of this course.

____
____
____
____
____
____
____

3. True or false: A homeostatic control mechanism maintains a physiological variable precisely at a fixed, ideal level called the operating point.

4. Homeostatic control systems achieve ____________________ of an internal environment variable by balancing ____________________ and ____________________.

5. Are positive feedback mechanisms important for homeostasis? Explain.

____
____
____
____
____

6. True or false: The more sensitive a homeostatic regulating system for a given variable, the smaller will be the magnitude of the error signal generated by environmental perturbations influencing the regulated variable.

7. How do feedforward systems interact with feedback systems to bring about homeostasis? ______________________________

____

____

____

8. Explain the statement, "Homeostatic control systems are inherited biological adaptations." ______________________________

____

____

____

9. True or false: A person who is acclimated to a hot environment will begin to react physiologically to a decreased environmental temperature faster than a person who is not.

10. Tenzing Norgay, a Sherpa from Nepal, was one of the first human beings to set foot on the summit of the world, Mt. Everest. Explain why he was physiologically better prepared for this feat than his partner from New Zealand, Sir Edmund Hillary, even though both men were in top physical condition. ______________________________

____

____

11. True or false: "Entrainment of biological rhythms" refers to the imposition of cycles of bodily activity by environmental cues.

12. John Doe, a San Francisco research physiologist, is attending a scientific meeting in London, which is in a time zone 8 h earlier than San Francisco. On his first night there, he awakens at 3 A.M. feeling warm and cannot go back to sleep. At noon the next day, just as he is about to start his presentation, he begins to feel uncontrollably drowsy.

Chart A is a record of Doe's normal sleep/wake cycle and circadian rhythm in body temperature. Chart B shows the shift to London time. Copy onto Chart B the body temperature curve shown on Chart A. Notice the time of day in London when the shifts take place. Use the data on Chart B to explain Doe's symptoms.

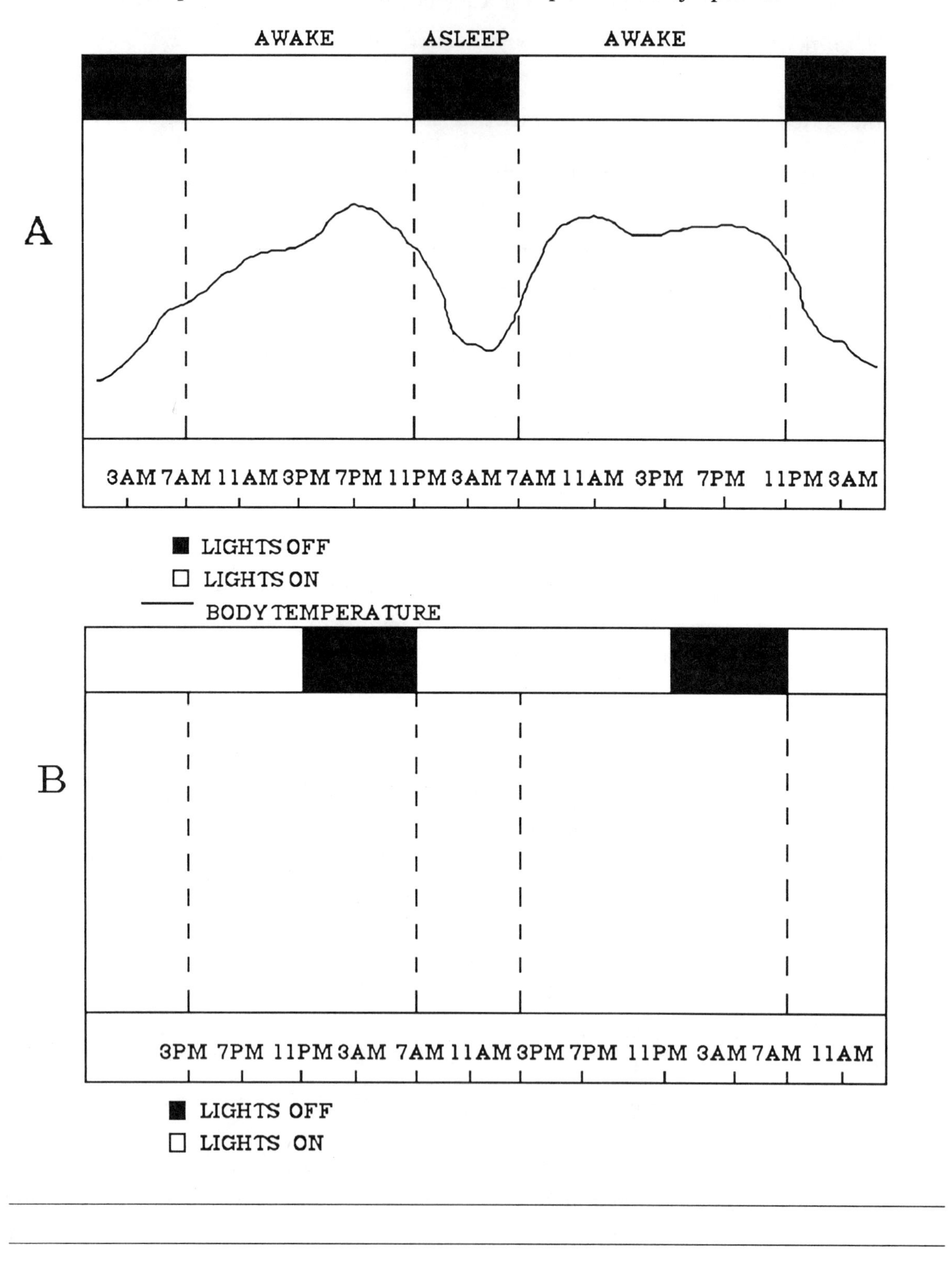

13. True or false: Gradual deterioration of homeostatic control processes is a common correlate of aging.

❐ The Balance Concept and Chemical Homeostasis (text pages 146-148)

14. That portion of the body's total amount of a given chemical substance readily available for use is called the ____________________ of that substance. This amount is often the same as the quantity of that substance found in the ____________________ fluid compartment. Three ways that such substances enter the body are ____________________, ____________________, and by ____________________. They are removed either by excretion or by ____________________.

15. Refer to text Figure 7-5, showing a change in sodium ingestion over a five-day period and subsequent changes in the amount of sodium excreted and in total body sodium. The data have been extracted to the table below. Fill in the appropriate word (positive, negative, or stable) for each line in the column marked "Balance."

Day	Ingested (g/day)	Excreted (g/day)	Body sodium content (g)	Balance
1	7	7	1.00	____________
2-3	15	10-13	1.00	____________
4-5	15	15	1.03	____________

❐ Components of Homeostatic Systems (text pages 148-152)

16. True or false: A learned reflex differs from an unlearned one in that premeditation is required for the former to occur.

17. Blood calcium concentration is closely regulated by homeostatic mechanisms. Calcium levels are monitored by receptors in the parathyroid, an endocrine gland that secretes parathyroid hormone (PTH). PTH acts to increase blood calcium levels by (among other actions) causing calcium to be reabsorbed from bone into the blood.

Using this information, draw a reflex arc depicting blood calcium regulation. Assume that the stimulus is a fall in blood calcium concentration.

18. The major classes of tissues that constitute reflex-arc effectors are ______________ and ______________.

19. Local homeostatic responses differ from reflex arcs in that the former lack ______________.

20. Can the same chemical messenger act as a hormone, a paracrine, and a neurotransmitter? Explain.

____

____

____

____

21. Match the description on the right with the substance on the left:

________ lipoxygenase

________ eicosanoids

________ cyclooxygenase

________ phospholipase A_2

a. membrane-bound enzyme that splits arachidonic acid from membrane phospholipid

b. enzyme important in synthesis of prostaglandins, prostacyclins, and thromboxanes from arachidonic acid

c. enzyme necessary for synthesis of leukotrienes from arachidonic acid

d. general name for molecules that act as local messengers and are derived from unsaturated fatty acids

❐ Receptors (text pages 152-155)

22. Receptors for chemical messengers are
    (a) always proteins.
    (b) found in cells.
    (c) found in the plasma membrane of cells.
    (d) found in the extracellular fluid.
    (e) responsible for conferring specificity to a chemical messenger.

23. Intercellular chemical messengers
    (a) are ligands.
    (b) bind only to proteins.
    (c) are proteins only.

24. True or false: Two cell types having the same type of receptor for a chemical messenger will respond to that messenger in the same way.

25. True or false: Competition for receptors is strictly a pharmacological phenomenon, since naturally occurring chemical messengers do not compete with each other for the same receptor site.

26. True or false: An agonist blocks the action of a chemical messenger by binding to its receptor site.

27. Stimulation of endocytosis by the binding of a messenger to its receptor is one cause of receptor down-regulation. Explain.

    ____

    ____

    ____

    ____

    ____

28. Diabetes mellitus is a disease in which the body's cells are unable to utilize glucose. One cause of this disease is a lack of the hormone insulin when the pancreas secretes little or none. Another cause is excess secretion of the same hormone. Explain how two opposite conditions might cause the same disease.

____

____

____

____

❐ Signal Transduction Mechanisms for Plasma-Membrane Receptors (text pages 155-163)

29. Activation of a receptor by a chemical messenger
    (a) occurs when the messenger binds to its receptor.
    (b) is the first step leading to the ultimate response of a cell to the messenger.
    (c) requires a change in messenger conformation.
    (d) requires a change in receptor conformation.

30. Second messengers
    (a) are necessary for all receptor signal-transduction mechanisms.
    (b) act in the cell cytoplasm.
    (c) act as intercellular messengers.
    (d) always function to activate enzymes.
    (e) are always proteins.

31. G proteins may
    (a) act as second messengers.
    (b) be stimulatory for second-messenger production.
    (c) be inhibitory for second-messenger production.
    (d) act as transducers for activated receptors by opening or closing ion channels.

32. True or false: Phosphorylation is a necessary component of any enzyme activation.

33. True or false: Cyclic AMP activates allosteric proteins.

34. The figure below is like text Figure 7-13, but it is missing a few labels. Write in the appropriate label on the lines below. Indicate where on the diagram amplification of the first messenger's signal takes place.

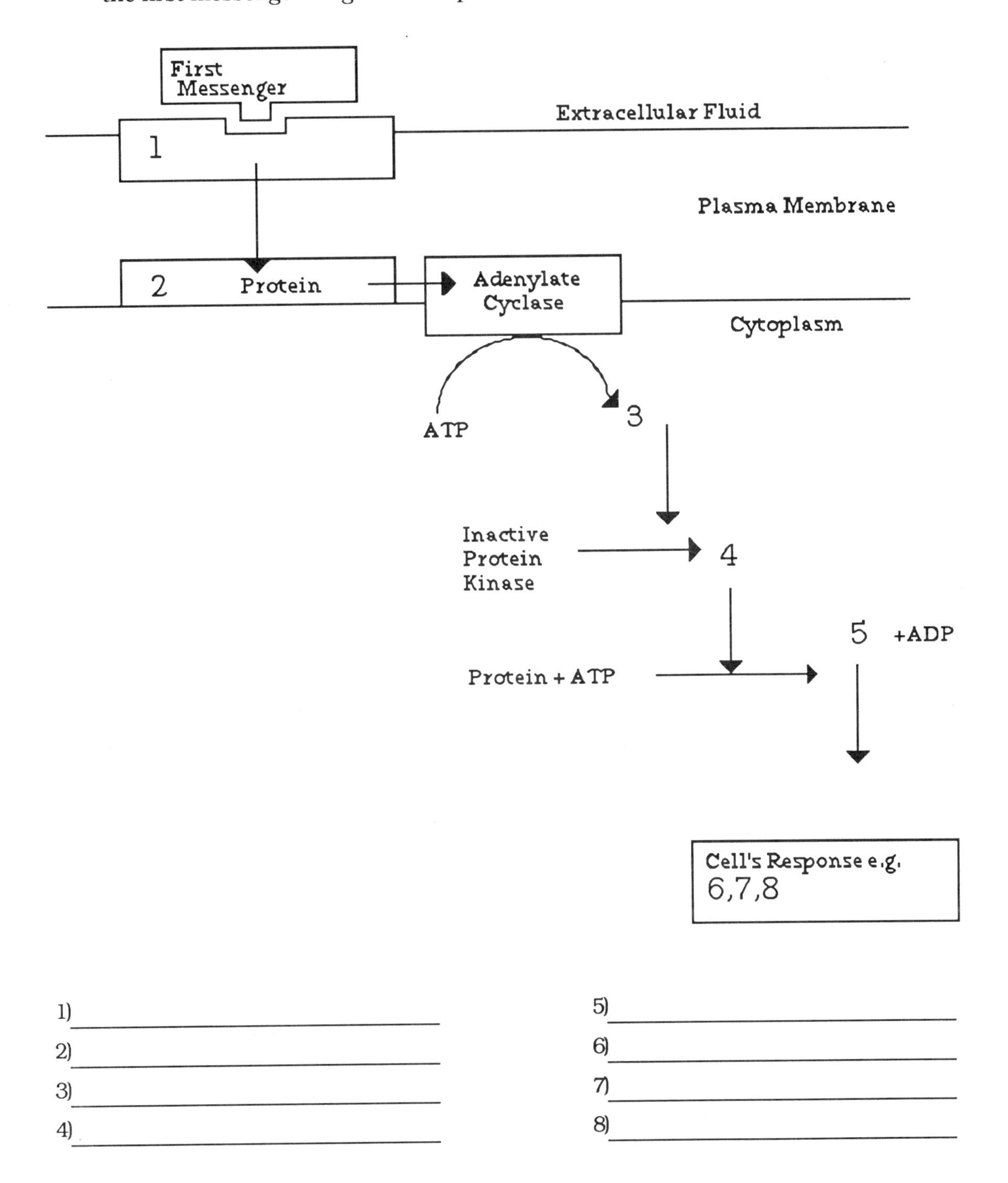

1) ______________________

2) ______________________

3) ______________________

4) ______________________

5) ______________________

6) ______________________

7) ______________________

8) ______________________

35. A fat cell responds to the presence of the hormone epinephrine by increasing cytosolic cyclic AMP production, which leads to the catabolism of both glycogen and fat. These responses can best be explained by assuming *one* of the following (in other words, all choices except one are incorrect). Explain all answers, both the correct one and the three incorrect ones:

(a) Epinephrine is binding to two types of receptors in the plasma membrane.

____

____

____

____

(b) The activated receptor complex stimulates production of two different second messengers. ______________________________

____

____

____

(c) Cyclic AMP can directly activate two kinds of enzymes.

____

____

____

____

(d) Cyclic-AMP-dependent protein kinase can activate two kinds of enzymes.

____

____

____

____

36. Calcium ion

(a) may act as a second messenger.

(b) concentration in the cytosol is increased by activation of phospholipase C.

(c) concentration in the cytosol is increased by activation of membrane calcium channels.

(d) concentration in the cytosol may be decreased by activation of adenylate cyclase.

37. What specific function do cyclic AMP, cyclic GMP, activated calmodulin, and diacylglycerol share?

ANSWERS

Boldface type indicates the answers you should have given. Words in medium-face (ordinary) type explain or expand upon the answer.

1. **"Homeostasis" refers to the stability of the conditions of the body's internal environment. A "homeostatic control system" is the mechanism that regulates the amount of some component in the body or some other variable so that the body's internal environment remains stable.**

2. **Body temperature**
   Blood pressure
   Blood volume
   Blood acidity
   Blood concentrations of electrolytes (sodium, potassium), glucose, water, and so forth.

3. **False.** A homeostatic control mechanism maintains a physiological variable ~~precisely at~~ **close to** a fixed, ideal level called the operating point.

4. Homeostatic control systems achieve **stability** of an internal environment variable by balancing **input** and **output.**

5. **Positive feedback mechanisms result in system instability, not stability, and thus by definition cannot themselves promote homeostasis.** However, they are important for physiological functions necessary for life, such as blood clotting following hemorrhage, reproduction (both ovulation and childbirth are dependent upon positive feedback signals), and even, as you will discover in Chapter 8, the functioning of nerves, which are required for much negative feedback regulation. Thus, positive feedback mechanisms are *indirectly* important for homeostasis. (This "trick" question is meant to ensure that you understand [1] the difference between negative and positive feedback mechanisms and [2] that each is both necessary and physiological.)

6. **True.** There are some variables, such as body core temperature and pH of the extra-cellular fluid, to which components of homeostatic regulatory systems are exquisitely sensitive. These variables are not allowed to fluctuate much from their operating point, so that the error signal generated by a perturbation of the variable need not be large to drive the system to correct it.

7. **A feedforward system is one in which a change in a homeostatic variable is predicted before it occurs; predicting such a change allows for faster compensation by response mechanisms to limit the magnitude of change.** (Recall the text's example of body temperature regulation on page 144: The regulated variable is body core temperature, not skin temperature, but temperature sensors in the skin "alert" the homeostatic system to changes in environmental temperature before there is any change in core temperature. Feedforward mechanisms are thus helpful in minimizing the error signal, as in Question 6.)

8. **A biological adaptation is defined as a mechanism whereby an organism can change or adapt to optimize its chances for survival in a given environment. Homeostatic systems maintain stable internal environmental conditions, a necessary requirement for life in a multicellular organism, and are genetically programmed. Thus, these systems can be said to be inherited biological adaptations.**

9. **False.** A person who is acclimatized to a ~~hot~~ **cold** environment will begin to react physiologically to a decreased environmental temperature faster than a person who is not.

   Alternatively, the answer below is also correct:

   False. A person who is acclimatized to a hot environment will begin to react physiologically to a decreased environmental temperature ~~faster~~ **more slowly** than a person who is not.

10. **Sherpa Tenzing had the advantage of developmental** (irreversible) **acclimatization to high altitudes, a condition that allowed him greater maximal capacity for high-altitude exercise than Hillary was capable of, even though Hillary did experience some acclimatization during the time he spent in Nepal before beginning his ascent of Mt. Everest.**

11. **False.** "Entrainment of biological rhythms" refers to the ~~imposition~~ **setting of the actual hours** of cycles of bodily activity by environmental cues.

12.

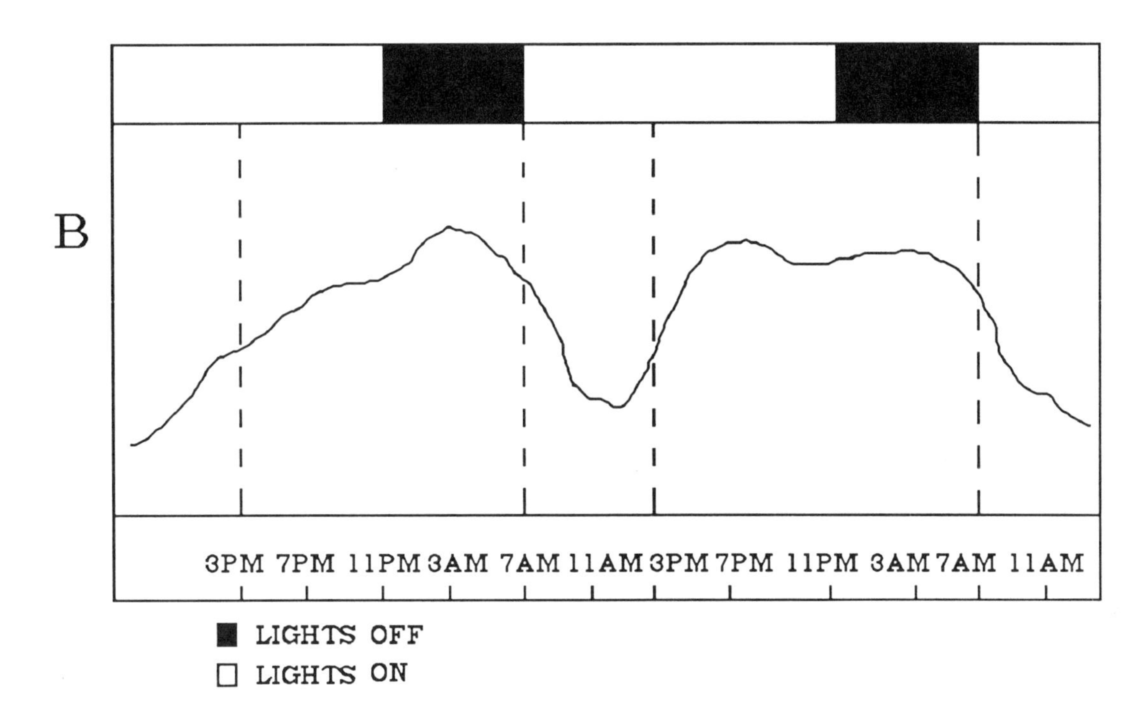

At 3 A.M. in London, Doe's body temperature is elevated, just as it is at 7 P.M. in San Francisco. However, it begins to fall after 7 A.M. London time and is at its lowest point at noon. Thus, even though environmental cues (lights on and off) tell Doe he should sleep at night and be awake during the day, his body temperature (and other rhythms not pictured here) **tell him, in effect, that the cues are misleading and he should "turn night into day."** Physiologists estimate that it takes one day to readjust a biological rhythm (at least the ones mentioned here) to compensate for one hour of time change. Therefore, Doe's body rhythms should be "in sync" with London time in eight days. Of course, by that time he will undoubtedly be heading back to San Francisco!

13. **True.**

14. That portion of the body's total amount of a given chemical substance readily available for use is called the **pool** of that substance. This amount is often the same as the quantity of that substance found in the **extracellular** fluid compartment. Three ways that such substances enter the body are **ingestion in food**, **uptake of air into lungs**, and by **synthesis in the body.** They are removed by excretion or by **metabolism.**

15. Day 1: **Stable**
Days 2-3: **Positive**
Days 4-5: **Stable** (Note that in this case body sodium is in balance even though the total body content is higher than it was on days 2 and 3.)

16. **False.** A learned reflex differs from an unlearned one in that ~~premeditation is required for the former to occur~~ **the latter is instinctive.** (Premeditation is required for acquiring a learned reflex but not necessarily for performing it.)

17.

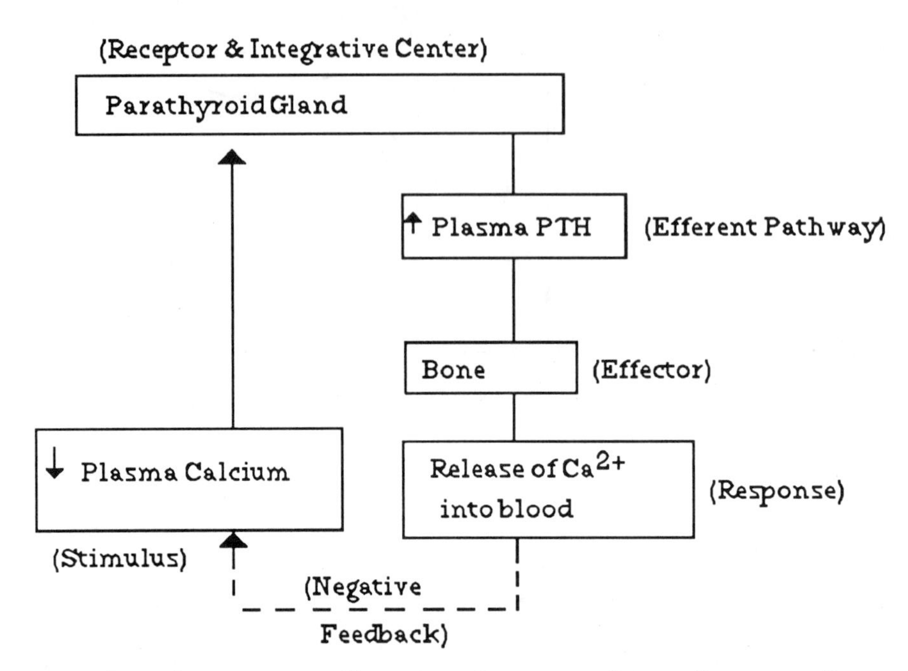

Note that there is no afferent pathway in this reflex arc; the receptors and the integrating center both reside in the parathyroid gland and may in fact be the same cells. Also note that a hormone, PTH, is the efferent "pathway." This can be a confusing concept, but remember that hormones travel through the blood, and one can think of them as the messengers in the blood "pathway." Similarly, in other examples given in the text, nerve cells themselves constitute the afferent and efferent pathways. The nature of the "messenger" in nerve cells is addressed in Chapter 8.

18. The major classes of tissues that constitute reflex-arc effectors are **muscle** and **gland cells**.

    A common error is the response "nerve" to this question. We tend to think of nerves as "doers"; in fact, they (and hormones) direct the activity of the actual effectors — muscles and glands.

19. Local homeostatic responses differ from reflex arcs in that the former lack **afferent and efferent pathways**.

20. **Yes, the same chemical can act as any type of messenger, depending solely on what type of cell secretes it, into what compartment it is secreted, and how quickly it is metabolized or inactivated.**

21. **c** lipoxygenase — **d** eicosanoids
b cyclooxygenase — **a** phospholipase A_2

Hint: Keep in mind that phospholipases are membrane-bound enzymes that catalyze phospholipid metabolism, as their name implies; "A_2" suggests arachidonic acid. (You will encounter another phospholipase in the next section.) The oxygenases can be differentiated by associating cycloxygenase with prostacyclin, which is related to prostaglandin; lipoxygenase is important for leukotriene synthesis.

Why is it important to know all this detail? For one thing, knowledge about various synthetic pathways for these ubiquitous molecules is important for pharmacological (drug-related) reasons. For example, aspirin and indomethacin specifically block the activity of cyclooxygenase, and steroidal anti-inflammatory drugs, such as cortisone, inhibit the activity of phospholipase A_2. Researchers are constantly trying to find ever more specific drugs to block or stimulate synthesis of these molecules because such drugs would be useful in treating patients whose synthesis rates are inappropriately high or low.

22. (a) **Correct.** (The messengers themselves are not always proteins, however.)
(b) **Correct.** (Receptors for steroid and other lipid-soluble hormones are found in the cytoplasm and nucleus of cells.)
(c) **Correct.** (Receptors for lipid-insoluble messengers are found in the plasma membrane of cells.)
(d) **Incorrect. By definition, receptors are binding sites in or on cells.**
(e) **Correct.**

23. (a) **Correct.** (Recall from Chapter 4 that ligands are molecules that bind to specific binding sites on proteins, a definition that fits intercellular chemical messengers.)
(b) **Correct.**
(c) **Incorrect. Many such messengers are not proteins.**

24. **False.** Two cell types having the same type of receptor for a chemical messenger will **not necessarily** respond to that messenger in the same way.

A cell's response to a given receptor-messenger complex depends upon the specific post-receptor pathways that cell possesses. A muscle cell has functions different from those of a gland cell, for example, but both may be stimulated (to do different things) by the same messenger-receptor complex.

25. **False.** Competition for receptors is **not** strictly a pharmacological phenomenon, since naturally occurring chemical messengers ~~do not~~ **may** compete with each other for the same receptor site.

26. **False.** An ~~agonist~~ **antagonist** blocks the action of a chemical messenger by binding to its receptor site.

27. **"Down-regulation of receptors" refers to the decrease in number of a specific kind of receptor. The number of receptors for a given messenger in the plasma membrane of a given cell can be reduced by endocytosis of the messenger-receptor complex.**

    (Note: The receptor in this case may be recycled back to the membrane. The rapidity with which this recycling occurs will influence changes in receptor number.)

28. **Insulin interacts with a plasma membrane receptor to promote glucose utilization. Diabetes can result from either too little insulin or too little receptor. The latter is a consequence of the down-regulation that results when the pancreas secretes excessively high levels of insulin over a long period of time.**

29. (a) **Correct.**
    (b) **Correct.**
    (c) **Incorrect. The chemical messenger is not changed by its interaction with the receptor.**
    (d) **Correct.** (That is what is meant by "receptor activation.")

30. (a) **Incorrect. Some signal-transduction mechanisms (e.g., receptor-operated channels) are confined to the plasma membrane and do not require second messengers.**
    (b) **Correct.**
    (c) **Incorrect. By definition, second messengers act intracellularly, unlike first messengers.**
    (d) **Incorrect. Some second messengers inactivate enzymes.**
    (e) **Incorrect. Most second messengers are not proteins.**

31. (a) **Incorrect. G proteins function only in the plasma membrane. Second messengers act in the cytoplasm.**
    (b) **Correct.** (These are known as "G_s" proteins.)
    (c) **Correct.** (These are known as "G_i" proteins.)
    (d) **Correct.**

32. **False.** Phosphorylation is **not** a necessary component of any enzyme activation. (Some enzymes are *in*activated by phosphorylation.)

33. **True.** (Recall from Chapter 4 that protein kinases are allosteric proteins.)

34. 1) **Receptor**
 2) **G_s**
 3) **Cyclic AMP**
 4) **Active protein kinase**
 5) **Phosphorylated protein**
 6) **Secretion**
 7) **Contraction**
 8) **Effects on membrane channels (Also: Changes in metabolic pathways, effects on the nucleus, etc.)**

 Amplification occurs with the generation of cyclic AMP, the conversion of inactive protein kinase to active protein kinase, and the phosphorylation of protein by protein kinase (see text Figure 7-15).

35. (a) **Incorrect.** **Although epinephrine can bind to more than one kind of receptor, only one kind** (the beta receptor, text Figure 7-21) **responds to that binding by increasing production of cyclic AMP. Therefore, this statement cannot account for the observations.**

 (b) **Incorrect.** **The cyclic AMP second-messenger system is not linked to any other** (unlike the phosphatidylinositol system).

 (c) **Incorrect.** **Cyclic AMP can activate only cyclic AMP-dependent protein kinase.**

 (d) **Correct.** **This answer explains the observation, because activated protein kinase phosphorylates many enzymes.** In general, phosphorylation activates enzymes in degradative pathways and inactivates enzymes in synthetic pathways. See text Figure 7-17 for an example of cyclic AMP mediating the stimulation of glycogen breakdown and the inhibition of glycogen synthesis by a liver cell in response to epinephrine.

36. (a) **Correct.**

 (b) **Correct.** Activation of phospholipase C leads to production of inositol triphosphate, which stimulates release of calcium ion from the endoplasmic reticulum into the cytosol.

 (c) **Correct.** Calcium ion concentration is higher extracellularly than intracellularly. Thus, calcium ion will enter the cell along its concentration difference when its membrane channels are activated.

 (d) **Correct.** (This is true in the case of certain smooth muscle cells. See text Figure 7-21.)

37. **Cyclic AMP, cyclic GMP, activated (calicium-bound) calmodulin, and diacylglycerol all activate different protein kinases.** (This question emphasizes the importance of protein phosphorylation in regulating cell activities.)

The number of second-messenger systems can appear bewildering at first. It is helpful to keep in mind functions they have in common, as well as differences. Of the four second messengers listed, calcium-bound calmodulin is the most versatile, in that it can bind to a number of different enzymes and affect their activity by changing their conformation.

You must also be aware of conventions regarding nomenclature (the naming of things) that can be confusing. For example, inositol triphosphate (IP_3) is considered a second messenger, but its function is limited to stimulating calcium ion release from the endoplasmic reticulum. You may wonder why the calcium released by IP_3 is not called a third messenger in this case. As a matter of convention and in an effort to maintain order in terminology, hormones, neurotransmitters, paracrines, and autocrines are all considered first messengers and all cytoplasmic mediators for these first messengers are called second messengers. Furthermore, enzymes are considered "effectors," not second messengers, even though they can be early links in the chain of events that constitutes a cell's response to a message.

PRACTICE

True or false (correct the false statements):

P1. A control system in which the response increases the strength of the original stimulus is known as negative feedback.

P2. Homeostatic control systems and acclimatization are examples of biological adaptations.

P3. Free-running biological rhythms exist in the absence of environmental time cues.

P4. Body changes associated with aging are independent of changes in lifestyle.

P5. When loss of a substance from the body exceeds gain, the body is said to be in positive balance for that substance.

P6. Eicosanoids are a family of ubiquitous, fatty-acid-derived, local chemical messengers.

Multiple choice (correct each incorrect choice):

P7-9. Homeostatic mechanisms

P7. result in an absolutely constant, unchanging internal environment.

P8. operate by positive feedback so that the response tends to minimize change in the variable being controlled.

P9. depend on receptors, inputs to integrating centers, and outputs to effectors.

P10-13. Feedforward regulatory processes

P10. work in anticipation of changes in regulated variables.

P11. work in conjunction with negative feedback processes.

P12. lead to instability of the regulated variable.

P13. maximize fluctuations in the regulated variable.

P14-16. The efferent pathway of a reflex arc

P14. is so named because it carries information to the integrating center.

P15. can be neural.

P16. can be hormonal.

P17-20. Some intercellular messengers act by

P17. affecting only one kind of cell.

P18. increasing the synthesis of an enzyme.

P19. converting inactive enzyme molecules to active ones.

P20. changing the permeability of a cell's plasma membrane.

Matching: For each numbered compound, identify the single best description from the column on the far right:

P21. ___Adenylate cyclase	P26. ___ATP	(a) first messenger
P22. ___Cyclic AMP	P27. ___Phospholipase A_2	(b) second messenger
P23. ___Phospholipase C	P28. ___Diacylglycerol	(c) membrane-bound enzyme
P24. ___Calcium ion	P29. ___Paracrine	(d) is converted to cyclic AMP
P25. ___Epinephrine	P30. ___Calmodulin	(e) binds calcium

8: NEURAL CONTROL MECHANISMS

Just as the explosion of knowledge from the study of molecular biology made that discipline *the* science of the 1980s, a similar explosion is rapidly hurtling neuroscience (the study of the nervous system) into the starring role for the 1990s. Although the fundamentals of how nerve cells process and transmit information have been known for decades, new and exciting information is being gleaned daily about how cells in the brain communicate with one another. In other words, we are learning the language of the mind. The Pulitzer-prize-winning journalist Jon Franklin in "Molecules of the Mind" compares this discovery to that of atomic energy:

> *"A thousand years hence, when our descendants look back on this time, it will not be the name of Albert Einstein that comes to their lips. For while the forces contained within the nucleus of the atom are truly powerful, and though the hydrogen fires may burn hot and bright, they pale when compared with the energy contained within the human mind."*

Instructions for answering questions in this Study Guide

1. True or false: Correct each false statement. Whenever you must correct a false statement, there will almost always be more than one way to do so. In the Guide, the answer requiring the least correction will generally be given, with the understanding that other correct answers are possible. It is usually not sufficient, in terms of demonstrating understanding, to simply insert a "no" or "not," however.
2. Multiple choice: Any, all, or none of the responses may be correct. Explain why each incorrect response is incorrect.
3. Fill in the blanks: Choose the best word or words to complete each statement.
4. Directions will be given for other types of questions as they appear.

❐ Introduction

1. The nervous system consists of two major divisions: 1) the ____________________ ____________________________________(abbreviated CNS), which is composed of the __________________ and the ___________________, and 2) the ____________________ __________________________, which is composed of ____________________________.

SECTION A: STRUCTURE OF THE NERVOUS SYSTEM

❐ Functional Anatomy of Neurons (text pages 167-172)

2. Label the structures lettered in the diagram.

(a)_______________

(b)_______________

(c)_______________

(d)_______________

(e)_______________

(f)_______________

(g)_______________

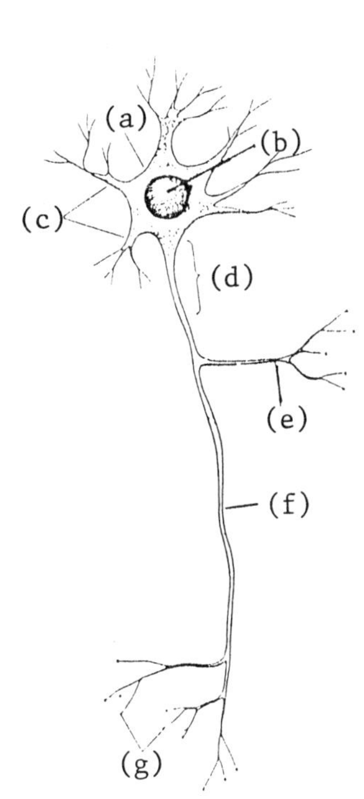

3. A neuron
   (a) is the basic unit of the nervous system.
   (b) is a nerve cell.
   (c) operates by generating electrical signals.
   (d) communicates with other neurons via chemical signals.

4. Myelin
   (a) is a fatty membranous sheath.
   (b) is formed from membranes of adjacent neurons.
   (c) covers the axons of all neurons.
   (d) influences the rate of conduction of the electrical signal down an axon.

5. True or false: A myelinated axon is shielded from direct contact with the extracellular fluid all along its length.

6. Axon transport
   (a) refers to the passage of materials from the cell body of a neuron to the axon terminals.
   (b) refers to the passage of materials from axon terminals to the cell body of a neuron.
   (c) refers to the transport of materials across the axonal membrane.
   (d) is especially important for maintaining the integrity of neurons with long axons.

7. Indicate whether the following are correct descriptions of: afferent neurons (A); efferent neurons (E); interneurons (I); any combination of them (e.g., A & I) or none (N).

   ________ lie entirely within the CNS
   ________ have no dendrites
   ________ most numerous of the nerve cells
   ________ have cell bodies outside the CNS
   ________ have cell bodies within the CNS but most of the axons are outside the CNS
   ________ are associated with effector cells
   ________ least numerous of the nerve cells
   ________ lie entirely outside the CNS
   ________ are associated with receptors (remember that there are two kinds of receptors, and the one meant here is the sensory receptor)

8. In a reflex arc involving nerve cells, ____________________ neurons constitute the ____________________ pathway, ____________________ neurons constitute the ____________________ pathway, and interneurons form the ____________________ ____________________

9. The junction between two neurons that are communicating with each other is called a ____________________. The most common type of such junction is one in which a chemical messenger called a ____________________ is released by ____________________ of the cell sending the message, called the ____________________ neuron. The cell receiving the message, called the ____________________ neuron, has specific receptors for the messenger. Receptors are most commonly found on the ____________________ and ____________________ of the nerve cells.

10. A given neuron can
    (a) be either a presynaptic neuron or a postsynaptic neuron, but not both.
    (b) receive information from more than one other neuron.
    (c) transmit information to more than one other neuron.

11. True or false: The most numerous cells in the CNS are interneurons.

12. List four functions of glial cells.
    (a) ____
    ____
    (b) ____
    ____
    (c) ____
    ____
    (d) ____
    ____

❐ Divisions of the Nervous System (text pages 172-184)

13. Match each structure on the left with the best description on the right.

_______ganglia	(a) clusters of neuronal cell bodies in the CNS
_______meninges	(b) group of nerve fibers running together in the CNS
_______nerve	(c) protective membranes covering nervous tissue
_______nuclei	(d) nonmyelinated tissue in the CNS
_______pathway	(e) clusters of glial cells
_______gray matter	(f) clusters of neuronal cell bodies in the peripheral nervous system
	(g) group of nerve fibers running together in the peripheral nervous system

14. True or false: Dorsal root ganglia contain the cell bodies of efferent neurons.

15. True or false: Spinal nerves are composed of the axons of both afferent and efferent neurons.

16. Complete the table:

SUBDIVISIONS OF THE BRAIN

(a) ______________ }
(b) ______________ } Forebrain ______________

Midbrain ______________ }
(c) ______________ } (e) ______________
(d) ______________ }

Cerebellum ______________

17. True or false: The hollow structures that contain cerebrospinal fluid are called cerebellar peduncles.

18. The reticular formation
(a) is part of the cerebellum.
(b) is essential for life.
(c) receives information from the spinal cord.
(d) sends relatively unprocessed information to the forebrain.

19. The cerebellum
(a) is important for coordinating body movement.
(b) has a cortex.
(c) has subcortical nuclei.

20. The cerebrum
(a) is important for controlling body movement.
(b) has a cortex.
(c) has subcortical nuclei.

21. True or false: Although nerve cells in each cerebral hemisphere make connections with other cells in the same hemisphere, there is no crossover of information between the two hemispheres.

22. Name the lobes of the cerebral cortex indicated on the diagram.

(a) ____________________

(b) ____________________

(c) ____________________

(d) ____________________

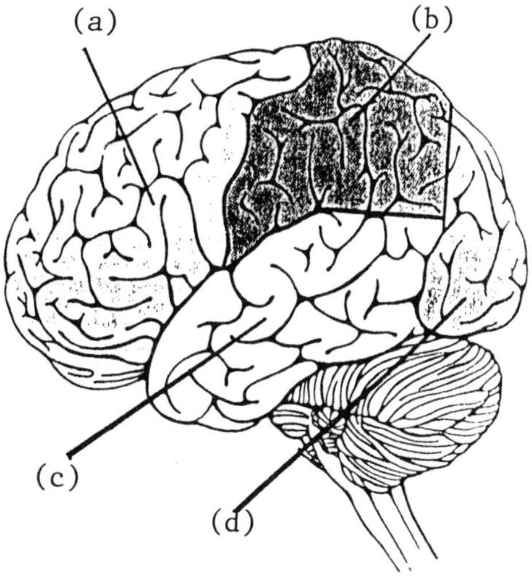

(a) (b) (c) (d)

23. True or false: The cerebrum consists only of gray matter.

24. Explain the significance of the extensive folding of the cerebral cortex.

____

____

____

____

25. True or false: The basal ganglia are important subcortical nuclei in the cerebrum.

26. The thalamus

(a) is a major part of the diencephalon.

(b) is a cluster of subcortical nuclei.

(c) sends information to the cerebral cortex.

27. The hypothalamus

(a) is a major part of the diencephalon.

(b) lies above the thalamus.

(c) is too small to be of much importance.

28. True or false: The thalamus is the single most important control area for regulating the internal environment.

29. The limbic system is
(a) composed of several parts of the diencephalon and cerebrum.
(b) composed of white matter and gray matter.
(c) associated with emotional responses and learning.

30. The peripheral nervous system consists of twelve pairs of ____________________ nerves and 31 pairs of ____________________ nerves. The former bring information primarily to and from the ____________________, while the latter do the same for the ____________________.

31. True or false: The efferent division of the peripheral nervous system consists of the somatic nervous system and the sympathetic nervous system.

32. The somatic portion of the peripheral nervous system
(a) refers to the mixed spinal nerves.
(b) consists of motor nerves.
(c) is multineuronal.

33. Indicate whether the descriptions below apply to the somatic nervous system (S), the autonomic nervous system (A), both (B), or neither (N):
(a) _____ innervates muscle
(b) _____ utilizes acetylcholine as a neurotransmitter
(c) _____ excites effector cells
(d) _____ innervates skeletal muscle
(e) _____ utilizes norepinephrine as a neurotransmitter
(f) _____ innervates smooth and cardiac muscle
(g) _____ inhibits effector cells
(h) _____ innervates endocrine glands
(i) _____ involves participation of ganglia
(j) _____ innervates exocrine glands

34. True or false: Preganglionic fibers of the parasympathetic division of the autonomic nervous system leave the CNS at the level of the brain and sacral portions of the spinal cord.

35. True or false: The efferent portion of vagus nerve is sympathetic.

36. True or false: Most of the parasympathetic ganglia lie in chains along the spinal cord called sympathetic trunks.

37. True or false: The sympathetic division of the autonomic nervous system is arranged so that it acts largely as a unit, whereas the components of the parasympathetic division generally act independently.

38. Indicate which of the descriptive terms on the right are true for each of the following:

___ preganglionic parasympathetic fibers	(a) cholinergic
___ postganglionic parasympathetic fibers	(b) adrenergic
___ preganglionic sympathetic fibers	(c) release epinephrine
___ postganglionic sympathetic fibers	(d) release norepinephrine
	(e) release acetylcholine

39. Nicotinic receptors are
(a) receptors for acetylcholine.
(b) receptors for nicotine.
(c) found on smooth muscle cells.

40. True or false: Alpha-adrenergic receptors are associated with a second-messenger signal transduction mechanism, whereas beta-adrenergic receptors are associated with opening ion channels.

41. True or false: "Dual innervation of effectors" refers to the innervation of the same effector organs by somatic and autonomic nerves.

❒ **Neural Growth and Regeneration** (text pages 184-185)

42. True or false: Neuron cell division and nervous system growth continue throughout life.

43. True or false: Damage to axons leads to inevitable loss of function of the innervated effector organ.

44. True or false: Neuronal cell death is caused only by injury and disease.

❐ Blood Supply, Blood-Brain Barrier Phenomena, and Cerebrospinal Fluid (text pages 185-186)

45. Give two reasons the brain is critically dependent upon a continuous flow of blood.

(a)____

____

____

(b)____

____

____

46. The most common form of brain damage is ____________, which is caused by ____.

47. The blood-brain barrier

(a) is formed by cells lining tiny blood vessels in the brain.

(b) is present in all parts of the brain.

(c) is ineffective against nonpolar substances.

(d) prevents entry of all lipid-insoluble molecules into the brain.

48. Morphine is a drug commonly given to patients in hospitals to relieve severe pain. Heroin, which also relieves pain, is not used for this purpose. Based on their solubility characteristics described in the text, explain why.

____

____

____

____

49. Cerebrospinal fluid
    (a) has the same composition as blood plasma.
    (b) acts as a cushion for the brain and spinal cord.
    (c) is secreted by cells lining the ventricles.
    (d) must be constantly reabsorbed.

SECTION B: MEMBRANE POTENTIALS

❐ Basic Principles of Electricity (text pages 186-187)

50. Extracellular fluid contains high concentrations of ______________ and ______________, relative to intracellular fluid, whereas intracellular fluid contains relatively more ______________ and ______________.

51. Substances with the same electrical charge are ______________ (attracted to/ repelled by) each other, and those with opposite charge are ______________ each other. Therefore, Na ions and Cl ions in solution are ______________ and Na and K ions are ______________.

52. The difference in electrical charge between two points
    (a) is called the potential difference between those points.
    (b) is called the electrical potential between those points.
    (c) is called the potential between those points.
    (d) depends upon charge flow between those points.

53. Ohm's law ($I = E/R$) states that
    (a) current increases with increasing voltage or increasing resistance.
    (b) current increases with decreasing voltage or decreasing resistance.
    (c) current increases with increasing voltage or decreasing resistance.
    (d) the electrical potential of a system can be determined by measuring the current and the resistance of the material the charge is moving in.

54. A solution of water and ions is a ______________ (good/poor) conductor of electricity and thus offers ______________ (low/high) resistance to charge flow. These electrical properties are due to the ions in the solution, which are called ______________.

55. True or false: The lipid portion of a cell's plasma membrane constitutes a barrier to current.

❐ The Resting Potential (text pages 187-191)

56. The resting membrane potential
    (a) occurs only in nerve and muscle cells.
    (b) is the same in all cells.
    (c) is oriented so that the cell's interior is positive with respect to the extracellular fluid.
    (d) requires the separation of most of the cell's charged particles.

57. List the two factors that determine the magnitude of the resting membrane potential.
    (a)____
    ____
    (b)____
    ____

58. True or false: Ions other than Na^+, K^+, and Cl^- play no role in generating the resting membrane potential of a cell.

59. True or false: A cell whose plasma membrane is more permeable to Na^+ than is the plasma membrane of a nerve cell will have a resting potential greater than that of the nerve cell.

60. Compartments A and B are separated by a membrane that is permeable to K^+ but not to Cl^-. At time zero, a solution of KCl is poured into compartment A and pure H_2O is poured into compartment B.

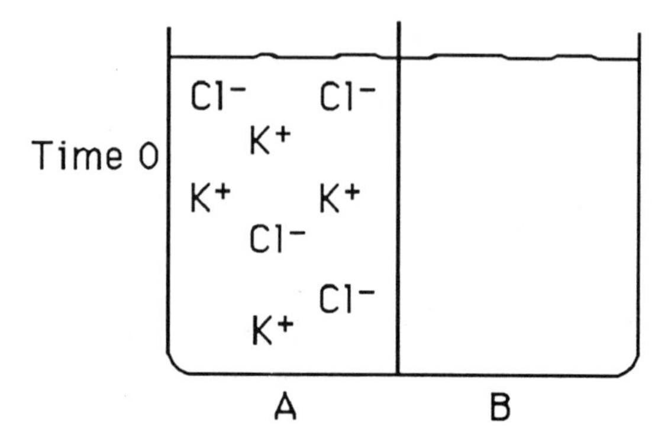

At equilibrium,

(a) the concentration of K^+ in A will be lower than it was at time zero.

(b) diffusion of K^+ from A to B will be equal to the diffusion of K^+ from B to A.

(c) compartment A will be electrically neutral and B will have an accumulation of positive charge.

(d) there will be a potential difference across the membrane, with side A negative relative to side B.

61. Now assume that the compartments in Question 60 are emptied and the membrane replaced with one that is permeable to both K^+ and Cl^-, but no other ions. At time zero, a solution of NaCl is poured into compartment A and a solution of KA (A is an anion other than Cl^-) is poured into compartment B.

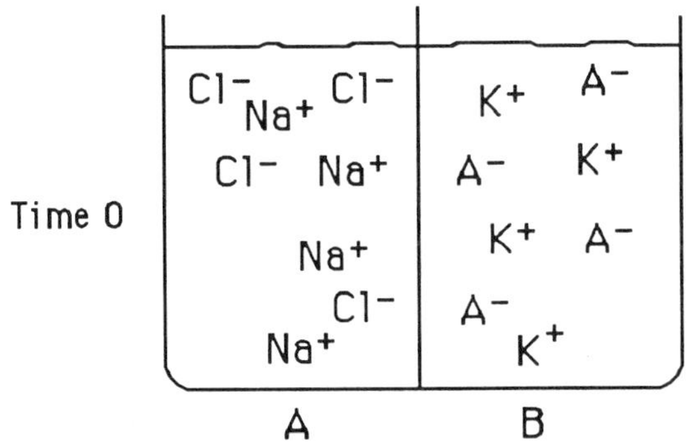

At equilibrium,

(a) compartment A will be negatively charged relative to B.

(b) there is a net flux of K^+ from B to A.

(c) there is a concentration gradient favoring Cl^- diffusion from A to B and this gradient is balanced by an electric force favoring diffusion from B to A.

(d) the potential difference across the membrane is the equilibrium potential of Cl^- in this example.

62. The equilibrium potential of an ion is the potential

(a) at which there is no movement of the ion across a cell membrane.

(b) at which there is no net movement of the ion across a cell membrane.

(c) difference across a membrane that creates an electric force favoring diffusion that is equal in magnitude and opposite in direction to the diffusion force provided by the concentration difference of the ion across the membrane.

63. The equilibrium potential for K ions in nerve cells is about -90 mV. The membrane potential for many neurons at rest is -70 mV. Therefore,
    (a) increasing the permeability of a resting neuronal membrane to K ions will decrease the membrane potential.
    (b) increasing the permeability of a resting neuronal membrane to K ions will increase the membrane potential.
    (c) in nerve cells at rest, there is net diffusion of K ions out of the cell.
    (d) increasing the membrane potential of a nerve cell would slow the diffusion of K ions out of it.

64. Neurons at rest are not at the equilibrium potential for K^+ because the neuronal membrane is
    (a) not as permeable to K^+ as to Na^+.
    (b) not as permeable to K^+ as to Cl^-.
    (c) slightly permeable to Na^+.

65. True or false: The maintenance of a resting potential in a neuron depends upon the functioning of the Na,K-ATPase pumps in the membrane.

66. What would be the effect of a poison such as cyanide (in other words, one that disrupts oxidative phosphorylation) on the resting potential of a neuron? Why?

    ____
    ____
    ____
    ____

67. Neuronal Na,K-ATPase pumps are said to be electrogenic. Explain.

    ____
    ____
    ____
    ____

68. True or false: The resting membrane potential of a neuron is constant because the components of the extracellular and intracellular fluid are in equilibrium.

❐ Graded Potentials and Action Potentials (text pages 191-200)

"The all-or-none transmission of a stimulus along a nerve fiber means a system even simpler than the Morse code. There are not even dots and dashes, just dots."

Anthony Smith, *The Body.*

69. Match the term on the left with the correct description on the right.

_____ depolarizing

_____ hyperpolarizing

_____ repolarizing

(a) increasing the magnitude of the membrane potential from its resting level to a more negative value
(b) returning the membrane potential to its resting level
(c) decreasing the magnitude of the membrane potential from its resting level toward or above zero
(d) maintaining a constant membrane potential

70. Graded potentials
(a) include various kinds of potential changes that are named for their location, their function, or how they are initiated.
(b) are always depolarizing.
(c) are related in magnitude to the magnitude of the stimulus initiating them.
(d) are unimportant for long-distance signalling in the nervous system.

71. True or false: A graded potential in a membrane results in an electric current along an adjacent area of membrane, and this current diminishes with distance from the site of the initial potential change.

72. True or false: During a hyperpolarizing graded potential, positively charged ions flow away from the site of the initial hyperpolarization on the outside of membrane, and toward this site on the inside.

73. Excitable plasma membranes are
(a) found in any cell that has a resting membrane potential.
(b) those that are capable of producing graded potentials.
(c) found only in nerve, muscle, and some gland cells.

74. Action potentials are
   (a) the mechanism for long-distance communication in the nervous system.
   (b) all-or-none in character.
   (c) conducted decrementally.
   (d) a form of graded potential.

75. Label the events of the action potential drawn below, in terms of depolarizing, hyperpolarizing, and repolarizing the membrane.

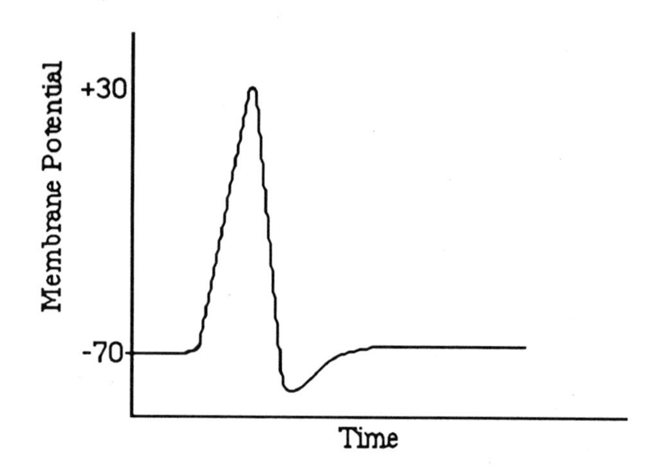

76. During the depolarization phase of the action potential,
   (a) channels for Na ions open.
   (b) the permeability of the membrane to Na ions becomes greater than that to K ions.
   (c) the membrane reaches the equilibrium potential for Na ions.

77. During the initial repolarization phase of the action potential,
   (a) Na^+ channels close.
   (b) K^+ channels open.
   (c) permeability to Na^+ is greater than it is when the membrane is at rest.

78. During an action potential, changes in the membrane's permeability to potassium
   (a) begin during depolarization.
   (b) speed repolarization.
   (c) are responsible for hyperpolarization.

79. True or false: Because so many Na ions move into a cell during an action potential and so many K ions move out, no further action potentials can be generated in a given membrane until the Na,K-ATPase pumps can restore the concentration gradient.

80. The permeability of an excitable membrane to Na ion increases with
   (a) membrane hyperpolarization.
   (b) membrane depolarization.
   (c) increasing permeability to Na ion.

81. True or false: The Na and K channels that open during an action potential are voltage-regulated, both responding to hyperpolarization of the membrane.

82. Explain the mechanism of action of a locally acting anesthetic such as Novocaine.

   ____

   ____

   ____

83. A threshold stimulus applied to an excitable membrane is one that is just sufficient to
   (a) trigger a graded potential in the membrane.
   (b) trigger an action potential in the membrane.
   (c) cause net inward movement of positive charge through the membrane.

84. True or false: The action potential elicited by a supra-threshold stimulus is larger than one elicited by a threshold stimulus.

85. True or false: The refractory period of an excitable membrane refers to the period of time during which no stimulus, however strong, will elicit a second action potential in the membrane.

86. True or false: The absolute refractory period of an excitable membrane roughly corresponds to the period when sodium channels are opening and closing.

87. An action potential generated in a membrane
   (a) travels decrementally down the membrane.
   (b) generates a new action potential in an adjacent area of membrane.
   (c) generates a local current that depolarizes adjacent membrane to threshold potential.
   (d) is the same size as the action potential ultimately generated at the end of the membrane.

88. The diagram below depicts a section of excitable membrane. At time zero, a threshold stimulus is applied to the membrane at point E, and this stimulus results in an action potential at E at time 1.

(a) Complete the diagram for time points 2, 3, and 4.

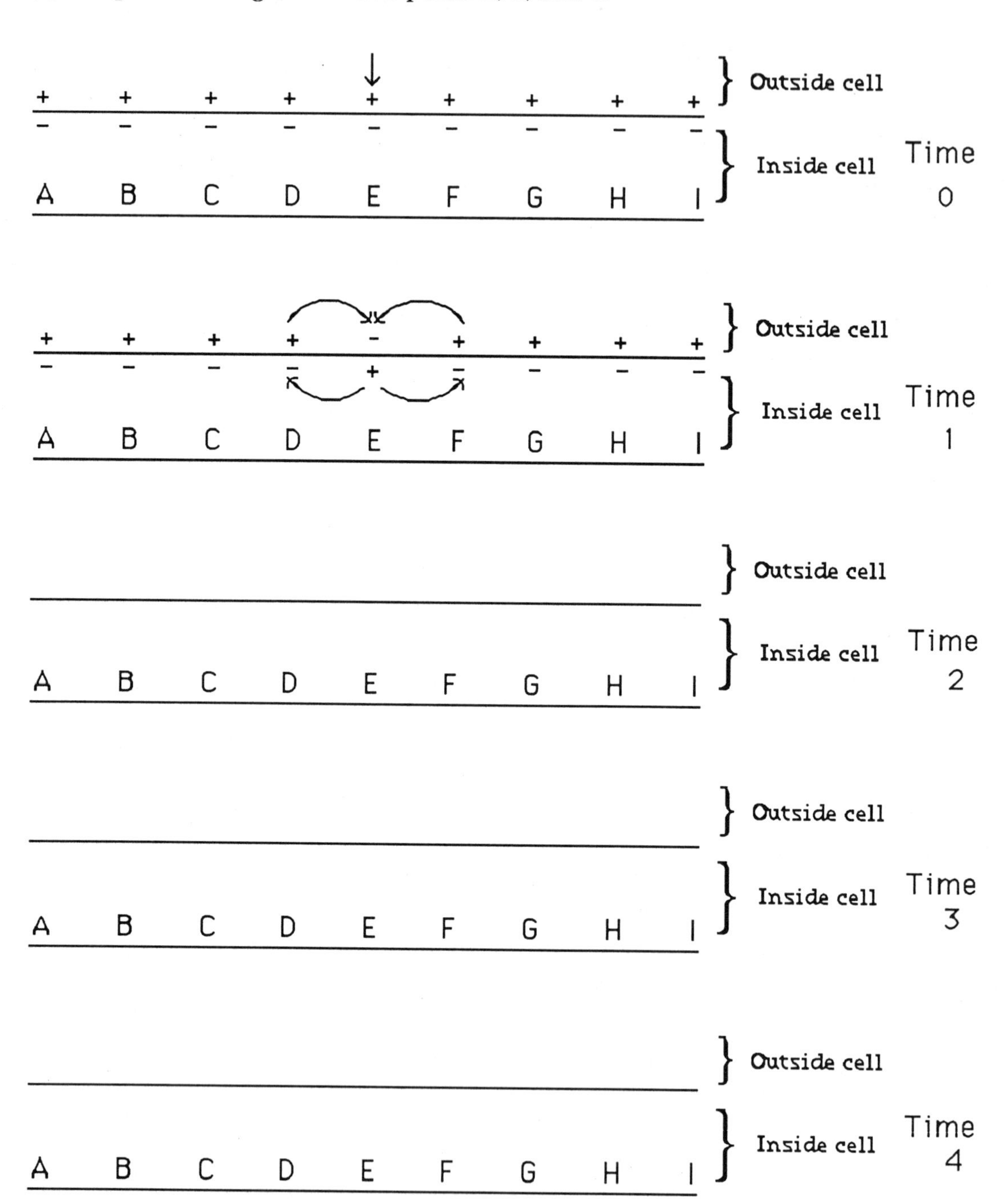

(b) Describe the direction of a local current. ____________________

____

____

(c) Describe the direction of action potential propagation. ____________________

____

(d) What would the direction of propagation be if the stimulus at time zero were given at point A? ____________________

89. Action-potential propagation down an axon is faster
    (a) for a strong stimulus than for a weak one.
    (b) in a small-diameter fiber than in a larger-diameter one.
    (c) in a myelinated fiber than in a nonmyelinated one.

90. Pacemaker potentials are
    (a) found in all excitable cells.
    (b) generated by a graded potential.
    (c) generated by a spontaneous change in ion channels in the membrane.

SECTION C: SYNAPSES

One of the most fascinating stories in the history of physiology is the discovery of neurotransmitters by the German scientist Otto Loewi, who was awarded the 1936 Nobel Prize for Physiology or Medicine for his discovery. Part of the fascination lies in the report that the idea for the critical experiment proving the existence of chemical messengers from nerves came to Loewi in a dream.

Loewi knew that stimulating the vagus nerve in frogs, as in humans, causes the heart rate to slow. The question was: how? Was the cause an electrical "spark" that inhibited the heart's pacemaker? Or was it a chemical messenger released by the nerve?

Loewi also knew that the frog heart can keep beating long after its removal from the animal if placed in an isotonic salt solution approximating extracellular fluid. Loewi performed a simple, yet brilliant, experiment: He placed one frog heart, with its vagus nerve still attached, into a dish of isotonic saline (dish A) and another heart without the vagus into another dish with the same saline (dish B). He then electrically stimulated the vagus nerve of the heart in dish A and watched the heart rate slow. After some time, he removed the stimulated heart and transferred the other heart from dish B to dish A — and observed that the beating of the second heart slowed to match that of the first one!

Something in the saline acted like a stimulated vagus nerve in affecting heart rate. The signal from nerve to muscle thus had to be a chemical.

The junction between a nerve and a muscle is called (not surprisingly) a neuromuscular junction, not a synapse. But scientists soon realized that most nerve cells "talk" to each other as well as to the heart and other muscle. This communication takes place by means of chemical messengers called neurotransmitters.

❐ **Functional Anatomy of Synapses** (text pages 200-205)

91. Interneurons are said to function as integrators of information because they
    (a) receive synaptic input from many other neurons.
    (b) sum excitatory and inhibitory synaptic input.
    (c) are presynaptic to many other neurons.

92. True or false: Two neurons joined by an electric synapse function as a single neuron.

93. At a chemical synapse
    (a) action potentials in the presynaptic neuron are transmitted across the synaptic cleft by local current.
    (b) neurotransmitter molecules released by the presynaptic cell diffuse across the synaptic cleft.
    (c) neurotransmitters released by the presynaptic cell bind to specific proteins in the subsynaptic membrane.

94. The first step in triggering neurotransmitter release by action potentials is the ____________________ of the axon terminal, a step that causes voltage-regulated channels for ________________ ion to open and consequent ________________ (influx/outflux) of that ion. This event results in binding of __________________ to specific sites on the axon terminal membrane, and such binding leads to exocytosis of the neurotransmitter.

95. True or false: The delay of approximately one millisecond between the time an action potential arrives at the axon terminal and the time the effects on the subsynaptic membrane occur is probably due to the time required for diffusion of the neurotransmitter across the synaptic cleft.

96. Following the binding of a neurotransmitter to a receptor on the subsynaptic membrane,
   (a) the activated receptor may open specific ion channels.
   (b) the neurotransmitter may be chemically modified.
   (c) the neurotransmitter may be actively transported into the postsynaptic cell.

97. True or false: A neurotransmitter that is excitatory at one synapse will be excitatory at all synapses where it is released.

98. Neurotransmitters at excitatory synapses
   (a) cause the postsynaptic cell membrane to become depolarized.
   (b) result in action potentials in the postsynaptic cell membrane.
   (c) result in graded potentials in the postsynaptic cell membrane.
   (d) result in EPSPs in the postsynaptic cell membrane.

99. An EPSP in a membrane
   (a) is usually a result of increased permeability to all small ions.
   (b) is usually a result of increased permeability to all small, positively charged ions.
   (c) is caused by net movement of positive charges into the cell.
   (d) spreads decrementally along the membrane.

100. An inhibitory postsynaptic potential is
   (a) a depolarizing potential.
   (b) a hyperpolarizing potential.
   (c) caused by increased membrane permeability to Na^+.
   (d) caused by increased membrane permeability to K^+ and/or Cl^-.

101. Explain how increasing a postsynaptic membrane's permeability to Cl ion will stabilize the membrane's potential at the resting level.________________________________

____

____

____

____

____

102. Activation of a postsynaptic cell
   (a) will generally result from one excitatory synaptic event, provided no inhibitory synapse on the cell is active simultaneously.
   (b) requires summation of EPSPs.
   (c) requires that excitatory synaptic activity predominate over inhibitory activity.

103. Temporal summation on a postsynaptic membrane refers
   (a) to the effect on the membrane of one (or more) synaptic event before the effects of a previous synaptic event have died away.
   (b) only to addition of EPSPs.
   (c) only to the effect of stimulating the same synapse repeatedly.

104. True or false: Because all parts of a neuronal cell body have the same threshold, no one synapse on the cell is more important than any other.

105. True or false: The frequency of action potentials in a postsynaptic cell is directly related to the degree of postsynaptic depolarization of the cell.

❐ **Synaptic Effectiveness** (text pages 205-206)

106. True or false: The effects of activation of any given synapse on a subsynaptic membrane are invariable.

107. During a high-frequency burst of action potentials down an axon, the amount of neurotransmitter released with each action potential increases (for a finite time). Explain.______________________________

____

____

____

____

108. A presynaptic synapse
   (a) is a synapse between axons.
   (b) may be stimulatory.
   (c) may be inhibitory.

109. True or false: Some synapses regulate their own activity.

110. Drug X interferes with the action of norepinephrine at synapses. Which of the following mechanisms could explain the effects of X?
   (a) X inhibits synthesis of norepinephrine at the axon terminal.
   (b) X inhibits norepinephrine release from the terminal.
   (c) X blocks reuptake of norepinephrine by the terminal.
   (d) X inhibits the synthesis of norepinephrine-degrading enzymes.
   (e) X is a norepinephrine agonist.
   (f) X is a norepinephrine antagonist.

❐ Neurotransmitters and Neuromodulators (text pages 206-210)

111. Neuromodulators may
   (a) influence synaptic effectiveness.
   (b) act through second messengers.
   (c) be released with neurotransmitters.
   (d) be hormones.
   (e) be the same molecules as neurotransmitters.

112. True or false: A major distinction between the action of neurotransmitters and that of neuromodulators is that the former is considerably slower than the latter.

113. The four major classes of neurotransmitters/neuromodulators are ________________, ________________, ________________, and ________________.

114. Acetylcholine
   (a) synthesis occurs in neuronal cell bodies.
   (b) synthesis requires acetyl coenzyme A.
   (c) is broken down by enzymes present in the postsynaptic cell membranes.
   (d) is important for peripheral nervous system function but relatively unimportant in the brain.

115. True or false: After the breakdown of acetylcholine by acetylcholinesterase, the acetate is actively transported back into the presynaptic cell.

116. The neurotransmitters dopamine and norepinephrine belong to a subclass of biogenic amines called ______________________. The rate-limiting enzyme in the synthesis of these amines is the one converting the amino acid ______________________ to ______________________. Another important neurotransmitter that is also a biogenic amine but is synthesized from a different amino acid is ______________________.

117. The two enzymes that catalyze breakdown of catecholamines are called ____ (MAO) and ____ (COMT). Catecholamine catabolism differs from the breakdown of acetylcholine in that the enzymes for the former are found primarily ______________________ (inside/outside) the cell and those for the latter are found ________________ (inside/outside).

118. True or false: Catecholamines are thought to act as important neuromodulators in the CNS.

119. Serotonin
   (a) metabolism resembles that of the catecholamines.
   (b) is a neuromodulator.
   (c) activity in the brain is highest during sleep.

120. Given: Serotonin inhibits sensory pathways in the brainstem; the hallucinatory drug psilocybin is chemically related to serotonin; the even more potent hallucinogen LSD inhibits the release of serotonin by serotoninergic neurons. Which of the following statements is/are likely to be correct? Explain.
   (a) Serotonin causes hallucinations.

____

____

____

   (b) Psilocybin may be a serotonin antagonist.

____

____

____

____

(c) Hallucinations are related to sensory "overload."

____

____

____

____

121. True or false: Catecholamines are the most abundant neurotransmitters in the CNS.

122. Gamma-aminobutyric acid (GABA) is
    (a) an amino acid neurotransmitter.
    (b) a major excitatory transmitter in the CNS.
    (c) synthesized from glutamic acid.

123. Neuropeptides are
    (a) synthesized by neurons.
    (b) composed of amino acids.
    (c) found only in the nervous system.

124. Synthesis of neuropeptides differs from that of other neurotransmitters because it
    (a) takes place in the axon terminals of neurons.
    (b) takes place in the cell bodies of neurons.
    (c) involves synthesis of precursor molecules.

125. True or false: Opiate drugs such as morphine are agonists of a class of neurotransmitters called endorphins.

126. Substance P is a neurotransmitter important for conveying the sense of ____________________.

❐ **Neuroeffector Communication** (text pages 210-211)

127. True or false: Neuroeffector communication is similar to synaptic communication except that, in the former, the effector cells are not neurons.

128. True or false: The most common neurotransmitters for neuroeffector communication are dopamine and acetylcholine.

SECTION D: RECEPTORS

❐ Introduction (text pages 211-212)

129. True or false: The process by which sensory receptors change various forms of energy into electrical energy is called translation.

130. Sensory receptors are
   (a) found only on nerve cells.
   (b) found on efferent neurons.
   (c) specific for only one form of energy.

131. True or false: According to the doctrine of specific nerve energies, only one kind of stimulus can activate any one kind of receptor.

❐ The Receptor Potential (text pages 212-213)

132. The receptor potential
   (a) is an action potential.
   (b) triggers action potentials.
   (c) varies in magnitude with stimulus strength.
   (d) is usually depolarizing.

133. The Pacinian corpuscle is an example of a ______________________ receptor, meaning that it responds to ______________________.

134. True or false: The greater the magnitude of the receptor potential generated by a stimulus, the greater the amplitude of the action potentials the receptor potential induces.

135. True or false: The first action potential in an afferent neuron is usually generated in the first node of Ranvier.

❐ **Magnitude of the Receptor Potential** (text page 213)

136. True or false: The rate of change of a stimulus may be important in determining receptor response.

137. Adaptation in sensory receptors
   (a) refers to the decrease or cessation of receptor potentials despite maintenance of a stimulus.
   (b) occurs to the same extent in all receptors.
   (c) may be rapid or slow, depending on the receptor.

ANSWERS

Boldface type indicates the answers you should have given. Words in medium-face (ordinary) type explain or expand upon the answer.

1. The nervous system consists of two major divisions: 1) the **central nervous system** (abbreviated CNS), which is composed of the **brain** and **spinal cord**, and 2) the **peripheral nervous system**, which is composed of **nerves leading to and from the CNS**.

2. (a) **Cell body**
   (b) **Nucleus**
   (c) **Dendrites**
   (d) **Initial segment of axon**
   (e) **Axon collateral**
   (f) **Axon**
   (g) **Axon terminals**

   Note: A "typical" neuron has about 1000 dendrites and from 1000 to 10,000 axon terminals.

3. (a) **Correct.**
   (b) **Correct.** (The terms are synonymous.)
   (c) **Correct.**
   (d) **Correct.** (There is some direct electrical signalling between nerve cells, as you will see later, but the primary means of neural communication is through chemical messengers — the neurotransmitters.)

4. (a) **Correct.**

   (b) **Incorrect. Myelin is formed by non-neural cells adjacent to the neurons.** (Myelin-forming cells in the peripheral NS are called Schwann cells; those in the CNS are oligodendroglia.)

   (c) **Incorrect. Some axons are not myelinated.**

   (d) **Correct.** (It increases the rate.)

5. **False.** A myelinated axon is shielded from direct contact with the extracellular fluid ~~all along its length~~ **except at the nodes of Ranvier** (which are the spaces between myelin-forming cells).

6. (a) **Correct.** (This is the most common definition.)

   (b) **Correct.** (Some nutritive material and growth-regulating factors are transported this way, as are some harmful agents such as toxins and viruses. This kind of transport is sometimes referred to as "retrograde" to distinguish it from transport in the other direction.)

   (c) **Incorrect. It refers to transport lengthwise, within the axon only.**

   (d) **Correct.** (The longer the axon, the farther it is from the support structures in the cell body — the nucleus, endoplasmic reticulum, and ribosomes. If an axon is cut so that it is separated from its cell body, it will degenerate.)

7. **I** lie entirely within the CNS

   A have no dendrites

   I most numerous of the nerve cells

   A have cell bodies outside the CNS

   E have cell bodies within the CNS but most of the axons are outside the CNS

   E are associated with effector cells

   A least numerous of the nerve cells

   N lie entirely outside the CNS

   A are associated with sensory receptors

8. In a reflex arc involving nerve cells, **afferent** neurons constitute the **afferent** pathway, **efferent** neurons constitute the **efferent** pathway, and interneurons form the **integrative center**.

9. The junction between two neurons that are communicating with each other is called a **synapse**. The most common type of such junction is one in which a chemical messenger called a **neurotransmitter** is released by **axon terminals** of the cell sending the message, called the **presynaptic** neuron. The cell receiving the message, called the

postsynaptic neuron, has specific receptors for the messenger. Receptors are most commonly found on the **dendrites** and **cell body** of the nerve cells. (The dendrites of most neurons have the great majority of synaptic input because their collective surface area is so much greater than that of the cell body.)

10. (a) **Incorrect. Interneurons are both pre- and postsynaptic.** (Most neural pathways are multineuronal, and many involve large numbers of interneurons.)

    Note: Efferent neurons use neurotransmitters to communicate with effector cells, but these junctions are called neuroeffector junctions, not synapses. You will learn more about these special neuron-effector cell junctions later in this chapter.

    (b) **Correct.** (The vast majority of neurons receive information from many other neurons. The average number of synapses per neuron is 1000, while the range is from 1 to more than 100,000.)

    (c) **Correct.** (Although this fact has not been stated explicitly in the text at this point, you should be able to infer it from the information that most neurons receive multiple inputs.)

11. **False.** The most numerous cells in the CNS are ~~interneurons~~ **glial cells.**

12. (a) **They form myelin for neurons in the CNS.** (oligodendroglia)

    (b) **They regulate the composition of the extracellular fluid in the CNS.**

    (c) **They deliver fuel molecules (glucose) to neurons and remove waste products of metabolism.**

    (d) **They are important for the development of the nervous system, acting as guides for neuron migration and stimulating neuronal growth.**

    (Functions 2, 3, and 4 are performed by astroglia.)

13. **f** ganglia
    c meninges
    g nerve
    a nuclei (do not confuse this type of nucleus with the organelle of the same name)
    b pathway (also called "tract")
    d gray matter

14. **False.** Dorsal root ganglia contain the cell bodies of ~~efferent~~ **afferent** neurons. (The cell bodies of efferent neurons are located in the gray matter of the spinal cord.)

15. **True.** (Spinal nerves are "mixed," with fibers containing information *to* the CNS running alongside fibers containing information *from* the CNS. There is no exchange of information between individual fibers in a nerve.)

16.

SUBDIVISIONS OF THE BRAIN

(a) **Cerebrum**
(b) **Diencephalon** } Forebrain

Midbrain
(c) **Pons** } (e) **Brainstem**
(d) **Medulla oblongata**

Cerebellum

17. **False.** The hollow structures that contain cerebrospinal fluid are called ~~cerebellar peduncles~~ **ventricles**.

Alternatively, the answer below is also correct:

False. The ~~hollow structures that contain cerebrospinal fluid~~ **pathways linking the cerebellum and the brainstem** are called cerebellar peduncles.

18. (a) **Incorrect. It is part of the brainstem.**
(b) **Correct.** (It contains nuclei that regulate heart rate, blood pressure, and breathing and is required for consciousness.)
(c) **Correct.** (And from all other parts of the CNS as well.)
(d) **Incorrect. It is an important processing and integrating center for neural information.** (Many pathways in the reticular formation have hundreds of synapses, allowing for such processing. Some relatively unprocessed information does reach the forebrain by way of the long neural pathways in other parts of the brainstem, however. See text Figure 8-9.)

19. (a) **Correct.** (Which you will learn about in Chapter 12.)
(b) **Correct.** (An outer shell, or "bark," of neuronal cell bodies; "cortex" means "bark.")
(c) **Correct.** (Cell bodies beneath the cortex.)

20. (a) **Correct.**
(b) **Correct.** (The cerebral cortex is the largest and most complex part of the human brain. It is responsible for the cognitive functions, including the ability to reason, that set humans apart from other animals.)
(c) **Correct.**

The purpose of asking Questions 19 and 20 was not to lead you to think that the cerebellum and cerebrum are essentially the same, for they are not. The cerebellum is specialized to help carry out and coordinate body movement only. The functions of the cerebrum, as indicated in Answer 20(b), are considerably more diverse. The question was intended to alert you that the words "cerebellum" and "cerebrum" are similar, as are aspects of their anatomy. You should be on guard against confusing them.

21. **False.** ~~Although~~ Nerve cells in each cerebral hemisphere make connections with other cells in the same hemisphere~~, there is no crossover of information between the two hemispheres~~ **and also with cells in the opposite hemisphere by way of tracts called commissures** (or the **corpus callosum**.)

22. (a) **Frontal lobe**
    (b) **Parietal lobe**
    (c) **Temporal lobe**
    (d) **Occipital lobe**

23. **False.** The cerebrum consists ~~only~~ of gray matter **and white matter**. (Even though we emphasize the gray matter — the cerebral cortex and subcortical nuclei — the white-matter fiber tracts, called association fibers, are critically important in getting information to and from the various processing and integrative centers.)

24. **The cortical layer of the cerebrum is extensively folded to allow for a larger surface area than would fit in the skull if this layer were not folded.**

    The human brain, which is dominated by the cerebral cortex, resembles an oversized walnut with its many folds and grooves and fissures. When unfolded, the surface area is more than 2000 cm^2. We should be grateful that our heads do not need to be large enough to carry that size brain in an unfolded state.

    (The amount of cerebral cortex present in an animal's brain, that is, the amount of folding one can observe, is an indication of the complexity of that brain. A rat brain, for example, has much less folding than a monkey brain, which in turn is less folded than a human brain. The cerebral cortex of a bird is almost devoid of folds — perhaps accounting for the epithet "bird-brain" for someone who is somewhat slow.)

25. **True.** (Note that the name is unfortunate, because "ganglia" refers to clusters of cell bodies *outside* the CNS, and the basal ganglia are very much within it.

26. (a) **Correct.**
    (b) **Incorrect. It is a cluster of nuclei in the diencephalon. The subcortical nuclei are in the cerebrum.**
    (c) **Correct.** (It is a major relay station for incoming messages from lower parts of the CNS and other forebrain structures.)

27. (a) **Correct.**
    (b) **Incorrect. It lies beneath the thalamus.** ("Hypo" means "under.")
    (c) **Incorrect. Even though small, the hypothalamus is critically important for maintenance of the internal environment** (and other "housekeeping" functions; see text Table 8-4).

28. **False.** The ~~thalamus~~ **hypothalamus** is the single most important control area for regulating the internal environment. (This question emphasizes the point made in Question 27. The British medical journal *Lancet* once put it, "Is there any pie in the vertebrate organization [that is, the physiology of vertebrate animals] into which the hypothalamus does not dip its finger?")

29. (a) **Correct.** (The thalamus and hypothalamus from the diencephalon, parts of the frontal lobe cortex and deep temporal lobes, and the fiber tracts interconnecting them.)
    (b) **Correct.** (The white matter, again, consists of the fiber tracts that connect the gray-matter nuclei.)
    (c) **Correct.** (The limbic system is evolutionarily very old and has not changed much as we have evolved, unlike the "newer" areas of the cortex. We shall learn more about this fascinating structure in Chapter 20.)

30. The peripheral nervous system consists of twelve pairs of **cranial** nerves and 31 pairs of **spinal** nerves. The former bring information primarily to and from the **brain**, while the latter do the same for the **spinal cord**.

    Look at text Table 8-3, which lists the cranial nerves. Note that the first and second, the olfactory and optic, are not true nerves but rather are fiber tracts. In other words, their fibers do not leave the CNS. The same is true for number VIII, the vestibulo-cochlear (which is also called "auditory" by many texts), although the fact that these fibers do not leave the CNS is not mentioned in the table. The cranial nerves, with one exception, innervate (send nerve fibers to and from) structures in the head and neck. The exception is number X, the vagus, which also innervates structures in the thorax (including the heart) and abdomen (including the digestive system). You will be encountering the vagus nerve in several chapters ahead.

31. **False.** The efferent division of the peripheral nervous system consists of the somatic nervous system and the ~~sympathetic~~ **autonomic** nervous system.

32. (a) **Incorrect. The somatic portion of the peripheral nervous system, like the autonomic portion, is efferent.**

    (b) **Correct.** ("Motor" refers to movement; these nerves innervate skeletal muscle and stimulate the contraction of muscles that move the body.)

    (c) **Incorrect. There is only one neuron between the CNS and the skeletal muscle effector.** (This one neuron is the motor neuron, which has its cell body in the CNS and most of its axon in the peripheral nervous system.)

33. (a) **B** innervates muscle

    (b) **B** utilizes acetylcholine as a neurotransmitter

    (c) **B** excites effector cells

    (d) **S** innervates skeletal muscle

    (e) **A** utilizes norepinephrine as a neurotransmitter

    (f) **A** innervates smooth and cardiac muscle

    (g) **A** inhibits effector cells

    (h) **A** innervates endocrine glands

    (i) **A** involves participation of ganglia

    (j) **A** innervates exocrine glands

34. **True.** (See text Figure 8-14.)

35. **False.** The efferent portion of vagus nerve is ~~sympathetic~~ **parasympathetic**.

36. **False.** Most of the ~~parasympathetic~~ **sympathetic** ganglia lie in chains along the spinal cord called sympathetic trunks.

    Alternatively, the answer below is also correct:

    False. ~~Most of~~ The parasympathetic ganglia lie ~~in chains along the spinal cord called sympathetic trunks~~ **within the effector organ**.

37. **True.** (As we shall see, the sympathetic division readies the body for action — what is called the "flight or fight response" — and must coordinate responses from the various effectors that enable the action, such as running away, to occur. See text Table 8-7.)

38. **a, e** preganglionic parasympathetic fibers

    a, e postganglionic parasympathetic fibers

    a, e preganglionic sympathetic fibers

    b, d postganglionic sympathetic fibers

Epinephrine is not released by postganglionic sympathetic fibers (or any of the other fibers mentioned) and so (c) is not a correct answer. As your text mentions, the adrenal medulla is part of the sympathetic nervous system and secretes epinephrine. However, the modified nerve cells of the adrenal medulla have no axons, and "fiber" means "axon" when one is referring to neurons, remember.

You may be wondering about the term "medulla," which you have encountered twice in this chapter. It refers to the middle, or "marrow," of a structure. The adrenal medulla is surrounded by a "bark" called the adrenal cortex, which secretes hormones unrelated to epinephrine and norepinephrine. The medulla oblongata of the brainstem received its name from its location between the spinal cord and the pons. When the word "medulla" is used alone, its meaning should be clear from the context in which it is used. The same is true for "cortex."

Another problem with nomenclature was mentioned by your text: Why do we speak of fibers that release norepinephrine as being "adrenergic"? The major hormone secreted by the adrenal medulla was discovered and characterized early in this century, essentially simultaneously by different investigators in the United States and Great Britain. Each group named the compound, and other scientists in each country retained the name assigned by their countrymen (epinephrine in the U.S. and adrenaline in Britain). As you are probably aware, however, most people in the U.S. are familiar with the term "adrenalin," associated with the "rush" one experiences (from the effects of epinephrine) in exciting situations. Although it would seem to make sense for U.S. scientists to adopt the British usage and call this substance adrenaline, such a change is very unlikely, and U.S. students will continue to have to memorize two names for the same hormone.

How does *nor*epinephrine differ from epinephrine? The difference in structure is that epinephrine has a methyl (CH_3) group that the "nor" compound lacks (see text Figure 8-48). Epinephrine acts at the same receptors as does norepinephrine. (These receptors are sometimes called noradrenergic.) Finally, keep in mind that both molecules are hormones (norepinephrine is also secreted by the adrenal medulla, but in smaller amounts) and both molecules are neurotransmitters, but epinephrine is used this latter way only in the CNS, and much less commonly than norepinephrine.

39. (a) **Correct.**

(b) **Correct.**

You may rightly be puzzled by the notion that a receptor exists for a compound that is not normally found in the body (nicotine is found in tobacco and some other plants). However, many drugs work by acting as agonists or antagonists of naturally occurring substances, as we learned in Chapter 7. These drugs are very useful tools for helping us identify subtle differences in receptors. Remember that receptors are proteins, and they evolve as other proteins do by undergoing slight mutations. With time, the modified receptor, which still binds its original ligand, can become associated with a signal transduction mechanism different from its original one, and the messenger-receptor complex will develop a new function. (The knowledge that nicotine acts as an agonist for ACh on some receptors explains some of the physical symptoms associated with cigarette smoking.)

(c) **Incorrect. They are found in skeletal muscle. Smooth muscle ACh receptors are muscarinic.**

(Memorization of the two classes of cholinergic receptors would be easier if all muscle-cell receptors were muscarinic. However, you may find it useful to remember that receptors for the ACh released by the postganglionic parasympathetic fibers are muscarinic, whereas those for ACh released by motor neurons are nicotinic. Then just remember which nerves do what!)

40. **False.** ~~Alpha~~ **Beta**-adrenergic receptors are associated with a second-messenger signal transduction mechanism, whereas ~~beta~~ **alpha**-adrenergic receptors are associated with opening ion channels. (This one is easy to remember because "beta" is the second letter of the Greek alphabet.)

41. **False.** "Dual innervation of effectors" refers to the innervation of the same effector organs by ~~somatic and autonomic~~ **sympathetic and parasympathetic** nerves. (Somatic nerves innervate skeletal muscles *only*. Autonomic nerves do not innervate skeletal muscles.)

42. **False.** Neuron cell division and nervous system growth ~~continue throughout life~~ **are essentially completed before birth.**

    Once neurons are formed from neuroblasts, the cells are too differentiated to divide again. Thus, no new neurons can be formed after fetal development. (Research with animals has shown that transplanted fetal brain tissue can "take" — that is, grow and function — in, for example, the brains of aging rats, restoring some mental function that is lost as the animals age. There is hope that similar procedures may be used for humans who suffer brain damage or disease.)

    Alternatively, the answer below is also correct:

    False. ~~Neuron cell division and nervous system growth~~ **Formation of synaptic connections between neurons** continue**s** throughout life. (Indeed, it is thought that learning, which we continue to do throughout life, is dependent upon forming new synapses and strengthening old ones.)

43. **False.** Damage to axons ~~leads~~ **outside the CNS may not lead** to inevitable loss of function of the innervated effector organ. (The rate of regrowth of peripheral axons is slow — about 2.5 cm a week.)

    Alternatively, the answer below is also correct:

    False. Damage to axons **within the CNS** leads to inevitable loss of function of the innervated effector organ. (An injury within the spinal cord to the axons of motor neurons innervating a leg muscle, say, would lead to permanent paralysis of that muscle, whereas injury to the same axons after they left the CNS might be followed by gradual return of function as the axons grow back.)

44. **False.** Neuronal cell death is **a normal aspect of development of the nervous system** (as well as being caused ~~only~~ by injury and disease).

45. (a) **The brain absolutely requires glucose for the production of the ATP it needs, and it stores essentially no glycogen. Therefore, the brain is completely dependent upon the glucose supplied by the blood.**

    (b) **The efficient production of ATP requires oxygen for oxidative phosphorylation. Oxygen is also carried by the blood.** (The brain requires relatively more ATP for its functions than do other tissues, for reasons that will become clear in Section B of this chapter.)

46. The most common form of brain damage is **stroke**, which is caused by **stoppage of blood flow through a region of the brain.**

47. (a) **Correct.** (These lining cells are the endothelial cells of brain capillaries. In most tissues, capillary endothelial cell junctions are quite "leaky" and permit ready passage of all but the largest proteins out of or into the blood.)

    (b) **Incorrect. Certain areas of the brain lack the barrier.** (Certain areas of the brain monitor the concentrations of molecules that do not cross the barrier, for example.)

    (c) **Correct.** (Recall that "nonpolar" is synonymous with "lipid-soluble.")

    (d) **Incorrect. Some lipid-insoluble molecules, such as glucose and amino acids, are selectively transported across the barrier by membrane transport systems.**

    As mentioned in the text, drugs that affect the brain obviously have to cross the blood-brain barrier (or else they affect the brain in those few areas that have no barrier). The brain has no defense against lipid-soluble drugs; lipid-insoluble ones may or may not cross, depending on whether they are transported across the endothelial cell membranes. Developers of drugs intended to treat conditions in the body (such as antihistamines for allergies) try to find formulas that will *not* cross the barrier, because these drugs may have unwanted side-effects in the brain. (In our example, antihistamines tend to induce drowsiness. An antihistamine that does not enter the brain has recently been developed by one company, which is good for allergy-sufferers *and* for that company's profits!)

48. **Heroin is more lipid-soluble than morphine and crosses the blood-brain barrier much more readily. Once across the membrane, heroin is quickly metabolized to morphine and is unable to diffuse out of the brain. Thus, it acts like a "super-strength, super-long-acting" version of morphine and is dangerously addictive.**

49. (a) **Incorrect. It is selectively secreted across a barrier and differs from plasma in its concentration of various ions and other molecules.**

    (b) **Correct.**

    (c) **Correct.** (mostly by the choroid plexuses)

    (d) **Correct.** (This is true because cerebrospinal fluid is being constantly secreted, and an increase in the volume of the fluid would cause a dangerous buildup of pressure on the brain.)

50. Extracellular fluid contains high concentrations of **sodium ion** and **chloride ion**, relative to intracellular fluid, whereas intracellular fluid contains relatively more **potassium ions** and **ionized organic molecules**.

51. Substances with the same electrical charge are **repelled by** each other, and those with opposite charge are **attracted to** each other. Therefore, Na ions and Cl ions in solution are **attracted to each other** and Na and K ions are **repelled**.

52. (a) **Correct.**

    (b) **Correct.**

    (c) **Correct.**

    It is important to recognize that these three terms — "potential difference," "electrical potential," and "potential" —are used synonymously. They refer to the potential energy that is stored in unlike charges separated over a distance. Recall the boulder example of potential energy from Chapter 5 in this Guide. Some of the energy used to hoist the boulder to the hilltop was stored as potential energy in the boulder. The potential energy of separated charges can be used to do work if the charges are allowed to come together. We are all familiar with the work that can be done by electricity.

    (d) **Incorrect. The electrical potential still exists whether or not electric charges can move from one point to the other.** (This movement of electrical charge is the meaning of current.)

53. (a) **Incorrect. Increasing resistance will decrease current.** (We say that there is an inverse or negative correlation between current and resistance.)

    (b) **Incorrect. Current decreases with decreasing voltage.** (We say that current and voltage are positively correlated.)

    (c) **Correct.**

    (d) **Correct.** This can be seen by rearranging the equation to $E = IR$.

54. A solution of water and ions is a **good** conductor of electricity and thus offers **low** resistance to charge flow. These electrical properties are due to the ions in the solution, which are called **electrolytes**. (This last term was introduced in Chapter 2. If you have forgotten it, perhaps a quick review of ion properties would be a good idea.)

55. **True.**

56. (a) **Incorrect. All living cells have a potential difference across their plasma membrane.**

    (b) **Incorrect. The magnitude of the potential difference varies from cell to cell.**

    (c) **Incorrect. In all resting cells, the potential is negative inside relative to outside.**

    (d) **Incorrect. Only a tiny percentage of the cell's charged particles must be separated.** (The magnitude of the potential differences in cells is very small — from -5 to -100 mV; a flashlight battery produces 1.5 V, or 1500 mV, of potential difference.)

    Note that reference is made to *resting* cells and the *resting* potential difference. This distinction is made because later we shall be dealing with membrane potentials in cells not at rest. Be sure you have a clear understanding of the concept of resting potentials and how they are produced and maintained before going on to nonresting conditions.

57. (a) **The difference in ion concentrations across the cell membrane.**

    (b) **The permeability of the membrane to the different kinds of ions.**

58. **False.** Ions other than Na^+, K^+, and Cl^- play ~~no~~ **a** (minor) role in generating the resting membrane potential of a cell.

    The purpose of this intentionally tricky question is to remind you that inorganic anions and some molecules, particularly amino acids and proteins found in intracellular fluid, play a role in a cell's resting membrane potential. These negatively charged ions and molecules counteract the positive charge created by the high intracellular concentrations of K^+. Otherwise, as you can see from adding up the charges in text Table 8-8, the inside of the cell would be positively charged relative to the extracellular fluid. Nevertheless, our discussion largely ignores these anions because their concentrations change relatively little and the cell's plasma membrane is essentially impermeable to them, especially the proteins. Therefore, the permeability of the plasma membrane to the ions that are in highest concentration — K^+, Na^+, and Cl^- — determines the magnitude of the resting potential.

59. **False.** A cell whose plasma membrane is more permeable to Na^+ than is the plasma membrane of a nerve cell will have a resting potential ~~greater~~ **less** than that of the nerve cell. (That is, the potential will be less negative inside with respect to outside, because Na ions have both a concentration gradient and an electrical force that impel

cell if the membrane is permeable to them. By convention, increasing the negativity of the inside of the cell increases the magnitude of the potential difference, while decreasing it or even making the inside of the cell positive with respect to the outside decreases the potential difference.

60. (a) **Correct.** (Thus, there was a net flux of K^+ from A to B.)
    (b) **Correct.** (At equilibrium, the two one-way fluxes are equal. Recall that movement of molecules is continuous and random.)
    (c) **Incorrect. Compartment A will have excess negative charge.**
    (d) **Correct.**

61. (a) **Incorrect. Compartment A will be positively charged relative to B** (from both the influx of K^+ and the outflux of Cl^-).
    (b) **Incorrect. At equilibrium, there is no net flux.** (The net flux occurred while equilibrium was being reached.)
    (c) **Correct.**
    (d) **Correct.** (Answer [c] is the definition of equilibrium potential. The potential difference is also the equilibrium potential for K^+ in this example.)

The situation in the two compartments in this example is similar, but not identical, to the composition of fluid in and around nerve cells, with compartment A being similar to extracellular fluid and compartment B similar to intracellular fluid. Nerve cells are more complicated than this example, of course.

62. (a) **Incorrect. There is no <u>net</u> movement at equilibrium.**
    (b) **Correct.**
    (c) **Correct.** (This is simply a restatement of the definition in Question 61-c.)

63. (a) **Incorrect. Increasing the permeability will increase the potential.** (K ions will diffuse more quickly out of the cell, driven by the concentration gradient, until the equilibrium potential is reached. At that point [-90 mV], the electric force will balance the concentration force, and no more net movement of K ions will occur.)
    (b) **Correct.**
    (c) **Correct.** (since the neuron potential difference is less than the equilibrium potential of K^+)
    (d) **Correct.**

64. (a) **Incorrect. The neuronal membrane is 50 to 75 times more permeable to K^+ than to Na^+.**

    (b) **Incorrect. Neurons at rest are at the equilibrium potential for Cl^-, meaning that the permeability of the membrane to that ion does not affect the membrane potential.** (Neither does the permeability to the ion affect its net flux; Cl^- is at equilibrium.)

    (c) **Correct.** Even though permeability to Na^+ is very much less than that to K^+ (because there are many fewer Na^+ channels open than K^+ channels), the combined concentration and electrical forces on the Na^+ push it into the neuron at an even higher rate than potassium leaves. (Note the relative length of the arrows in text Figure 8-25.)

65. **True.** (Without the membrane pumps, Na ions would leak into the cell continuously and K ions would leak out, decreasing the potential until it eventually reaches zero.)

66. **The poison would cause the resting potential to fall to zero because the Na,K-ATPase pumps require ATP, and neurons depend almost entirely upon oxidative phosphorylation for generation of ATP.**

67. **The Na,K-ATPase pumps of neurons are electrogenic because they pump three sodium ions out of the cell for every two potassium ions pumped in.** (Thus, these pumps contribute slightly to the neurons' resting membrane potential.)

68. **False.** The resting membrane potential of a neuron is constant because the components of the extracellular and intracellular fluid are in ~~equilibrium~~ **a steady-state condition.** (This distinction is an important one because it emphasizes that the resting membrane potential is being actively maintained by the Na,K-ATPase pumps. If neurons at rest were completely impermeable to Na ions, then a true equilibrium could be achieved, as in our simple example in Question 61.

69. <u>c</u> depolarizing

    <u>a</u> hyperpolarizing

    <u>b</u> repolarizing

70. (a) **Correct.**

    (b) **Incorrect. They may be either depolarizing or hyperpolarizing.**

    (c) **Correct.**

    (d) **Incorrect. Although graded potentials are not carried long distances, they initiate the long-distance signal, the action potential.**

71. **True.** If, for example, the potential is depolarizing, the current causes depolarization of adjacent segments of membrane. This depolarization would continue *along* the membrane for long distances (indefinitely if the intra- and extracellular fluids were perfect conductors of charge) if it were not for the leakage of current (charged molecules) *across* the membrane. This leakage decreases the depolarization as it travels and causes it to fade away after a few millimeters. Thus, the propagation of graded potentials is said to be decremental.

72. **True.** (See text Figure 8-28 for an example of charge flow after a depolarizing graded potential, and reverse the flow patterns. Keep in mind that like charges repel, while unlike ones attract.)

73. (a) **Incorrect. All cells have resting membrane potentials, but all cells do not have excitable membranes.**
    (b) **Incorrect. They are those capable of producing action potentials.** (Graded potentials can be induced in nonexcitable membranes by various means, as we shall soon see.)
    (c) **Correct.**

74. (a) **Correct.**
    (b) **Correct.**
    (c) **Incorrect. They are conducted without decrement.**
    (d) **Incorrect. Graded potentials are not all-or-none** (they are, as their name suggests, of variable magnitude), **and they are conducted decrementally, unlike actions potentials.** (You may also have mentioned other differences described in text Table 8-10.)

75.

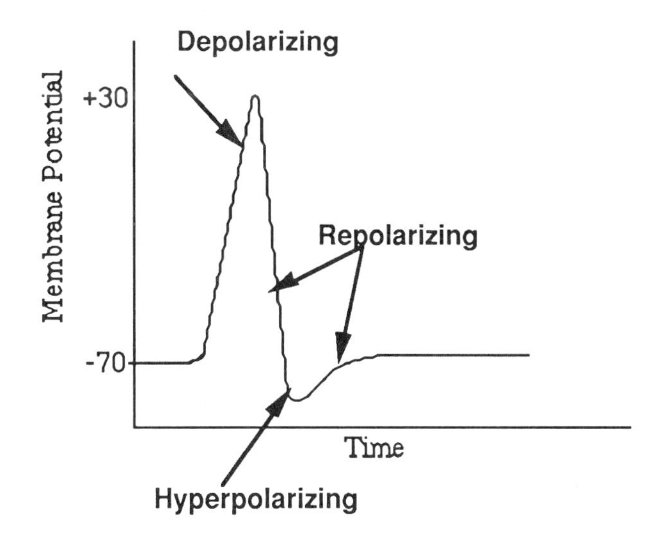

(The hyperpolarization is described in text Table 8-10 as an "overshoot." Note that two separate repolarizations occur, one from the original depolarization and one from the overshoot.)

76. (a) **Correct.**

    (b) **Correct.** (Look at the scale on the right side of text Figure 8-30. The permeability [P] for Na goes from 1 to 600, while that for K is at about 75. Note, too, that P_K begins to increase during the depolarization phase.)

    (c) **Incorrect. The membrane reaches about +40 mV, and the equilibrium potential for Na ions is +60 mV.**

77. (a) **Correct.**

    (b) **Correct.** (Notice in text Figure 8-30 that P_K increases during the repolarization.)

    (c) **Correct.** You may have been tempted to say that the permeability is less than at rest because the membrane potential is becoming more negative (repolarizing). But note from the P_{Na} curve in text Figure 8-30 that Na permeability does not reach the resting value of 1 until repolarization is finished, and never goes below that value.

78. (a) **Correct.** (See note at end of Answer 76-b.)

    (b) **Correct.** (Without these changes, the flux of K ions out of the cell would still be increased compared to their flux during rest, because the accumulation of positive charges — depolarization — inside the cell creates an electrical force considerably stronger than the one present during resting conditions. This increased electrical force stimulates the K^+ flux. However, simultaneously increasing the permeability to K ions increases the rate of diffusion and speeds repolarization.)

    (c) **Correct.** (Unlike repolarization, hyperpolarization would not occur if P_K did not increase during the action potential.)

    The purpose of asking Questions 76 to 78 is in part to make certain you understand the difference between permeability (the number of channels open) for a given ion and the forces — concentration gradient and electrical — influencing the rate of diffusion of that ion (flux) across the membrane. Again, changes in permeability mean nothing if diffusional forces are not present (the case for chloride ion when a neuron is at rest).

79. **False.** Because so ~~many~~ **few** Na ions move into a cell during an action potential and so ~~many~~ **few** K ions move out, ~~no further~~ **many** (thousands) **more** action potentials can be generated in a given membrane ~~until~~ **before** the Na,K-ATPase pumps ~~can~~ **are required to** restore the concentration gradient. (If the statement were true as it read originally, the capacity of nerve membranes to convey information about a stimulus using action potentials would be very limited.)

80. (a) **Incorrect. Na^+ permeability decreases with membrane hyperpolarization.** However, the *flux* of Na ions through the Na channels still open *increases* with membrane hyperpolarization because the electrical force "pushing" them increases. (See explanation after Answer 78.)

(b) **Correct.**

(c) **Correct.** This "circular" kind of statement describes a positive feedback loop. (Recall that we hinted at the existence of this mechanism in Chapter 7.) Now, what calls a halt to the cycle? The Na channels close, but why? The nature of the proteins forming the Na channels is such that they can remain in the open position for only a brief time (about one millisecond) before changes in their conformation cause them to again become barriers to the passage of Na ions into the cell. A good analogy to help you grasp how this kind of barrier might work is a turnstile, which stops between open phases.

81. **False.** The Na and K channels that open during an action potential are voltage-regulated, both responding to ~~hyperpolarization~~ **depolarization** of the membrane. (It may be tempting to think of the channels for the two ions as being regulated by different kinds of potentials, depolarizing ones for Na and hyperpolarizing ones for K, but this situation is clearly not the case. If you missed this question, go back to look at text Figure 8-30 again.

82. **Local anesthetics such as Novocaine keep Na channels in afferent neuronal membranes from responding to depolarization and prevent the generation of action potentials. Thus, the pain "message" does not reach the brain.**

83. (a) **Incorrect. A threshold stimulus must be sufficient to trigger an *action* potential.** (One just sufficient to trigger a graded potential would be subthreshold.)

(b) **Correct.**

(c) **Correct.** (The key word here is "net" — where sodium influx is greater than potassium outflux. Keep in mind that a depolarization stimulates outflow of potassium ions because it makes the cell's interior less electrically attractive to the positively charged ion.)

84. **False.** The action potential elicited by a supra-threshold stimulus is ~~larger than~~ **the same size as** one elicited by a threshold stimulus.

85. **False.** The **absolute** refractory period of an excitable membrane refers to the period of time during which no stimulus, however strong, will elicit a second action potential in the membrane.

86. **True.**

87. (a) **Incorrect. Action potentials are propagated** (they do not "travel") **without decrement.**

(b) **Correct.**

(c) **Correct.** (That is how the action potential can generate another action potential, and why the propagation is nondecremental. The key words here are "to threshold.")

(d) **Correct.** (Because of the nondecremental propagation.)

88. (a)

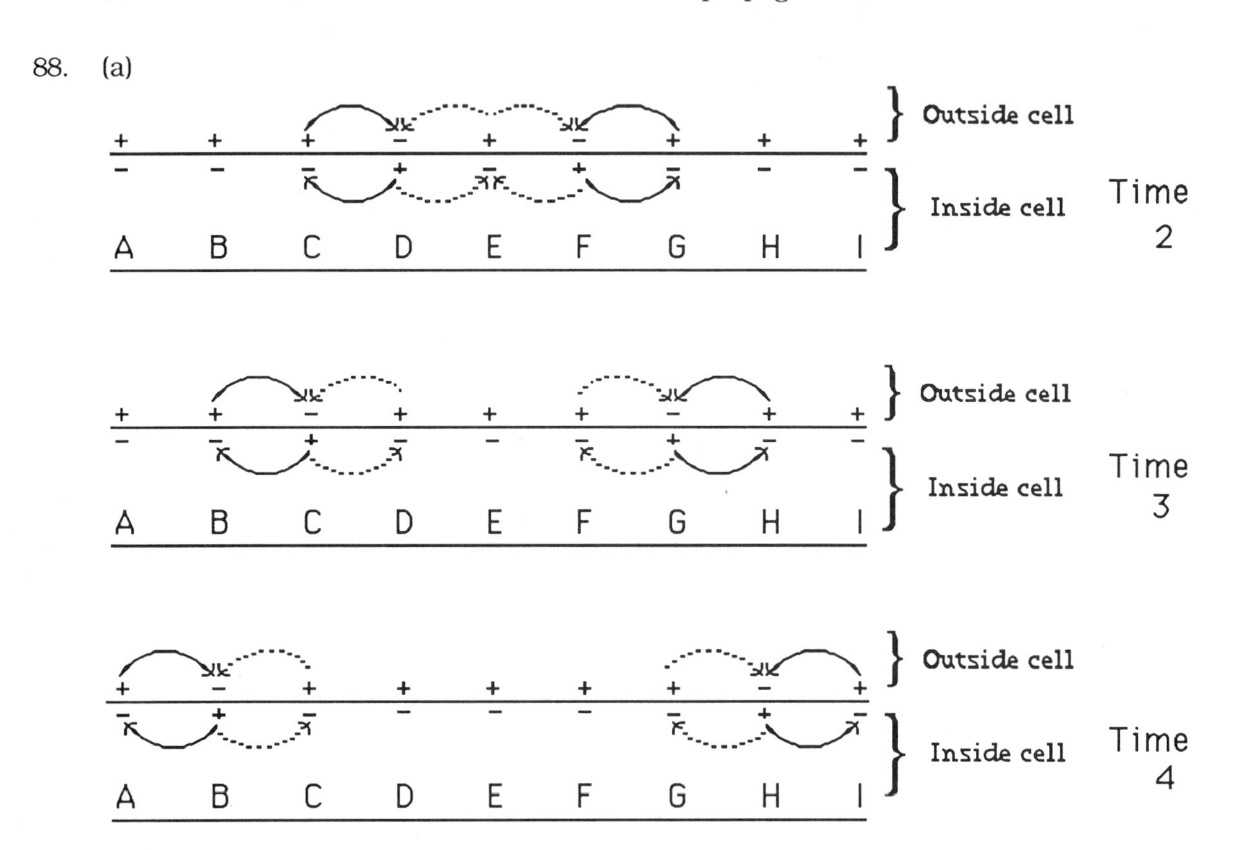

(b) **It would be in two opposite directions simultaneously: The first direction** (at time 1) **is circularly from E to D inside the cell and from D to E outside the cell; then from D to C and so on to A. The opposite direction is circularly from E to F on the inside and so on to I.** (There would be some current back from D and F to E inside the cell at time 2, and likewise from C to D and G to F at time 3 and so on, but this backcurrent doesn't concern us because it occurs during the absolute refractory period and dies away.)

(c) **From E to A and from E to I.**

(d) **From A to I.**

This exercise illustrates that action potentials can be propagated in two directions along an excitable membrane if the stimulus is in the middle of the membrane. As the text points out, this two-way transmission is common in skeletal muscle but not in nerve cells, because of the differences in where nerve cells and muscle cells are normally stimulated to begin generating action potentials.

89. (a) **Incorrect. In any given axon, the velocity of action-potential propagation is the same regardless of stimulus strength.** (The text mentions that frequency of action potentials codes for the strength of the stimulus. This topic is covered in Chapter 9.)

(b) **Incorrect. The opposite is true.** (There is less resistance to current in axons of large diameter.)

(c) **Correct.**

It is easy to understand why myelin makes a difference in the velocity of action-potential conduction when you consider that the time it takes to generate an action potential is greater than the time it takes for charge to flow along the axon. In an unmyelinated fiber, action potentials are generated at regular small intervals along the membrane, whereas these potentials are generated at the more widely spaced nodes of Ranvier in the myelinated fibers. For example, with the sequence

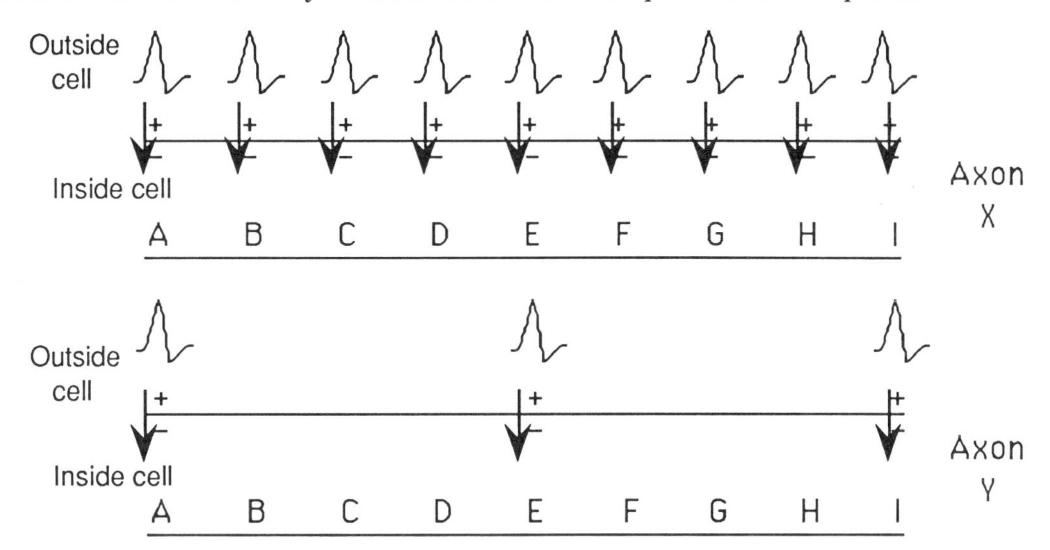

axon X is unmyelinated and axon Y has myelin. In X, action potentials would be generated at points A, B, C, D, E, and so on. In Y, they would jump from A to E to I. (Note that action potentials can be generated only where there are voltage-regulated sodium channels. In unmyelinated fibers these channels are distributed evenly along the axon [110 per square micrometer], whereas they are bunched at the nodes [2,000 - 12,000 per square micrometer] in myelinated ones.)

The text mentions that the largest myelinated fibers conduct action potentials up to 400 times faster than the smallest unmyelinated ones. So which is more important, size or myelination? Myelin wins. Given two fibers of the same diameter, the action potential message is conducted fifty times faster in the myelinated fiber. If we had unmyelinated fibers in the tracts of our spinal cord instead of myelinated ones, the cord would have to be as large as a medium-sized tree trunk in order for neural signals to be conveyed as quickly as our bodies need them.

This information about how myelin affects the speed of nerve impulse conduction, coupled with the knowledge that myelinated fibers in the brain convey information to and from the cerebral cortex and other brain centers and the spinal cord, explains why diseases involving myelin breakdown can be profoundly debilitating. The most important of these diseases is multiple sclerosis, which involves progressive degeneration of myelin in the brain. MS is thought to be an autoimmune disease — one in which the body's immune system mistakenly identifies parts of the body as foreign and attacks them. There is no cure as yet for MS, although drugs that selectively suppress the immune system can often lessen the symptoms. Current research has found an animal model (a strain of rat that develops an MS-like disease) that can be used to study the disease's onset and to test potentially helpful drugs.

90. (a) **Incorrect. They are found only in special neurons and muscle cells.**
    (b) **Incorrect. They occur spontaneously.**
    (c) **Correct.**

91. (a) **Correct.** (This is the property called convergence.)
    (b) **Correct.**
    (c) **Incorrect. While it is true that most interneurons are presynaptic to many other neurons, their role as integrators is a postsynaptic one.** (This question was intended to remind you that interneurons are both postsynaptic and presynaptic.)

92. **True.** (When any two cells are connected by gap junctions, they function as a unit. Electrical synapses do not allow for integration of signals.)

93. (a) **Incorrect. There is no electric signal in the synaptic cleft.** (This was the "spark" theory disproved by Loewi.)
    (b) **Correct.**
    (c) **Correct.** (The proteins are receptors, of course. The part of the postsynaptic cell membrane beneath the axon terminal or axon varicosity is termed the "subsynaptic" membrane — see text Figure 8-38.)

94. The first step in triggering neurotransmitter release by action potentials is the **depolarization** of the axon terminal, a step that causes voltage-regulated channels for **calcium** ion to open and consequent **influx** of that ion. This event results in binding

of **synaptic vesicles** to specific sites on the axon terminal membrane, and such binding leads to exocytosis of the neurotransmitter.

95. **False.** The delay of approximately one millisecond between the time an action potential arrives at the axon terminal and the time the effects on the subsynaptic membrane occur is probably due to ~~the time required for diffusion of the neurotransmitter across the synaptic cleft~~ **events in the presynaptic cell prior to release of the neurotransmitter.** (The significance of synaptic delay is that conduction of a message along a pathway with many synapses in it — a multineuronal pathway — is measurably slower than conduction of a message along a pathway with few synapses.)

96. (a) **Correct.**
    (b) **Correct.** (Chemical modification is one mechanism for inactivating the neurotransmitter.)
    (c) **Incorrect. It may be actively transported into the presynaptic cell** (a way of recycling the transmitter).

97. **False.** A neurotransmitter that is excitatory at one synapse will **not necessarily** be excitatory at all synapses where it is released. (The response of any cell to any messenger depends upon the cell's receptor-transducer mechanism, remember.)

98. (a) **Correct.**
    (b) **Incorrect. They do not necessarily result in action potentials.** (They cannot if they are released in insufficient amount or if inhibitory synapses are simultaneously activated.)
    (c) **Correct.** (They invariably cause depolarization of the membrane directly under the synapse. Their effect can be cancelled out by inhibitory neurotransmitters, but only downstream from the synapse.)
    (d) **Correct.** (Excitatory postsynaptic potentials are the kind of graded potential caused by excitatory neurotransmitters.)

99. (a) **Incorrect. Negatively charged ions are usually not involved.**
    (b) **Correct.** (This increased permeability increases the flux of Na ions into and K ions out of the cell.)
    (c) **Correct.** (More Na ions will diffuse into the cell than K ions will diffuse out, because the forces driving the Na ions are stronger.)
    (d) **Correct.** (This is one of the characteristics of all graded potentials.)

100. (a) **Incorrect. A depolarizing potential is excitatory.**
     (b) **Correct.**

(c) **Incorrect. Increasing permeability to Na^+ would lead to depolarization.**

(d) **Correct.**

Increasing the Cl^- permeability of a membrane that actively transports Cl^- out of the cell will result in hyperpolarization, because the equilibrium potential of Cl^- in those cells is greater than the resting membrane potential in neurons — -80 mV versus -70 mV. In most cells, that is, those that do not actively transport Cl ions, increasing permeability only to Cl ions would not result in a hyperpolarizing IPSP, but instead would stabilize the resting membrane potential. (See answer to Question 101.)

101. **Increasing a postsynaptic membrane's permeability to Cl ions will not, by itself, change the membrane's resting potential. However, such an increase will allow Cl ions to move into the cell more rapidly to counter the effects of an influx of Na ions. Thus, the resting potential is stabilized, and activation of more excitatory synapses would be required to depolarize it.**

102. (a) **Incorrect. One EPSP will provide only 1/30th to 1/50th of the amount of depolarization necessary to bring the membrane to threshold.**

(b) **Correct.** (Since one EPSP cannot do the job alone, obviously many must add together to activate the cell.)

(c) **Correct.** (The grand total of EPSPs and IPSPs must be a depolarizing potential of threshold level [usually 15 to 25 mV].)

103. (a) **Correct.**

(b) **Incorrect. It refers to summation of EPSPs and IPSPs** (in any combination).

(c) **Incorrect. It refers to any summation of synaptic input on a postsynaptic cell that is not simultaneous.** (In other words, much temporal summation is also spatial. These terms are not mutually exclusive.)

104. **False.** Because ~~all parts of a neuronal cell body have the same threshold, no one synapse on the cell~~ **the initial segment of an axon has a lower threshold than other parts of a neuronal cell body, a synapse close to the initial segment** is more important than any other.

105. **True.** The greater the depolarization (that is, the greater the amount of excitatory synaptic activity), the greater the number of action potentials per unit time propagated down the postsynaptic cell's axon. In turn, the higher the frequency of action potentials propagated down an axon, the greater the synaptic or neuroeffector junction activity at that axon's terminals. Action-potential frequency is obviously an important signal in the nervous system.

106. **False.** The effects of activation of any given synapse on a subsynaptic membrane are ~~invariable~~ **influenced by presynaptic and postsynaptic factors.**

107. **During a high-frequency burst of action potentials, there is not sufficient time between each action potential to pump out all the calcium ion that rushed in with the membrane depolarization, and the calcium concentration rises in the terminal. The more calcium present, the greater the number of synaptic vesicles that can bind to the terminal membrane and the greater the amount of neurotransmitter released.** (This increased efficiency of synaptic stimulation cannot go on indefinitely, of course, because eventually the vesicle population will become depleted.)

108. (a) **Correct.** (This sort of synapse requires the axon terminal to have receptors for neurotransmitter on its membrane. See text Figure 8-45.)
   (b) **Correct.** (In this case, the presynaptic axon facilitates the release of excitatory neurotransmitter from the terminal with which it synapses. This sort of presynaptic facilitation may involve mediation by a second messenger and effects on, for example, ion channel proteins.)
   (c) **Correct.** (As the text mentions, presynaptic inhibition is particularly common in the afferent neuronal pathways.)

109. **True.** (This is the case when a terminal has receptors, called autoreceptors, for its own transmitter or cotransmitter. Thus, a neuron can exercise negative-feedback regulation of its own activity.)

110. **Responses a, b, and f are correct. All the others would enhance the action of norepinephrine.**

111. (a) **Correct.** (The example given in the answer to 108-b is a neuromodulator.)
   (b) **Correct.**
   (c) **Correct.**
   (d) **Correct.**
   (e) **Correct.** (Again, the function of a messenger is determined by the receptor for that messenger and by the receptor-transduction mechanisms in a particular cell.)

112. **False.** A major distinction between the action of neurotransmitters and that of neuromodulators is that the former is considerably ~~slower~~ **faster** than the latter.

113. The four major classes of neurotransmitters/neuromodulators are **acetylcholine**, **biogenic amines**, **amino acids**, and **neuropeptides**.

114. (a) **Incorrect. It takes place in the axonal synaptic terminals** (close to its site of action).
     (b) **Correct.**
     (c) **Correct.** (The presynaptic membranes also contain acetylcholinesterase. These membrane-bound enzymes quickly reduce the molecule to choline and acetate.)
     (d) **Incorrect. ACh is very important for both efferent peripheral nervous system function and brain function.** Alzheimer's disease, which results from destruction of ACh-releasing or ACh-sensitive neurons in the brain, is one of the most important and tragic diseases involving the CNS.

115. **False.** After the breakdown of acetylcholine by acetylcholinesterase, the ~~acetate~~ **choline** is actively transported back into the presynaptic cell. (Choline is recycled; acetyl coenzyme A is constantly being generated from glycolysis.)

116. The neurotransmitters dopamine and norepinephrine belong to a subclass of biogenic amines called **catecholamines**. The rate-limiting enzyme in the synthesis of these amines is the one converting the amino acid **tyrosine** to **L-DOPA**. Another important neurotransmitter that is also a biogenic amine but is synthesized from a different amino acid is **serotonin**.

117. The two enzymes that catalyze breakdown of catecholamines are called **monoamine oxidase** (MAO) and **catecholamine O methyl transferase** (COMT). Catecholamine catabolism differs from the breakdown of acetylcholine in that the enzymes for the former are found primarily **inside** the cell and those for the latter are found **outside**.

118. **True.** (This action is in addition to the important role of norepinephrine in the peripheral nervous system.)

119. (a) **Correct.** (It is catabolized by MAO.)
     (b) **Correct.** (in addition to other functions)
     (c) **Incorrect. It is lowest during sleep.**

120. (a) **Incorrect. LSD inhibits serotonin release and is a potent hallucinogen. Therefore, serotonin is probably not hallucinogenic.**
     (b) **Correct. This assumption would explain how a drug that is chemically related to serotonin (and so could presumably bind to serotonin's receptors) can be hallucinogenic if serotonin is not.**
     (c) **Correct. This makes sense because serotonin normally inhibits incoming sensory information, and inhibition of serotonin induces hallucinations.**

Serotonin secretion, and the inhibition of sensory input to the brain, cease during dreaming. Consequently, the hallucinations induced by LSD have been compared to dreaming while awake. Unfortunately, some of those dreams are nightmares.

121. **False.** ~~Catecholamines~~ **Amino acids** are the most abundant neurotransmitters in the CNS.

122. (a) **Correct.**

(b) **Incorrect. It is the major *inhibitory* transmitter.** (GABA is also a common mediator of presynaptic inhibition.)

(c) **Correct.** (This fact is particularly interesting because the ionized form of glutamic acid, glutamate, is one of the most common of the excitatory amino acid neurotransmitters.)

The importance of inhibition in the CNS can be illustrated by the fact that GABA agonists, such as Valium, are among the most commonly prescribed drugs in the U.S. (These drugs are prescribed for conditions related to feelings of anxiety. The need for such drugs tells us something about our lifestyle.)

123. (a) **Correct.**

(b) **Correct.** (more than one, linked by peptide bonds)

(c) **Incorrect. Many are found throughout the body** (where they function as either hormones or paracrines).

Answers (a) and (c) may generate some confusion. Many of the neuropeptide molecules are also synthesized by cells that are not neurons. For example, the neuropeptide insulin is synthesized by epithelial cells in the pancreas, which secretes it into the blood. Hormonal insulin is not a neurotransmitter, but it may still be referred to as a neuropeptide because it is the same molecule as the one synthesized by neurons. This discussion emphasizes the need to distinguish between the structure and function(s) of molecules.

124. (a) **Incorrect. Other kinds of neurotransmitters are synthesized in axon terminals.**

(b) **Correct.**

(c) **Correct.** (There is one exception to this generalization. The tripeptide TRH, which you will meet in Chapter 10, is not derived from a larger precursor.)

Notice that the text refers to the precursors of the neuropeptides as "prohormones" and "preprohormones." This usage is a result of the fact that the precursor nature of peptide messenger synthesis was first discovered by scientists studying hormones secreted by the pituitary gland. Moreover, many of the neuropeptides are neurohormones — they are secreted by axon terminals into the blood (as we learned in Chapter 7).

125. **True.** The word "endorphin" actually means "endogenous (which in turns means 'found in the body') morphine."

126. Substance P is a neurotransmitter important for conveying the sense of **pain**. (Remembering that substance P is associated with pain is easy because of the accidental coincidence of its naming. The name was given to it as one of a series of substances of unknown function being studied by scientists several years ago.)

127. **True.**

128. **False.** The most common neurotransmitters for neuroeffector communication are ~~dopamine~~ **norepinephrine** and acetylcholine.

129. **False.** The process by which sensory receptors change various forms of energy into electrical energy is called ~~translation~~ **transduction**. (Recall that this word is also used to denote the relay mechanisms that convey information about cellular receptors activated by messenger molecules. This repeat usage is useful because it does not require learning a new word. But be on guard that you keep clear in your mind what kind of receptor and what kind of transduction are meant in any discussion.)

130. (a) **Incorrect. They are sometimes separate cells near nerve cells.**
     (b) **Incorrect. They are found on afferent neurons.**
     (c) **Correct.**

131. **False.** According to the doctrine of specific nerve energies, only one kind of ~~stimulus can activate~~ **sensation can be evoked by stimulation of** any one kind of receptor. (Think of the eye. Rubbing your eyes or even squeezing your eyelids tightly closed produces the sensation of light.)

132. (a) **Incorrect. It is a graded potential.**
     (b) **Correct.** (if it is large enough)
     (c) **Correct.**
     (d) **Correct.**

133. The Pacinian corpuscle is an example of a **mechano**receptor, meaning that it responds to **mechanical deformation**. (Mechanoreceptors are responsible for our sense of touch.)

134. **False.** The greater the magnitude of the receptor potential generated by a stimulus, the greater the ~~amplitude~~ **frequency** of the action potentials the receptor potential induces.

135. **True.**

136. **True.**

137. (a) **Correct.**
     (b) **Incorrect. Some receptors adapt completely, while others adapt very little.**
     (c) **Correct.**

PRACTICE

True or false (correct the false statements):

P1. The neurotransmitter released by stimulation of a preganglionic sympathetic fiber is acetylcholine.

P2. In general, the sympathetic and parasympathetic branches of the autonomic nervous system produce opposite effects in the organs they innervate.

P3. One function of cerebrospinal fluid is to cushion the brain.

P4. Transient changes in the resting potential of a neuronal membrane that are characterized by variable magnitude and duration are referred to as action potentials.

P5. The period following an action potential during which an excitable membrane will respond only to a suprathreshold stimulus is called the relative refractory period.

P6. Action potentials are propagated by means of local membrane currents causing depolarization of adjacent regions of membrane.

P7. Neural activity can be transmitted in either direction at any given chemical synapse.

P8. The integration of information in the CNS is primarily the result of summation of IPSPs and EPSPs in postsynaptic neurons.

P9. The transmission of information at a synapse occurs at about the same rate as the conduction of the action potential along the axon.

P10. In general, one action potential arriving at an excitatory presynaptic terminal will lead to the generation of one action potential in the postsynaptic neuron.

P11. Summation of postsynaptic potentials due to the successive stimulation of the same presynaptic fiber is called spatial summation.

P12. In any given postsynaptic cell, each excitatory synapse plays an equal role with every other synapse in determining whether the cell will become depolarized to threshold.

P13. Some neuromodulators are hormones.

P14. A drug that inhibits monoamine oxidase synthesis would be likely to inhibit the activity of norepinephrine in the brain.

P15. Alzheimer's disease is thought to involve loss of neurons that secrete and respond to dopamine.

P16. Receptors on afferent neurons are found in the CNS.

P17. One kind of graded potential is the receptor potential.

P18. Increasing the intensity of a stimulus applied to a receptor will increase the frequency of action potentials produced by the receptor.

Multiple choice (correct each incorrect choice):

P19-21. The cerebral cortex

P19. consists primarily of white matter.
P20. is part of the brainstem.
P21. is the "thinking" part of the brain.

P22-24. The ventricles of the brain contain

P22. air.
P23. blood.
P24. cerebrospinal fluid.

P25-27. The equilibrium potential of a given ion across a membrane is

P25. a function of the concentration of that ion on both sides of the membrane.

P26. the potential at which there is no net movement of that ion across the membrane due to electrochemical forces.

P27. the resting potential of the membrane.

P28-30. The resting potential of a nerve cell

P28. depends ultimately on the ability of the cell to actively transport Na and K ions across the membrane.

P29. will decrease if its plasma membrane permeability to K ion increases.

P30. is closer to the equilibrium potential of Na ion than to that of K ion.

P31-33. At some time during an action potential in a nerve cell

P31. the neuron is absolutely refractory to a second stimulus.

P32. the membrane permeability to K^+ increases.

P33. positive feedback increases Na^+ permeability.

P34-36. The propagation of an action potential along an axon is faster

P34. in large-diameter axons than in small-diameter ones.

P35. in myelinated nerve fibers than in nonmyelinated ones.

P36. for a strong stimulus than for a weak one.

Fill in the blanks:

P37. The most abundant cells in the CNS are __________ cells.

P38-41. Groups of cell bodies in the CNS are called ______________, while similar cell groups in the peripheral NS are referred to as ___________. Similarly, axon fibers travelling together in the two divisions of the nervous system are called ____________________ in the CNS and ____________________ in the peripheral NS.

P42. A neuron will not generate an action potential unless excited beyond its _________________________ potential.

P43-48. During an action potential in an axon, the membrane potential changes from __________ mV to _______ mV. During this time, __________ ions rush into the cell as a result of a change in permeability. The membrane then becomes more permeable to _______________ ions, which then ___________(enter/leave) the cell, returning the potential to its resting level. The absolute refractory period corresponds to the period of __________ ion permeability changes.

P49. Transmission at a chemical synapse requires activation of voltage-regulated _____ channels.

P50. The decrease or cessation of action potentials in a sensory neuron despite continued stimulation of the receptor is called _______________.

9: THE SENSORY SYSTEMS

"At the heart of our experience of life there is a deep and subtle paradox. Most of us consider the world of physical matter to be the trustworthy world of true reality. We think of rocks and rivers and tables and chairs and our own flesh and bones as objects composed of solid certainty. The world of consciousness and thought, on the other hand, is regarded with more suspicion as an abstract and ghostly place in which there are no certainties and great doubt about the true nature of its reality. The physical world is the world we feel comfortable with, while the world of the mind seems a domain of mystery and even magic. The paradox is that the world of the mind is the only one we have direct experience of, and the only one each of us can be certain exists. Our experience of the physical world comes to us indirectly, along nerves which convert the effects of that physical world into sensations, thoughts and ideas in our brain. Each of us can be certain that our own thoughts exist, but that is the only thing we can be absolutely certain about."

Andrew Scott, "Vital Principles"

This quote, taken from a highly readable (and recommended) book by a British science writer and broadcaster, introduces the subject of this chapter: how information from the world around (and within) us reaches our brain.

Instructions for answering questions in this Study Guide

1. True or false: Correct each false statement. Whenever you must correct a false statement, there will almost always be more than one way to do so. In the Guide, the answer requiring the least correction will generally be given, with the understanding that other correct answers are possible. It is usually not sufficient, in terms of demonstrating understanding, to simply insert a "no" or "not," however.

2. Multiple choice: Any, all, or none of the responses may be correct. Explain why each incorrect response is incorrect.

3. Fill in the blanks: Choose the best word or words to complete each statement.

4. Directions will be given for other types of questions as they appear.

❐ Introduction (text pages 219-220)

1. In a sensory system, structures called ______________________ receive information in the form of ______________________ from the internal or external environment. This information is conducted in the form of ______________________ along ______________________ nerve fibers and is processed by ______________________ in the brain.

2. True or false: "Sensation" is sensory information that reaches the brain.

3. True or false: Perceptions are derived from higher-order processing of sensory information.

4. True or false: The transmission of information in a sensory system is analogous to the transmission of sound in a telephone system except that, unlike the telephone system, the sensory system does not re-translate the electrical signaling code back to the specific energy of the stimulus.

❐ Neural Pathways in Sensory Systems (text pages 220-222)

5. A sensory unit is
   (a) all the sensory receptors in a given area of the body that respond to the same stimulus.
   (b) a single receptor ending and its afferent nerve fiber.
   (c) a single afferent neuron and all its receptor endings.

6. The receptive field of an afferent neuron
   (a) refers to the number of interneurons with which the central process of the afferent neuron makes synaptic contact via divergence.
   (b) refers to the body surface that, when stimulated by a specific stimulus, gives rise to generator potentials in the receptors of that neuron.
   (c) is determined by the number and length of branches of the peripheral terminals of that neuron.

7. True or false: The minimum number of synapses in a specific ascending pathway is four.

8. The primary cortical receiving areas for sensory information are
   (a) composed of the first neurons in the cerebral cortex to receive sensory information from specific ascending pathways.
   (b) the only cerebral-cortex neurons that process specific sensory information.
   (c) located in anatomically specific areas of the brain according to stimulus type.

9. True or false: "Somatosensory" refers to the part of the cerebral cortex that receives synaptic input from specific ascending pathways originating only with receptors for touch.

10. Match the primary receiving area for specific sensory information in the left column with the correct brain location in the right column.

____ Somatosensory	(a) Frontal lobe
____ Visual	(b) Temporal lobe
____ Auditory	(c) Occipital lobe
____ Taste	(d) Limbic system
____ Olfactory	(e) Parietal lobe

11. Polymodal neurons are
    (a) afferent neurons.
    (b) interneurons that receive synaptic input from different kinds of sensory units.
    (c) part of specific ascending sensory pathways.
    (d) unimportant because they do not convey specific sensory information.

❒ **Basic Characteristics of Sensory Coding** (text pages 222-224)

12. The three kinds of information specific sensory pathways provide to higher brain centers are the ________________________, the ________________________, and the ____________________ of a stimulus.

13. Which of the following is/are important for the determination of stimulus type?
    (a) the relative sensitivity of different receptors to different stimulus energies
    (b) the presence of polymodal neurons in the sensory pathway
    (c) the location of the fourth-order (or higher) neuron in the ascending pathway

14. Which of the following is/are correct and important for determining stimulus intensity? Explain each answer.
    (a) The magnitude of receptor potentials increases with increasing stimulus intensity. ______________________________

    ____
    (b) The frequency of receptor potentials increases with increasing stimulus intensity. ______________________________

    ____
    (c) The magnitude of action potentials increases with increasing stimulus intensity.

    ____

    ____
    (d) The frequency of action potentials in second-order neurons in the ascending pathway increases as a result of recruitment of nearby receptor units.

    ____

    ____

15. True or false: Information about the location of a given stimulus on or in the body is conveyed by the same mechanisms that convey information about stimulus intensity.

16. True or false: The density of receptors in a receptive field is usually greatest in the periphery of the field.

17. The precision of locating a stimulus is
    (a) greater in areas of the body that have large, nonoverlapping receptive fields than in areas with small, overlapping fields.
    (b) about the same for all parts of the skin.
    (c) greater for the skin than for internal organs.
    (d) greater for pathways lacking lateral inhibition than for those with extensive lateral inhibition.

❐ **Somatic Sensation** (text pages 224-230)

18. The specific ascending pathways for somatic sensation on the right side of the body cross to the left side of the CNS at the level of the ____________________ or ____________________. Information from somatic receptors on the right side, therefore, is carried to the ____________________ cortex on the ____________________ side of the brain.

19. True or false: In the somatosensory cortex, neuronal representation of body parts is proportional to their size.

20. Examine text Figures 9-12 and 9-13 (and their legends) carefully. Now tell whether each of the following statements is correct, and explain your reasoning.

    (a) The axons of second-order neurons in the ventral (anterior) spinothalamic tract cross from one side of the spinal cord to the other at the same level of the cord (that is, at the same distance from the brain) as the synapses with their afferent neurons.____________________

    ____

    ____

    ____

    (b) A person with spinal cord damage to the left lateral spinothalamic tract just above the level where afferent information from the foot enters the cord would be wise not to test the temperature of his or her bath water with his or her right great toe.

    ____

    ____

    ____

    ____

    (c) A person with damage to the dorsal (posterior) columns on the right side of the cord at the same level as in Question 20-b would have difficulty distinguishing the touch of one pencil point from that of two pencil points applied simultaneously to the skin of the right foot.

    ____

    ____

    ____

    ____

21. The general term for receptors that respond to mechanical stimulation is ________________________. In the skin, these receptors are grouped into two categories — rapid and slow — according to how quickly they ____________________ to stimuli. Receptors in the rapid category include those that give rise to sensations of ______________________, ______________________, ______________________, and ________________________. Receptors in the slow category mediate the sensation of ________________________. Mechanical stimulation of receptors in skeletal muscles, joints, tendons, ligaments, and skin produces information regarding the sense of the body's position in space, which is termed ________________________, and sensations of joint movement, termed ________________________.

22. The sense of effort differs from other somatic senses in that it requires information from ______________________________ pathways.

23. True or false: At normal body temperature, internal warm and cold receptors are activated approximately equally.

24. True or false: The afferent pathways for pain differ from those for other somatic sensations in that information about the pain stimulus is transmitted to the thalamus.

25. True or false: A common neurotransmitter for afferent neurons carrying information about pain is prostaglandin.

26. True or false: The phenomenon of referred pain may occur because the ascending pathways for pain are not completely specific with respect to localization of the pain receptor.

27. True or false: "Phantom limb" pain is imaginary pain because it is "felt" in an amputated limb.

28. True or false: The reduction of pain by electrical stimulation of inhibitory neurons in the pain pathway is called stimulation-produced kinesthesia.

29. Describe the *rationale* (this term means "the reason it works") of TENS therapy.

____

____

____

____

30. For each somatic sense, summarize the properties of (1) the receptors, (2) the afferent neurons, and (3) the second-order neurons by matching the letters of the indicated senses with each appropriate description:

Receptors

_______ mechanoreceptors
_______ nociceptors
_______ complex structures
_______ naked nerve endings
_______ respond to chemicals

Afferent Neurons

_______ have thick, myelinated processes
_______ have thin, unmyelinated or thinly myelinated processes
_______ have central processes that travel in the dorsal horns

Second-order Neurons

_______ have axons that travel in lateral spinothalamic tract
_______ have axons that travel in ventral spinothalamic tract

Sense

(a) touch
(b) vibration
(c) pain
(d) heat
(e) pressure
(f) proprioception and kinesthesia

❐ Vision (text pages 230-241)

31. The wavelengths of visible light
    (a) are longer than those of radio waves.
    (b) are longer than those of X-rays.
    (c) increase as the frequency decreases.

32. True or false: The wavelength of blue light is shorter than that of red light.

33. Label the parts of the eye indicated on the diagram below.

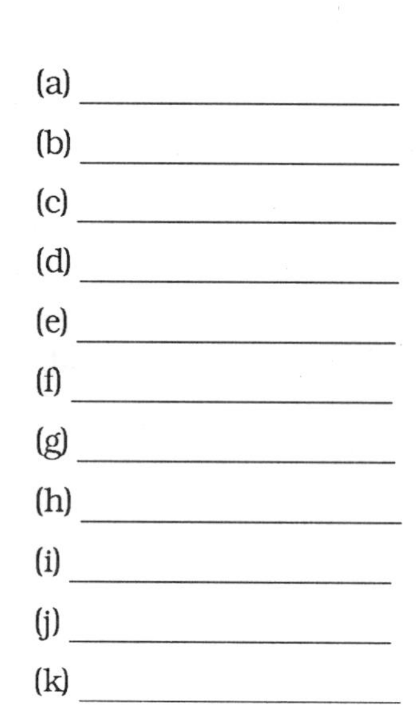

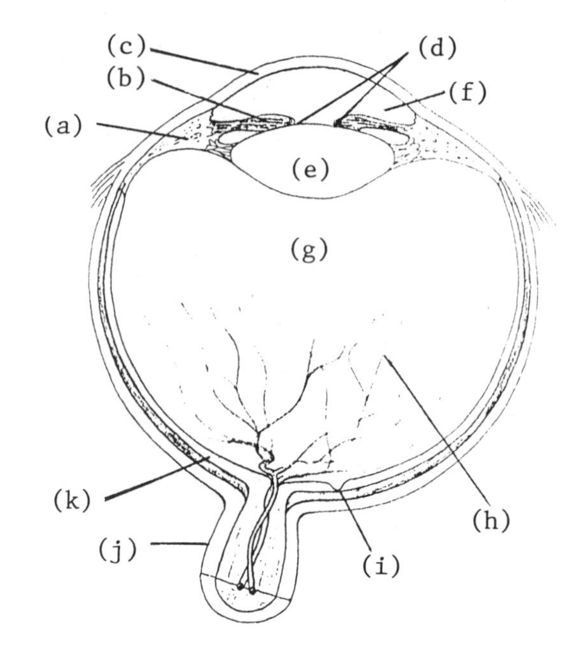

34. Describe the functions of the cornea and lens in vision.

35. When viewing a distant object, the
   (a) firing of parasympathetic nerves to ciliary muscles increases.
   (b) ciliary muscles relax.
   (c) zonular fibers relax.
   (d) lens flattens.

36. Presbyopia is
   (a) caused by having an eyeball that is too short for the lens size.
   (b) caused by coloration of the lens.
   (c) caused by stiffening of the lens.
   (d) correctable with a convex lens.

37. True or false: Myopia is corrected by a lens that increases the angle of light striking the cornea.

38. True or false: Increased pressure in the eye caused by an accumulation of aqueous humor is called astigmatism.

39. The iris of the eye
    (a) contains smooth muscle.
    (b) is pigmented.
    (c) is innervated by both sympathetic and parasympathetic nerves.

40. During a thorough eye examination, ophthalmologists administer to the eye drops of atropine, a blocker of muscarinic acetylcholine receptors. The drug causes the pupil to dilate and allows close examination of the retina. For one to two hours afterward, while the effects of the atropine are wearing off, bright light is painful and reading or other focusing on near objects is difficult. Explain:
    (a) the rationale for using atropine. ____________________

    ____

    ____

    ____

    ____

    (b) why focusing on near objects is difficult while the atropine is present.

    ____

    ____

    ____

    ____

    ____

41. Receptors in the visual system are
    (a) cells.
    (b) located on the surface of the retina that is bathed by the vitreous humor.
    (c) neurons.

42. Which of the following statements about photoreceptors is/are true?
    (a) There are four types of photoreceptors in the eye.
    (b) There are four types of photopigments in the eye.
    (c) There are four types of chromophores in the eye.
    (d) There are four types of opsin in the eye.

43. Vitamin A deficiency can lead to blindness. Explain.

____

____

____

____

____

44. Fill in the missing terms or arrows:

photon ⇒ retinal ⇒ change of __________ of retinal

⇓

____(↑/↓) concentration of cyclic GMP in photoreceptor cytoplasm ⇐ alteration of __________

⇓

__________(opening/closing) of membrane Na^+ channels ⇒ ____(↑/↓) P_{Na^+}

⇓

____(↑/↓) neurotransmitter release by photoreceptors ⇐ __________(depolarization/hyperpolarization) of photoreceptor

45. True or false: The optic nerve fibers from each eye meet at the optic chiasm, where they cross. Thus, visual information from the right eye is received by the left side of the brain.

46. (a) List five characteristics of visual images that are conveyed to the brain by the visual system. __________ __________ __________ __________ __________

(b) True or false: These characteristics of the visual system are first sorted out by cells in the lateral geniculate nucleus of the thalamus.

47. The receptive field of a ganglion cell
- (a) is the area of the retina that, when stimulated by light, causes a change in the rate of action-potential firing in that cell.
- (b) consists of a group of photoreceptor cells adjacent to one another.
- (c) is usually in the shape of a line.
- (d) is usually circular.

48. An off-center ganglion cell
    (a) increases its rate of action-potential firing when light strikes the periphery of its receptive field.
    (b) increases its rate of action-potential firing when light strikes the center of its receptive field.
    (c) does not fire unless light strikes its receptive field.

49. The lateral geniculate nucleus of the thalamus
    (a) relays information about different characteristics of the visual world to different regions of the primary visual cortex.
    (b) receives input only from the retina.
    (c) is important for synchronizing biological clocks.

50. True or false: The conscious sensation of sight is a result of integration of information that is processed by cortical areas of the occipital, temporal, parietal, and frontal lobes of the brain.

51. True or false: An object that absorbs the longer wavelengths of visible light while reflecting the shortest ones will be perceived as yellow.

52. Picture in your mind's eye a caravan of camels and riders crossing the Sahara. The riders are all wearing long white robes and head coverings. Explain why white clothing is a good choice in a hot, sunny environment.

    ____

    ____

    ____

53. Opponent color ganglion cells that respond to blue light by increasing their rate of firing will respond to red light by ____________ (increasing/decreasing) their rate of firing. White light will cause a/an ____________ (increase/decrease/no change) in the firing rate of these cells.

54. True or false: The greatest concentration of rods is in the fovea of the retina.

55. True or false: One function of saccades is to prevent adaptation of photoreceptors to a visual image.

❐ **Hearing** (text pages 241-246)

56. Sound waves consist of alternating areas of ______________________ and ______________________ of air molecules. The pitch of a sound is determined by the ______________________ of the sound source.

57. True or false: A sound with a frequency of 4000 Hz has a lower pitch than a sound with a frequency of 1000 Hz.

58. True or false: The tympanic membrane separates an air-filled chamber from a fluid-filled chamber.

59. True or false: The vibration of the tympanic membrane varies in magnitude according to the pitch of the sound it receives.

60. True or false: The function of the tympanic membrane is to amplify sound waves on their way to the inner ear.

61. When we experience a sudden change in atmospheric pressure, such as when driving in the mountains or taking off and landing in an airplane, pressure can build up in the middle ear, sometimes to a painful level. A natural response to this feeling of pressure is to yawn or swallow. Why is yawning or swallowing helpful in these circumstances? ______________________

____

____

____

62. Some of the structures in the ear are given Latin names corresponding to common objects they resemble. Examine the diagrams in the text (e.g., Figures 9-34 and 9-35) and match the names in columns A and B.

A	B
____ cochlea	(a) hammer
____ incus	(b) sickle
____ malleus	(c) snail
____ stapes	(d) anvil
	(e) stirrup
	(f) saddle

63. An inherited disease called *otosclerosis* causes the bones of the middle ear, particularly the stapes, to become soft. This results in a type of deafness referred to as *conduction deafness.* Explain why softened inner-ear bones can cause deafness.

____

____

____

____

____

64. In the cochlea, pressure transmitted to the ______________________ from the oval window causes the ______________________ of the cochlear duct to vibrate. This vibration causes stimulation of mechanoreceptors, called ______________. During stimulation of these receptors, bundles of actin myofilaments, called ______________, are bent against the overhanging ______________. This deformation causes channels for Na^+ and Ca^{2+} to ______________ (open/close) and consequent ______________________ (depolarization/hyperpolarization) of the cells. The region of the cochlear duct that contains the receptor cells is called ______________________.

65. True or false: A tone caused by striking one of the lowest notes on a piano keyboard will cause vibration of the basilar membrane at a point closer to the helicotrema than to the oval window.

66. True or false: Unlike information from the eyes, information from the ears bypasses the thalamus on its way to the cerebral cortex.

❐ **Vestibular System** (text pages 246-248)

67. The vestibular apparatus
   (a) includes the cochlea.
   (b) is filled with air.
   (c) consists of five receptor organs.
   (d) lies in the two temporal bones, one on each side of the head.

68. True or false: Hair cells in the semicircular canals detect changes in the rate of angular motion of the head, while the same kinds of cells in the utricle and saccule detect changes in the head's rate of linear motion.

69. True or false: When you turn your head to the left, to look over your left shoulder, the hairs in the cupula of the horizontal semicircular canals will be bent to the left.

70. Information from the vestibular system
    (a) is conveyed in the form of action potentials in the optic nerve.
    (b) is conveyed to nerves controlling eye movements.
    (c) does not reach consciousness.

71. True or false: An overproduction of inner ear fluid on one side of the head can cause symptoms of dizziness and loss of hearing.

❒ **Chemical Senses** (text pages 249-250)

72. The receptors for the senses of smell and taste belong to the general category of ______________________. The sense organs for taste are the ______________________, which are traditionally divided into four categories — those sensitive to ______________, ______________, ______________, and ______________ tastes. Information regarding taste is conveyed to the ______________ cortex in the ______________ lobe of the brain. Receptors for the sense of smell, called ______________________, are located in the ______________________ in the roof of the nasal cavity. Information regarding smell is conveyed directly to the ______________________, which are part of the ______________ system of the brain.

73. True or false: Olfactory receptors are neurons.

74. True or false: Detection of an odorous substance requires binding of odorant molecules to specific odorant-binding proteins on the receptor cells.

75. A person who is blindfolded and holding her or his nose will have difficulty tasting the difference between a piece of apple and a piece of onion. Explain.

____

____

____

____

____

❒ **Association Cortex and Perceptual Processing** (text pages 250-252)

76. Cortical association areas
   (a) may receive sensory information directly from primary sensory areas.
   (b) may receive sensory information directly from the thalamus.
   (c) may integrate two or more kinds of sensory information.
   (d) are found in each lobe of the cerebral cortex.

77. True or false: Highly processed sensory information is invested with emotional significance by neurons in the association cortex of the temporal lobes.

ANSWERS

Boldface type indicates the answers you should have given. Words in medium-face (ordinary) type explain or expand upon the answer.

1. In a sensory system, structures called **receptors** receive information in the form of **stimuli** from the internal or external environment. This information is conducted in the form of **action potentials** along **afferent** nerve fibers and is processed by **interneurons** in the brain.

2. **False.** "Sensation" is sensory information that reaches ~~the brain~~ **consciousness**. (As will become clear later in this chapter, not all sensory information reaches the brain, and not all that reaches the brain is consciously recognized as sensation. However, such information *must* reach the brain in order to reach consciousness, since consciousness resides in the brain.)

3. **True.** (The answer to this question is intuitive at this point. The text defines "perception" as the understanding of sensation and states that perceptions are acquired by neural processing of information. The "higher-order" refers to integration of several kinds of input in the "thinking" part of the brain — the cerebral cortex.)

4. **True.**

5. (a) **Incorrect. See answer (c).**
   (b) **Incorrect. See answer (c).**
   (c) **Correct.**

6. (a) **Incorrect. See answer (b).**
   (b) **Correct.** (Note that "body surface" includes internal structures as well as external ones.)
   (c) **Correct.** (A given receptive field may be either relatively large or relatively small, depending on the length of the branches. We shall discover the significance of receptive field size in the next few pages of this chapter.)

7. **False.** The minimum number of synapses in a specific ascending pathway is ~~four~~ **three**. (First-order neuron synapsing on second-order neuron, second-order on third-order, and third-order on fourth-order. Text Figure 9-4 does not show the last synapse — the one in the cerebral cortex.)

8. (a) **Correct.**
   (b) **Incorrect. Many higher-order neurons process this information.**
   (c) **Correct.**

9. **False.** "Somatosensory" refers to the part of the cerebral cortex that receives synaptic input from specific ascending pathways originating ~~only~~ with receptors for ~~touch~~ **somatic sensation** (which includes touch and other types of sensation, as we shall see later in this chapter).

10. __e__ Somatosensory
    __c__ Visual
    __b__ Auditory
    __e__ Taste
    __d__ Olfactory

11. (a) **Incorrect. They are interneurons.**

    (b) **Correct.**

    (c) **Incorrect. They are part of nonspecific ascending pathways.**

    (d) **Incorrect. They are important even though they do not convey specific sensory information.** (The information that "something is happening" is important for arousing higher brain centers, for example.)

12. The three kinds of information specific sensory pathways provide to higher brain centers are the **type**, the **intensity**, and the **location** of a stimulus.

13. (a) **Correct.**

    (b) **Incorrect. The *absence* of polymodal neurons in the pathway is important.**

    (c) **Correct.**

    Recall that there are specific areas of the cerebral cortex (and limbic system) that are primary receiving areas for the various senses. Information received by the primary auditory cortex will be perceived (if it is perceived at all) as sound. If one's "wires" (nerve fibers) became crossed so that information from a receptor for touch, for example, would go to a second-order or higher neuron in the temperature pathway, then touch stimuli would be perceived as temperature sensations. The primary receiving areas determine sensation.

14. (a) **Correct. This increase in magnitude of the receptor potential** (increased depolarization) **increases the frequency of action potentials in the afferent nerve fiber, which is the first step in signalling increased stimulus strength**. (Recall from Chapter 8 that raising the receptor potential to suprathreshold levels allows firing of action potentials during the relative refractory period — the period when K^+ permeability is increased — and so the number of action potentials per second is increased. The upper limit of action-potential frequency is determined by the absolute refractory period — the period of increased permeability to Na^+.)

    (b) **Incorrect. Receptor potentials, like other graded potentials, are not associated with a frequency.** (These potentials are present in the receptor and change in parallel with changes in stimulus intensity.)

    (c) **Incorrect. The magnitude of action potentials does not vary with stimulus intensity. Action potentials are always all-or-none**.

    (d) **Correct. Recruitment of other receptor units increases the magnitude of EPSPs in the second-order neuron** (by spatial and temporal summation) **and increases action-potential frequency in the ascending pathway.**

15. **False.** Information about the location of a given stimulus on or in the body is conveyed by the same mechanisms that convey information about stimulus ~~intensity~~ **type**. (That is, which of the specific ascending pathways is activated is the most important "locater" of a stimulus. Higher-order neurons compare two things — [1] action potential frequency in the pathway beginning with the afferent neuron whose receptors are closest to the stimulus with [2] the frequency in adjacent afferent pathways — in order to "fine-tune" the localization, as the text explains. See text Figure 9-8.)

16. **False.** The density of receptors in a receptive field is usually greatest in the ~~periphery~~ **center** of the field.

17. (a) **Incorrect.** **The reverse is correct.**

    (b) **Incorrect.** **Different areas of the skin vary in the number and amount of overlap of receptive fields and thus in the precision of localization.**

    (c) **Correct.** (The relative lack of precision of locating a painful stimulus in the internal organs causes problems in diagnosing some illnesses.)

    (d) **Incorrect.** **The reverse is correct.** Some students have difficulty reconciling "recruitment" with "lateral inhibition." How can both phenomena occur? Simply put, when a stimulus is strong enough to stimulate nearby receptor units strongly, then lateral inhibition is less effective. Thus, for a very intense stimulus, information about precise location is sacrificed for information about intensity.

18. The specific ascending pathways for somatic sensation on the right side of the body cross to the left side of the CNS at the level of the **spinal cord** or **brainstem**. Information from somatic receptors on the right side, therefore, is carried to the **somatosensory** cortex on the **left** side of the brain.

19. **False.** In the somatosensory cortex, neuronal representation of body parts is proportional to **the density of** their ~~size~~ **sensory innervation**.

20. (a) **Correct.** **From text Figure 9-12A and its legend, we learn that the ventral spinothalamic tract conveys information from receptors that respond to a light touch. From text Figure 9-13, we see that second-order neurons conveying information about a light touch cross the cord at the level of entry of afferent neurons into the cord.**

    (b) **Correct.** **This person would not be able to perceive either temperature or pain with that toe (or foot). From text Figure 9-12B we learn that the lateral spinothalamic tract carries information from receptors for temperature and pain, and also that the second-order axons in this**

tract cross the cord at the level of the synapses with their first-order neurons.

(c) **Correct.** **From text Figure 9-12C we learn that the nerve fibers forming the dorsal columns are the central processes of afferent neurons associated with receptors that mediate two-point discrimination. These fibers do not cross to the left side of the cord; thus, two-point discrimination in the right foot is lost because the right dorsal columns are damaged.**

Note that crossing over to the opposite side is always done by second-order nerve fibers, not first-order ones. In the case of afferent neurons that carry information about vibration, joint position, and two-point discrimination, synapses with second-order neurons occur in nuclei of the brainstem. Third-order neurons in the specific ascending pathways for all the somatic senses are located in the thalamus.

Second note: When one refers to human anatomy, "dorsal" (back side) is synonymous with "posterior" and "ventral" (belly side) with "anterior." Some authors favor use of dorsal/ventral, while others prefer posterior/anterior. Thus, you must recognize both.

21. The general term for receptors that respond to mechanical stimulation is **mechanoreceptors**. In the skin, these receptors are grouped into two categories — rapid and slow — according to how quickly they **adapt** to stimuli. Receptors in the rapid category include those that give rise to sensations of **touch**, **movement**, **vibration**, and **tickle**. Receptors in the slow category mediate the sensation of **pressure**. Mechanical stimulation of receptors in skeletal muscles, joints, tendons, ligaments, and skin produces information regarding the sense of the body's position in space, which is termed **proprioception**, and sensations of joint movement, termed **kinesthesia**.

22. The sense of effort differs from other somatic senses in that it requires information from **descending motor** pathways.

23. **False.** At normal body temperature, internal warm ~~and cold~~ receptors are activated ~~approximately equally~~ **and cold receptors are not**. (Recall that normal body core temperature is 37°. Warm receptors respond to this temperature, but cold receptors do not start to respond until the temperature drops to 35°.)

24. **False.** The afferent pathways for pain differ from those for other somatic sensations in that information about the pain stimulus is transmitted to the **hypo**thalamus. (All sensory information that reaches the cerebral cortex is relayed by the thalamus.)

25. **False.** A common neurotransmitter for afferent neurons carrying information about pain is ~~prostaglandin~~ **substance P**. (Prostaglandins are potent stimuli for nociceptors, however. You may recall from Chapter 7 that aspirin and indomethacin block the activity of cyclooxygenase, the enzyme necessary for prostaglandin synthesis from arachidonic acid. This enzyme-blocking activity explains the analgesic action of these drugs.)

26. **True.**

27. **False.** "Phantom limb" pain is ~~imaginary~~ **real** pain ~~because~~ **even though** it is "felt" in an amputated limb. (If you corrected the sentence to include mention of the cause of the pain — stimulation of any of the ascending fibers in the pain pathway that had originated in the lost limb — good for you!)

28. **False.** The reduction of pain by electrical stimulation of inhibitory neurons in the pain pathway is called stimulation-produced ~~kinesthesia~~ **analgesia**.

29. **Transcutaneous electric nerve stimulation (TENS) is effective in alleviating some cases of pain because some afferent neurons in nonpain pathways, which are activated by the mild electrical stimulation, inhibit transmission of pain information.** (Recall that the peripheral and central processes of afferent neurons conveying information about tactile stimuli are large in diameter and myelinated, while those serving pain receptors are small and poorly myelinated. Therefore, the inhibitory signals travel faster than the pain signals do.)

30. Receptors

a,b,e,f mechanoreceptors
c nociceptors
a,b,e,f complex structure
c,d naked nerve endings
c respond to chemicals

Afferent Neurons

a,b,e,f have thick, myelinated processes
c,d have thin, unmyelinated or thinly myelinated processes
b,f have central processes that travel in the dorsal horns

Second-order Neurons

c,d have axons that travel in lateral spinothalamic tract
a,e have axons that travel in ventral spinothalamic tract

31. (a) **Incorrect.** **They are smaller.** The wavelengths of visible light range from 400 to 700 nanometers ("nano-" refers to one-billionth of, in this case, a meter), whereas the wavelengths of radio waves are in the hundreds of meters.

(b) **Correct.**

(c) **Correct.** This relationship holds for all electromagnetic radiation and is inherent in the description of any cyclic phenomenon, such as sound waves. Wavelength is the more useful term for describing light, whereas frequency is more useful for describing sound, as we shall see in the next section. The unit of frequency is the hertz (Hz), which means "cycles per second." Note that neither wavelength nor frequency describes the intensity of the stimulus.

Note: Human beings can sense electromagnetic radiation only in the range of visible light. We are "blind" to the rest of the spectrum, even though we have learned how to use radar and radio waves for communication, for example. Information communicated in these ways must be translated to visible light or sound in order for us to detect it, however.

32. **True.** (See text Figure 9-16.)

33. (a) **Ciliary muscle**
    (b) **Iris**
    (c) **Cornea**
    (d) **Pupil**
    (e) **Lens**
    (f) **Aqueous humor**
    (g) **Vitreous body**
    (h) **Blood vessels**
    (i) **Fovea**
    (j) **Optic nerve**
    (k) **Retina**

34. **Both structures bend light rays to allow light from an object being viewed to be focused on the retina. The cornea does most of the bending and thus most of the focusing, but the lens is able to change shape and so can vary the amount of bending it does. This variability of the lens is responsible for *accommodation* — the ability to focus on objects at different distances.**

    Note: The importance of the cornea for vision can be appreciated by the fact that a defective cornea can result in blindness. Many people who die of accidents or diseases not affecting the eye donate their corneas to be used as transplants for people with eye disease, allowing those people the joy of seeing the world they live in.

35. (a) **Incorrect. It decreases.**
    (b) **Correct.**
    (c) **Incorrect. Relaxation of the ciliary muscles puts tension on the zonular fibers.**

(d) **Correct.** (This lens flattening results in less bending of the light rays, which arrive at a less sharp angle than rays from a nearby object.)

36. (a) **Incorrect. The too-short eyeball describes the farsighted, or hyperopic, eye. Both conditions result in difficulty in focusing on nearby objects. However, presbyopia is due to a change in the elasticity of the lens rather than to a change in eyeball length.**

(b) **Incorrect. Lens coloration causes cataract.**

(c) **Correct.**

(d) **Correct.** (Reading glasses are generally slightly convex. Some people correct mild presbyopia with a magnifying glass, which allows one to see small objects from farther away.)

Note: Nearsighted, or myopic, people also develop presbyopia. These people can generally see near objects clearly without corrective lenses, but they retain their inability to focus on distant objects without correction.

37. **True.** (See text Figure 9-23.)

38. **False.** Increased pressure in the eye caused by an accumulation of aqueous humor is called ~~astigmatism~~ **glaucoma**. (If untreated, glaucoma can cause blindness by injuring the delicate receptors in the retina.)

39. (a) **Correct.**

(b) **Correct.**

(c) **Correct.**

40. (a) **Parasympathetic stimulation of the iris (a smooth muscle) causes narrowing of the pupil. Using atropine to block the action of acetylcholine released by parasympathetic nerves allows the pupil to dilate.** (The same result would be achieved by administration of an adrenergic *agonist.*)

(b) **Activation of parasympathetic nerves is necessary for accommodation for near vision. Blocking the effects of acetylcholine on the ciliary muscles controlling the shape of the lens inhibits accommodation and thus makes focusing on near objects difficult.**

41. (a) **Correct.**

(b) **Incorrect. They are located on the surface closest to the back of the eye.**

(c) **Correct.**

42. (a) **True.** (You may have answered, "two: rods and cones." There are three types of cones, however, for a total of four types of photoreceptors.)

    (b) **True.**

    (c) **False. There is just one kind — retinal.**

    (d) **True.**

43. **The chromophore retinal is derived from vitamin A. Lack of vitamin A in the diet leads to breakdown of photopigments and eventual blindness.** (Vegetables such as carrots, which are good sources of vitamin A, really are good for your eyes.)

44. photon ⇒ retinal ⇒ change of **shape** of retinal ⇒ alteration of **opsin**

⇓

$\downarrow P_{Na^+}$ ⇐ **closing** of membrane Na^+ channels ⇐ ↓ concentration of cyclic GMP in photoreceptor cytoplasm

⇓

hyperpolarization of photoreceptor ⇒ ↓ neurotransmitter release by photoreceptors

Because photoreceptor stimulation results in hyperpolarization of the receptor and decreased neurotransmitter release at the synapse with the second-order neuron, you might expect that signaling from the eye to the brain would be totally inhibitory — in other words, that the presence of light on the retina is signaled by a decrease in action potentials in third- and fourth-order neurons. This is not the case. Although the interactions among the photoreceptors and other cells in the retina are complicated, in general the following takes place:

light → hyperpolarized photoreceptor → hyperpolarized bipolar cell → depolarized ganglion cell

↓

action potentials in ganglion cells

Thus, bipolar cells are themselves inhibitory; that is, when depolarized, they release a hyperpolarizing neurotransmitter that inhibits the firing of action potentials in ganglion cells. Inhibition of neurotransmitter release from an inhibitory cell leads to excitation of the neurons with which it makes synaptic contact and thus to an increase in the firing of action potentials in these neurons.

A further complication in the retina is that other cells, such as horizontal and amacrine cells (see text Figure 9-25), can stimulate or suppress activity in nearby bipolar and ganglion cells. These interactions account in part for the characteristic "on-off" responses of the visual system.

45. **False.** The optic nerve fibers from each eye meet at the optic chiasm, where ~~they~~ **some of them** cross. Thus, visual information from ~~the right eye~~ **both eyes** is received by ~~the left side~~ **both sides** of the brain.

The crossing of optic nerve fibers is not random, but has a pattern. Fibers from the ganglion cells in the nasal half of each retina (the half closest to the nose) cross over, while those from the temporal side do not. When you focus your eyes straight ahead, everything you can see to the left of your focus point (without moving your eyes) constitutes your *left visual field,* and everything you can see to the right of your focus point is your *right visual field.* Information from the left visual field strikes the nasal half of the left retina, whose fibers cross at the optic chiasm, and the temporal half of the right retina, whose fibers do not cross. Thus, information about the left visual field goes to the right side of the brain, and vice versa.

46. (a) **color** **movement** **depth/distance** **texture** **form**

(b) **False.** These characteristics of the visual system are first sorted out by cells in the ~~lateral geniculate nucleus of the thalamus~~ **retina** (by ganglion cells, and probably by bipolar cells as well). (Parallel processing of visual information begins very early in the visual pathway and continues at each higher step.)

47. (a) **Correct.**

(b) **Correct.**

(c) **Incorrect. It is circular.**

(d) **Correct.**

48. (a) **Correct.**

(b) **Incorrect. Its firing rate decreases.**

(c) **Incorrect. There is spontaneous action-potential activity in all ganglion cells.** (The information capacity of the signaling system is increased by the cell's ability to increase or to decrease firing in response to a visual stimulus.)

49. (a) **Correct.**

(b) **Incorrect. It receives information from the reticular formation and the visual cortex as well.**

(c) **Incorrect. Synchronizing of biological clocks is a function of the suprachiasmatic nucleus.**

50. **True.** (The regions of cerebral cortex other than the occipital-lobe primary visual cortex involved in processing visual information are referred to as "association areas" and are described in more detail at the end of this chapter and in Chapter 20.)

51. **False.** An object that absorbs the longer wavelengths of visible light while reflecting the shortest ones will be perceived as ~~yellow~~ **blue or violet**.

52. **White clothing reflects all wavelengths of light and thus keeps its wearers cooler than colored clothing, which absorbs light** (and heat).

53. Opponent color ganglion cells that respond to blue light by increasing their rate of firing will respond to red light by **decreasing** their rate of firing. White light will cause a **decrease** in the firing rate of these cells. (See text Figure 9-31.) We are capable of distinguishing a vast array of colors and hues because these opponent color cells can efficiently integrate input from the three types of cones.

    What about the rods? We have ignored them for most of this discussion. Rhodopsin (the photopigment in rods) is exquisitely sensitive to light and is responsible for our ability to see objects in dim light. In bright light, all the rhodopsin is "bleached" (the opsin is separated from retinal) and the rods contribute nothing to our vision. When we go from bright light to dim light, at first we are essentially blind because the light is insufficient for cone vision. After a few minutes, rhodopsin is regenerated and the rods begin to function. Rod vision is five to six orders of magnitude more sensitive to light than is cone vision.

    Why can't we see colors in dim light? Rhodopsin is maximally sensitive to light at a wavelength of 505 nm (between the blue and green cone photopigments). However, more than one photopigment is necessary to view color, and no color can be detected in dim light because there is only one photopigment that functions at that degree of luminance — rhodopsin. (If only one photopigment is present, the photoreceptor response is perceived as brightness, not color.) We can distinguish colors in bright light because of the three cone photopigments and the opponent color cells in the visual pathway. If there were only one cone photopigment, we could not see colors in bright light either. (There are people, called monochromats, who lack two of the cone pigments and can distinguish only white, black, and shades of grey. A much more common problem is red-green color blindness, which results in a defective photopigment in either the red- or green-detecting cones but not both. This condition is much more common in men than in women, for reasons that will become clear after you have studied Chapter 18.)

54. **False.** The greatest concentration of ~~rods~~ **cones** is in the fovea of the retina.
    Alternatively, the answer below is also correct:
    False. The greatest concentration of rods is in the ~~fovea~~ **periphery** of the retina.

55. **True.**

56. Sound waves consist of alternating areas of **compression** and **rarefaction** of air molecules. The pitch of a sound is determined by the **vibration frequency** of the sound source.

57. **False.** A sound with a frequency of 4000 Hz has a ~~lower~~ **higher** pitch than a sound with a frequency of 1000 Hz. (Recall that Hz means "cycles per second." The higher the number, the greater the frequency and therefore the higher the pitch.)

58. **False.** The tympanic membrane separates ~~an~~ **two** air-filled ~~chamber~~ **chambers** ~~from a fluid-filled chamber~~.

    Alternatively, the answer below is also correct:

    False. The ~~tympanic~~ **oval** membrane separates an air-filled chamber from a fluid-filled chamber.

59. **False.** The vibration of the tympanic membrane varies in magnitude according to the ~~pitch~~ **amplitude** (or **loudness**) of the sound it receives.

    Alternatively, the answer below is also correct:

    False. The vibration of the tympanic membrane varies in ~~magnitude~~ **frequency** according to the pitch of the sound it receives.

    The amplitude of sound waves is commonly expressed as decibels (dB). A whisper has a value of about 20 dB, while the sound from a jet airplane is about 160 dB. The level when sound becomes painful for a normal ear is about 140 dB.

60. **False.** The function of the ~~tympanic membrane~~ **bones of the middle ear** is to amplify sound waves on their way to the inner ear.

61. **Yawning and swallowing open the tube leading from the middle ear to the pharynx (back of the mouth) and allow the pressure in the middle ear to equilibrate with atmospheric pressure.**

62. **c** cochlea
    d incus
    a malleus
    e stapes

63. **Conduction deafness refers to problems with the conduction of sound waves to the inner ear. The middle ear bones function as a piston to amplify the tympanic-membrane vibration that is transferred to the oval window. Softened bones cannot perform as an adequate piston.**

    Fortunately, otosclerosis can be cured by a delicate surgical technique in which the stapes is removed and replaced with a platinum pin, resulting in a sort of "bionic ear."

64. In the cochlea, pressure transmitted to the **scala vestibuli** from the oval window causes the **basilar membrane** of the cochlear duct to vibrate. This vibration causes stimulation of mechanoreceptors, called **hair cells.** During stimulation of these receptors, bundles of actin myofilaments, called **stereocilia**, are bent against the

overhanging **tectonic membrane**. This deformation causes channels for Na^+ and Ca^{2+} to **open** and consequent **depolarization** of the cells. The region of the cochlear duct that contains the receptor cells is called **the organ of Corti.**

65. **True.**

66. **False.** ~~Un~~Like information from the eyes, information from the ears ~~bypasses~~ **is transmitted to** the thalamus on its way to the cerebral cortex. (Recall that all sensory information that reaches the cortex is first processed in the thalamus. The nucleus of the thalamus that is concerned with sound stimuli is the medial geniculate.)

67. (a) **Incorrect.** **It is connected to the cochlea.** (The inner ear includes the entire duct system, the cochlea, and the vestibular apparatus.)
    (b) **Incorrect.** **It is filled with fluid.**
    (c) **Correct.** (You may have answered "three," thinking that the three semicircular canals constitute one organ; however, each of the three detects a different direction of motion. The utricle and saccule are the fourth and fifth receptor organs of the vestibular system.)
    (d) **Correct.**

68. **True.** If the concept of "linear motion" for the head seems peculiar, think about walking straight ahead — such motion is horizontal linear motion and is detected primarily by the utricles. Jumping up and down is vertical motion and is primarily detected by the saccules. We also experience linear acceleration in automobiles, airplanes, and other conveyances and when swimming and diving. People who have vestibular disorders are warned not to dive in murky water because their ability to tell which way is up is impaired.

    Note: The utricles and saccules also give us information about static equilibrium — that is, about the orientation of the head with respect to the ground — whether or not the head is moving.

69. **False.** When you turn your head to the left, to look over your left shoulder, the hairs in the cupula of the horizontal semicircular canals will be bent to the ~~left~~ **right**.

70. (a) **Incorrect.** **It is conveyed by cranial nerve VIII** (the vestibulocochlear, also called auditory, nerve).
    (b) **Correct.**
    (c) **Incorrect.** **It gives us conscious awareness of position and movement of the head.**

71. **True.** (This may be a cause of Ménière's disease.)

72. The receptors for the senses of smell and taste belong to the general category of **chemoreceptors**. The sense organs for taste are the **taste buds**, which are traditionally divided into four categories — those sensitive to **sweet**, **salty**, **sour** (acid), and **bitter** tastes. Information regarding taste is conveyed to the **somatosensory** cortex in the **parietal** lobe of the brain. Receptors for the sense of smell, called **olfactory cells**, are located in the **olfactory mucosa** in the roof of the nasal cavity. Information regarding smell is conveyed directly to the **olfactory bulbs**, which are part of the **limbic** system of the brain.

73. **True.**

74. **False.** Detection of an odorous substance requires binding of odorant molecules to specific ~~odorant-binding proteins~~ **receptors** on the receptor cells. (The odorant-binding proteins may, as the text explains, facilitate binding of the odorant molecules to the receptors, but they are not themselves part of the receptor cells. This question illustrates another example of the two meanings of the word "receptor.")

75. **Neither an apple nor an onion is characteristically salty, sweet, bitter, or sour. Much of the "taste" of either one is detected by *olfactory* receptors (which are essentially bypassed when holding one's nose), not the taste buds. Furthermore, the texture of apples and onions is similar.**

    Note: A blindfolded, nose-holding person would have no trouble distinguishing between a bell pepper and a chile pepper. Chiles and other "hot" foods get their characteristic taste from molecules that stimulate pain receptors in the mouth.

76. (a) **Correct.**

    (b) **Incorrect. The thalamus relays sensory information directly to the primary sensory cortical areas only.**

    (c) **Correct.**

    (d) **Correct.**

77. **False.** Highly processed sensory information is invested with emotional significance by neurons in the association cortex of the ~~temporal~~ **frontal** lobes. (Recall that the frontal lobes are the link between the cerebral cortex and the limbic system, which is the "emotional" brain.)

    The text mentions that removal of part of the limbic association area may abolish the emotional response to a painful stimulus. The procedure known as "prefrontal lobotomy," in which the connections between the frontal lobes and the limbic system are cut, is sometimes performed on people who have intractable pain caused by cancer. After the operation, pain is still perceived but it doesn't bother the patient. This type of surgery has profound effects on the personality, however, and is used only in extreme cases.

PRACTICE

True or false (correct the false statements):

P1. Information on the type of a given stimulus is provided by the specific sensitivity of individual receptors and by the specific pathways conveying information to the primary sensory areas.

P2. In general, the larger the receptive field of a given sensory neuron and the lesser the degree of overlap among adjacent sensory fields, the greater the precision of locating the stimulus.

P3. The ability to discriminate between two adjacent stimuli is greatest in the back, thigh, and forearm.

P4. A somatosensory map of the left side of the body is present in the frontal lobe of the right side of the brain.

P5. The dorsal columns of the spinal cord carry information about joint position and vibration to the somatosensory cortex.

P6. The conduction of information regarding a painful stimulus can be inhibited by electrical stimulation of certain areas of the CNS that release endorphins.

P7. The majority of the fibers in the optic nerve project to the suprachiasmatic nucleus, which relays information from the retina to the visual cortex.

P8. The two parts of the eye most important for focusing an image on the retina are the lens and the iris.

P9. An important function of the bones of the inner ear is to amplify sound waves.

P10. Information on the loudness of a sound is conveyed by the frequency of action potentials in specific afferent fibers of the cochlear nerve.

P11. The receptors for both sound and changes in head position are hair cells.

P12. The primary sensory area of the brain for the sense of taste is found in the limbic system.

P13. Processing of sensory information becomes more complex as the information is conveyed from one primary sensory area of the cortex to another.

Indicate whether the following are mechanoreceptors (M), chemoreceptors (C), both (B), or neither (N):

P14. ___ Pacinian corpuscle

P15. ___ Hair cell in organ of Corti

P16. ___ Cone

P17. ___ Receptor cell in taste bud

P18. ___ Nociceptor

P19. ___ Hair cell in semicircular canal

P20. ___ Thermoreceptor

P21. ___ Olfactory receptor

Match the description on the left with the correct word on the right:

P22.	___	eyeball too long for lens	(a) astigmatism
P23.	___	eyeball too short for lens	(b) hyperopia
P24.	___	caused by stiffening of lens	(c) cataract
P25.	___	caused by accumulation of aqueous humor	(d) presbyopia
			(e) glaucoma
P26.	___	correctable by concave lens	(f) myopia

Fill in the blanks:

P27-30. The light-sensitive chemicals in the photoreceptors are molecules called ____________________. There are ________(how many?) different kinds of these molecules in the human eye. Each of these molecules is composed of an identical light-sensitive ____________________ molecule and a surrounding protein called ________________, which varies from one kind of photoreceptor to another.

10: HORMONAL CONTROL MECHANISMS

The nervous system, as we learned in Chapters 8 and 9, is a communication system dependent upon "wires," much as our homes have telephone wires connecting them with a central system and thus the means to communicate with other homes. In contrast, the endocrine system is more comparable to a highly efficient bulk mail system, in which chemical "parcels" are constantly delivered to each home via freeway, street, and lane and are addressed "To Whom It May Concern." Those who are concerned — that is, the cells that have receptors for a given hormone — respond to the message by changing some aspect of their activities. Some hormones, such as those from the thyroid gland, are necessary for the proper functioning of nearly all cells at some stages of their lives, while others, such as the pituitary hormone that stimulates thyroid hormone synthesis, are quite specific for one cell type. The specificity is "decided" by the addressee.

Instructions for answering questions in this Study Guide

1. True or false: Correct each false statement. Whenever you must correct a false statement, there will almost always be more than one way to do so. In the Guide, the answer requiring the least correction will generally be given, with the understanding that other correct answers are possible. It is usually not sufficient, in terms of demonstrating understanding, to simply insert a "no" or "not," however.

2. Multiple choice: Any, all, or none of the responses may be correct. Explain why each incorrect response is incorrect.

3. Fill in the blanks: Choose the best word or words to complete each statement.

4. Directions will be given for other types of questions as they appear.

❒ Introduction (text pages 255-258)

1. The endocrine system
   (a) is one of two major communication systems of the body.
   (b) is composed of glands that secrete chemical messengers into the blood.
   (c) is an important regulator of homeostatic mechanisms.
   (d) works essentially independently of the nervous system.

2. Examine text Table 10-1 and indicate which of the following statements is/are correct.
   (a) Most glands secrete more than one hormone.
   (b) Many glands have functions other than as endocrine organs.
   (c) Many hormones have multiple functions.

❒ Hormone Structures and Synthesis (text pages 258-262)

3. The three general chemical classes of hormones are the ______________________, the ______________________, and the ______________________. Four hormones derived from the amino acid tyrosine are ______________________ and ______________________ from the thyroid gland, and ______________________ and ______________________ from the adrenal medulla. One of these hormones, ______________________, also functions as an important neurotransmitter. The thyroid hormones are unusual molecules in that they contain atoms of ______________________.

4. In most if not all target cells for thyroid hormones (TH), T_4 is converted to T_3 before it interacts with the TH receptor. Which of the following statements is/are likely to be true?

   (a) T_4 may be considered a "precursor" or "prohormone" form of T_3.
   (b) T_4 is probably more active than T_3.
   (c) The TH receptor is probably on the inside of the cell.
   (d) Thyroid hormones are either lipid-soluble or actively transported across target cell membranes.

5. Peptide hormones
   (a) are derived from larger molecules called prohormones.
   (b) are packaged into secretory vesicles in the endoplasmic reticulum of the cells that synthesize them.
   (c) are secreted in response to a stimulus that causes depletion of intracellular calcium levels in the secreting cell.
   (d) may be secreted as a result of action potentials in the secretory cell.

6. Steroid hormones belong to the chemical class of ______________________. They are ______________________ (polar/nonpolar) molecules derived from ______________________. These hormones are produced by the following glands: the ______________________, the ______________________, the ______________________, and the ______________________.

7. True or false: Synthesis of steroid hormones is dependent upon protein synthesis.

8. True or false: Steroid hormones are stored in the cells that synthesize them until a stimulus is received that provokes their secretion.

9. True or false: The adrenal cortex secretes androgens.

10. Aldosterone is called a "mineralocorticoid." Cortisol is a "glucocorticoid." Explain these terms. ____

____

____

____

11. The major steroid hormone secreted by the male gonad is ______________________. The female gonad secretes two steroid hormones, ______________________ and ______________________.

12. True or false: Ovaries synthesize as much testosterone as they do estradiol.

❒ Hormone Transport in the Blood (text page 262)

13. True or false: In general, amine and peptide hormones circulate in the plasma as free hormones whereas steroid hormones are mostly bound to circulating binding proteins.

14. Carlos and Sally each have a routine physical examination that includes measurement of thyroid hormone levels in plasma.

	Total plasma TH	Bound TH
Carlos	8 μg/dL	7.998 μg/dL
Sally	12 μg/dL	11.998 μg/dL

Neither shows symptoms of having too much or too little TH. Explain.

____

____

____

____

❒ Hormone Metabolism and Excretion (text pages 262-263)

15. A hormone may be
    (a) inactivated by its target cell.
    (b) activated by its target cell.
    (c) inactivated by nontarget cells.
    (d) excreted before it has a chance to act on a target cell.

16. True or false: In general, the metabolism and excretion rates of steroid hormones are higher than those of peptide hormones.

17. In Table 10-1, find the hormone that is really an enzyme. The name of this enzyme is ________________. It is secreted by the ________________ and catalyzes the formation of a hormone called ________________, which has effects on the secretion of the mineralocorticoid ________________.

❐ **Mechanisms of Hormone Action** (text pages 263-264)

18. The receptors for ________________ and ________________ hormones are proteins located inside cells, whereas those for ________________ and ________________ hormones are proteins on the cell membrane.

19. True or false: The concentration of a hormone in the blood can affect the number of receptors for that hormone on or in a target cell.

20. True or false: The effect of thyroid hormones on the number of epinephrine receptors on epinephrine's target cells is an example of down-regulation.

21. True or false: The mechanism of action of lipid-soluble hormones is the stimulation of specific DNA-dependent protein synthesis.

22. True or false: Binding of lipid-insoluble hormones to their receptors usually results in the formation of one or more second messengers that lead to alterations of the cell's activity.

23. As a review of material from Chapter 7
    (a) list the five second messengers known to mediate the responses to peptide and adrenal medullary hormones:

    ________________ ________________

    ________________ ________________

    ____

    (b) name two other signal transduction mechanisms used by peptide hormones:

    ________________ ________________

24. For further review of Chapter 7: If a particular hormone causes a response in a target cell by activating adenylate cyclase, then
    (a) binding of the hormone to its receptor should result in increased levels of cyclic AMP inside the cell.
    (b) binding of the hormone to its receptor should result in increased levels of phosphorylated proteins inside the cell.
    (c) binding of the hormone to its receptor should result in increased levels of cyclic GMP inside the cell.

(d) treating the cell with a phosphodiesterase inhibitor should increase the cell's response to the hormone.

25. True or false: Pharmacological effects of hormones can be seen only when one is given large doses of the hormone as a drug.

❒ Types of Inputs That Control Hormone Secretion (text pages 264-267)

26. The four general types of inputs that alter rates of hormone secretion are changes in plasma concentrations of ______________________ or ______________________ and the interactions of ______________________ or ______________________ with endocrine cells.

27. True or false: Unlike neurons, endocrine cells generally do not act as integrators of various kinds of positive or negative inputs.

28. The two parts of the central nervous system that directly affect hormone secretion are the ______________________ and the ______________________.

29. True or false: In cases where the plasma concentration of an ion or nutrient affects the secretion rate of a hormone, the affected hormone is a regulator of the homeostasis of that ion or nutrient.

30. True or false: A general name for a hormone that regulates mineral ion homeostasis is "tropic hormone."

❒ Control Systems Involving the Hypothalamus and Pituitary (text pages 267-277)

31. The pituitary gland
    (a) is composed of two lobes.
    (b) is attached to the hypothalamus.
    (c) contains both neural and glandular tissues.

32. The posterior pituitary
    (a) synthesizes oxytocin and vasopressin.
    (b) is neural tissue.
    (c) secretes hormones when stimulated by action potentials generated in the hypothalamus.

33. True or false: The effect of some tropic hormones is to stimulate release of other tropic hormones.

34. True or false: The fragment of pro-opiomelanocortin with established endocrine function is ß-lipotropin.

35. True or false: The primary function of thyrotropin is to stimulate secretion of thyroid hormones.

36. The gonadotropic hormones
    (a) stimulate hormone secretion by the gonads.
    (b) are secreted by the anterior pituitary gland of both males and females.
    (c) are LH and prolactin.

37. Growth hormone
    (a) is a tropic hormone.
    (b) has metabolic effects in many types of cells.
    (c) secretion has a circadian rhythm.

38. Describe the unusual properties of the blood vessels that connect the hypothalamus with the anterior pituitary.

____

____

____

____

____

____

39. A "releasing" hormone does not always stimulate *release* of the hormone whose secretion it affects. Explain.

____

____

____

40. The hypothalamic releasing hormones
    (a) are all peptides.
    (b) are all specific for one pituitary cell type.
    (c) can have releasing activity or release-inhibiting activity, depending on the cells to which they bind.

41. Text Figure 10-21 shows one example of how neural input to the hypothalamus can influence secretion of a hypothalamic hormone, which in turn influences the secretion of a pituitary hormone. (Ignore for the moment the effects of the target-gland hormone on secretion of the first two hormones.) During lactation (breast feeding), the stimulus of an infant suckling on the nipple causes an increase of prolactin secretion by the anterior pituitary. Using text Figures 10-20 and 10-21 as guides, draw a scheme that can explain this observation.

42. In cases in which one adrenal cortex is destroyed, the other gland undergoes what is called compensatory *hypertrophy* (overgrowth). Explain.

____

____

____

____

43. During pregnancy, blood levels of estrogens are elevated, leading to, among many other effects, increased secretion of TH-binding proteins by the liver. What effect does this increased secretion of binding proteins have on the secretion of TH? Complete the diagram below to illustrate your answer. (Hint: Refer to Question 14.)

↑ plasma estrogens → ↑ liver secretion of TH-binding proteins → ↑ % bound TH in blood → ↓ free TH in blood

↓

______________________ (b) ← ______________________ (a)

↓

(c) ______________________ → (d) ______________________

44. Study text Figures 10-23 and 10-24 carefully, and then tell which of the following statements is/are correct.

(a) The negative effect of increased plasma IGF-I levels on GHRH secretion (or positive effect on SS secretion) is an example of long-loop negative feedback.

(b) The negative effect of increased plasma IGF-I levels on pituitary sensitivity to GHRH is an example of short-loop negative feedback.

(c) The negative effect of increased plasma GH levels on GHRH secretion (or positive effect on SS secretion) is an example of short-loop negative feedback.

(d) During exercise, stress, fasting, and sleep, a rise in the plasma levels of GH and/or IGF-I over basal levels has physiological adaptive value.

❒ **Candidate Hormones** (text page 278)

45. The candidate hormone from the pineal gland

(a) is a prostaglandin.

(b) is not known to be secreted into the blood.

(c) has no known function in animals.

46. Two candidate hormones from the anterior pituitary are ______________________ and ______________________.

❐ Types of Endocrine Disorders (text pages 278-279)

47. You are a physician specializing in the diagnosis and treatment of endocrine disorders. One day three patients come to you complaining of identical symptoms: mental sluggishness, physical lethargy, lack of appetite, and difficulty keeping warm. These are all symptoms of too little thyroid hormone (*hypo*thyroidism).

    (a) Patient A has an enlarged thyroid gland and low plasma TH levels. What is the most likely diagnosis?

    ____

    ____

    What treatment would you prescribe?

    ____

    (b) Patient B has low plasma concentrations of both TH and TSH. What test might you perform to determine which gland(s) is/are at fault in this case? Discuss the different possible results, and how you would arrive at your diagnosis. (In order not to overcomplicate this discussion, ignore pituitary hormones other than TSH and hypothalamic hormones other than TRH.)

    ____

    ____

    ____

    ____

    ____

    ____

    ____

    ____

    c) Patient C is a real puzzle. He has *elevated* TH and TSH levels, but all the symptoms of *hypo*thyroidism. What defect or defects might account for this disease?

    ____

    ____

    ____

    ____

    ____

ANSWERS

Boldface type indicates the answers you should have given. Words in medium-face (ordinary) type explain or expand upon the answer.

1. (a) **Correct.** (The nervous system is the other.)
   (b) **Correct.**
   (c) **Correct.** (This is one of the most important functions of the body's communication systems.)
   (d) **Incorrect. The two systems are interdependent.** (You already know a little about how hormones affect the nervous system by acting as neuromodulators. In this chapter you will learn about the major role the brain — especially the hypothalamus — plays in regulating hormone secretion and other endocrine functions.)

2. (a) **Correct.**
   (b) **Correct.**
   (c) **Correct.**

3. The three general chemical classes of hormones are the **amines**, the **peptides**, and the **steroids**. Four hormones derived from the amino acid tyrosine are **thyroxine (T_4)** and **triiodothyronine (T_3)** from the thyroid gland, and **epinephrine (E)** and **norepinephrine (NE)** from the adrenal medulla. One of these hormones, **NE**, also functions as an important neurotransmitter. The thyroid hormones are unusual molecules in that they contain atoms of **iodine**.

   (Note: There is another amine hormone, dopamine, you will learn about later in this chapter when we discuss hormonal control of pituitary hormone secretion.)

4. (a) **Correct.** (However, the "prohormone" designation is usually reserved for describing the synthesis of the peptide hormones. As the text mentions, both T_4 and T_3 are first synthesized as part of a large protein, thyroglobulin.)
   (b) **Incorrect. Cells would be expected to convert a less active form to a more active one, not the other way around.**
   (c) **Correct.** (As you will see in a later section, TH receptors are located in the nucleus.)
   (d) **Correct.** If 4-c is true, then one or the other of these options must be true as well. In fact, TH are lipid-soluble (nonpolar). This is *not* true for the other amine hormones — the catecholamines — however. Catecholamine

hormones are not lipid-soluble and have receptors on the cell membrane.

5. (a) **Correct.** Synthesis of large proteins followed by cleavage into smaller ones seems to be "easier" for cells to accomplish than, say, synthesis of a peptide twenty or thirty amino acids in length. One possible explanation for this is that short fragments of mRNA may not be as stable in the cell, or may not initiate protein synthesis on the ribosomes as efficiently as longer ones. In any case, the process of synthesizing large molecules, then packaging (and secreting) them as groups of smaller ones allows for the possibility of having "spare" molecules in the circulation for which a function might be found in the course of evolution.

   (b) **Incorrect. Secretory vesicles are packaged in the Golgi apparatus.**

   (c) **Incorrect. The stimulus for the exocytosis of secretion granules** (whether containing hormones or neurotransmitters or, presumably, any other molecules that are secreted in granules) **is always an increase in intracellular calcium.**

   (d) **Correct.** Action potentials trigger secretion in all neurons that secrete neurohormones. Action-potential activity has been observed in some non-neural endocrine cells (epithelial cells) as well. Depolarizing any cell generally activates it.

6. Steroid hormones belong to the chemical class of **lipids**. They are **nonpolar** molecules derived from **cholesterol**. These hormones are produced by the following glands: the **adrenal cortex**, the **testes**, the **ovaries**, and the **placenta**.

   Note: There are two adrenal glands, just as there are two ovaries and two testes. To be consistent, thc answer should have read "adrenal cortices."

7. **True.** (The synthesis of steroid hormones is mediated by enzymes, which are proteins.)

8. **False.** ~~Steroid~~ **Peptide** hormones are stored in the cells that synthesize them until a stimulus is received that provokes their secretion.

   Alternatively, the answer below is also correct:

   False. Steroid hormones ~~are stored in~~ **diffuse out of** the cells that synthesize them ~~until a stimulus is received that provokes their secretion~~ **as soon as they are synthesized**. (In the case of steroids, the stimulus that promotes secretion first stimulates their synthesis from cholesterol.)

9. **True.** ("Androgen" is the term used for any molecule that has masculinizing effects. As the text explains, adrenal androgens are important for women but not for men, whose testes secrete the much more potent androgen testosterone.)

10. **A mineralocorticoid is a steroid from the adrenal cortex (thus, corticoid) that is important for salt (mineral) and water homeostasis. Cortisol is a corticoid that is important for glucose metabolism.** (Cortisol and aldosterone are also referred to collectively as "corticosteroids" to distinguish them from the steroids secreted by the gonads, the gonadal steroids.)

11. The major steroid hormone secreted by the male gonad is **testosterone**. The female gonad secretes two steroid hormones, **estradiol** and **progesterone**.

    Note: Just as testosterone is one type of the generic hormone androgen, estradiol is one of several generic *estrogens*, or feminizing hormones. It is important to understand that both males and females make and secrete both androgens and estrogens. The amounts and the activities of the hormones secreted by the different sexes do differ, however.

    Progesterone has importance as a *hormone*, that is, a messenger that is secreted, primarily during pregnancy. However, it is a very important precursor for other steroid hormones, as is shown for the adrenal cortex in text Figure 10-4. Progesterone is also important for the synthetic pathways in the testis and ovary, which are not shown.

12. **True.** (Testosterone is the precursor of estradiol, the major estrogen secreted by the ovary.)

13. **False.** In general, ~~amine~~ **catecholamine (adrenal medullary)** and peptide hormones circulate in the plasma as free hormones, whereas steroid **and thyroid** hormones are mostly bound to circulating binding proteins. (This question is another reminder that the two kinds of amine hormones have different chemical properties: Catecholamines dissolve readily in plasma and so do not require carrier proteins, whereas the relatively insoluble TH do require such carriers.)

14. **Both Carlos and Sally have the same concentration of free (unbound) TH in their plasma — 0.002 μg/dL. Only free hormone can bind to receptors and exert biological effects.** (This sort of difference in the amount of bound TH and also of bound corticosteroid hormones would be expected if Sally were pregnant. During pregnancy, the levels of binding proteins in blood rise and the secretion of the hormones that bind to these proteins increases in parallel fashion, so that the amount of free hormone stays the same. The mechanism that provides for this parallel rise in hormone secretion is discussed later in this chapter.)

15. (a) **Correct.** (Inactivation occurs after the hormone binds to its receptor.)
    (b) **Correct.**
    (c) **Correct.** (Liver and kidney cells metabolize many hormones that have no function on those cells — at least as far as we know. Many hormones do act on liver and kidney cells, however, and these hormones are inactivated by those cells, as in 15-a.)
    (d) **Correct.**

16. **False.** In general, the metabolism and excretion rates of steroid hormones are ~~higher~~ **lower** than those of peptide hormones.

17. The name of this enzyme is **renin**. It is secreted by the **kidneys** and catalyzes the formation of a hormone called **angiotensin II**, which has effects on the secretion of the mineralocorticoid **aldosterone**. (You will learn where the precursor to angiotensin II comes from and the significance and functions of renin, angiotensin II, and aldosterone in Chapter 15.)

18. The receptors for **steroid** and **thyroid** hormones are proteins located inside cells, whereas those for **peptide** and **catecholamine** (**adrenal medullary**) hormones are proteins on the cell membrane. This question makes the same point as Question 13: Hormones that are soluble in lipid but not in water can easily penetrate the lipid barrier of the cell membrane, while hormones that are lipid-insoluble cannot. Similarly, hormones that are soluble in water can dissolve in the blood plasma and be carried freely in it, while those that are water-insoluble must be transported bound to molecules that are water-soluble. The location of a hormone's receptor is within the cell if the hormone is lipid-soluble and on the cell's surface if the hormone is not lipid-soluble. Thus, although there are three *chemical* classes of hormones, the amine group can be split and its members put into the peptide (catecholamine = adrenal medullary hormones) and steroid (thyroid hormones) categories for discussions of transport in blood and general mechanisms of action. (See text Table 10-2.)

19. **True.**

20. **False.** The effect of thyroid hormones on the number of epinephrine receptors on epinephrine's target cells is an example of ~~down~~ **up**-regulation. (This up-regulation is the mechanism by which thyroid hormones have permissive effects on the *actions* of epinephrine. If you crossed out "down-regulation" and wrote "permissiveness," that would not be quite correct because the question asked about receptors, not actions.)

21. **True.**

22. **True.**

23. (a) **Cyclic AMP**, **cyclic GMP**, **calcium**, **inositol triphosphate**, **diacylglycerol**

    (b) **Receptor-operated channels**, **receptor tyrosine kinases** (See text page 162.)

24. (a) **Correct.**

    (b) **Correct.**

    (c) **Incorrect. This would require an increase in guanylate cyclase, not adenylate cyclase.** (No hormone-receptor complex stimulates adenylate cyclase and guanylate cyclase simultaneously. Cyclic AMP and cyclic GMP usually have opposite activities.)

    (d) **Correct.** (See text page 156.)

25. **False.** Pharmacological effects of hormones can be seen ~~only~~ when one is given large doses of the hormone as a drug **or when excessive amounts of hormones are secreted in disease states.**

26. The four general types of inputs that alter rates of hormone secretion are changes in plasma concentrations of **mineral ions** or **organic nutrients** and the interactions of **neurotransmitters** or **other hormones** with endocrine cells.

27. **False.** ~~Unlike~~ **Like** neurons, endocrine cells generally ~~do not~~ act as integrators of various kinds of positive or negative inputs.

28. The two parts of the central nervous system that directly affect hormone secretion are the **hypothalamus** and the **autonomic nervous system**. (See text Figure 10-12.)

29. **True.**

30. **False.** A general name for a hormone that ~~regulates mineral ion homeostasis~~ **stimulates secretion of another hormone** is "tropic hormone." (There is no general name for hormones that regulate mineral ion homeostasis.)

31. (a) **Correct.**

    (b) **Correct.**

    (c) **Correct.**

32. (a) **Incorrect. These hormones are synthesized in the hypothalamus and stored in the posterior pituitary.**

    (b) **Correct.**

    (c) **Correct.** (As the text mentions, each hypothalamic cell that synthesizes posterior pituitary hormones makes only one kind of hormone. Therefore, stimulation of secretion of one or the other hormone is discretely controlled by stimulation of one kind of neuron or another. You may appreciate that, with target organs as different as those for vasopressin and oxytocin (kidney cells and uterine muscle, for example), secretion of both hormones at the same time would rarely be appropriate.)

33. **True.** (The hypothalamic hormones that stimulate secretion of the tropic anterior pituitary hormones are themselves tropic hormones.)

34. **False.** The fragment of pro-opiomelanocortin with established endocrine function is ~~ß-lipotropin~~ ACTH.

35. **True.** (Thyrotropin does have another, related function: It stimulates growth of the thyroid gland.)

36. (a) **Correct.**

    (b) **Correct.**

    (c) **Incorrect. They are LH** (luteinizing hormone) **and FSH** (follicle-stimulating hormone). (These hormones are named for actions they have in the ovary, even though they have quite different effects in the testis. You will learn about both kinds of effects in Chapter 18.)

    Note: As the text mentions, prolactin does have a role in reproductive function in both the male and the female, but these effects do not accord it the status of a gonadotropin. Prolactin probably is not a tropic hormone for the gonads, as "gonadotropin" implies, but recent studies have implicated prolactin in having a tropic influence on insulin secretion and also on certain growth factors in the liver. This information is not intended to burden you with "one more thing to memorize" but rather to alert you to the fact that new knowledge, in endocrinology as in other aspects of physiology, is constantly being gathered. Further, you should bear in mind that proof that a hormone has an effect on a system is easier to come by than proof that it doesn't. Negative studies may not be measuring the right parameter, for example.

37. (a) **Correct.**
    (b) **Correct.**
    (c) **Correct.**

38. **The hypothalamo-pituitary portal vessels connect two capillary beds, one in the median eminence of the hypothalamus and the other in the anterior pituitary gland. This connection provides for delivery of hormones secreted by the hypothalamus directly to the pituitary cells they stimulate without being diluted by the whole blood supply.**

    Note: The significance of this direct-delivery system is that the hypothalamic releasing hormones reach the pituitary cells in sufficient concentrations to have effects on the secretory activities of those cells, even though the neurohormones are secreted by a relatively few neurons. The efficiency of this arrangement is shown by the fact that the hypothalamic hormones are not detectable in the peripheral (general body) circulation by most methods of measurement, whereas they are readily detectable in samples of portal blood.

    The fact that these hormones are present in relatively tiny amounts in the brains of animals made the discovery of their function and the identification of their structure a particularly difficult task. Roger Guillemin and Andrew Schalley were awarded the Nobel Prize in Physiology or Medicine in 1981 for their efforts in this area.

39. **Two of the hypothalamic releasing hormones are actually release-inhibiting hormones: somatostatin, which inhibits GH and TSH secretion, and PIH, which inhibits prolactin release.**

40. (a) **Incorrect. One hypothalamic releasing hormone, PIH (dopamine), is an amine, not a peptide.**
    (b) **Incorrect. SS inhibits secretion of hormones from cells that contain TSH as well as from cells that contain GH. TRH stimulates secretion of hormones from cells that contain prolactin and from those that contain TSH.**
    (c) **Incorrect. SS and TRH are consistent in their effects, as noted in 40-b. There are no other known releasing hormones that act on more than one pituitary cell type.**

41.

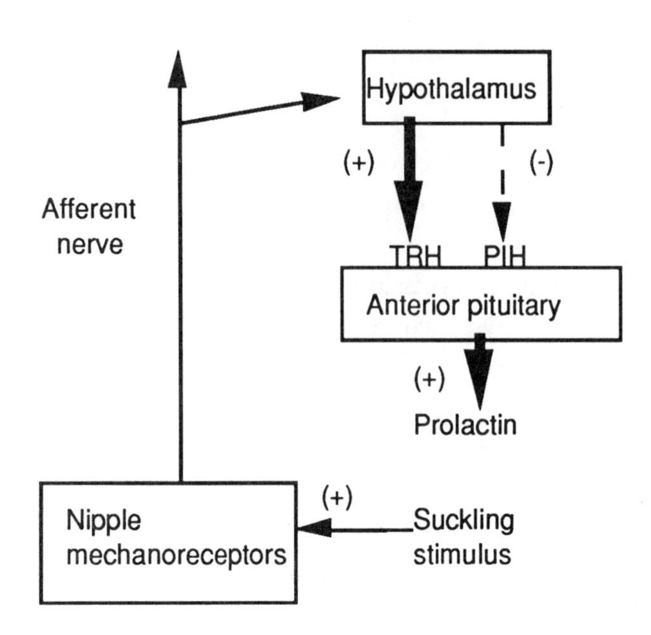

42. **As depicted in text Figure 10-22, the fall in cortisol secretion that accompanies removal of one adrenal gland leads to a decline in plasma cortisol concentrations. This decline is then detected by cells in the hypothalamus and anterior pituitary. Hypothalamic cells increase their secretion of CRH, and anterior pituitary cells increase their responsiveness to CRH. Both of these changes lead to increased ACTH release by the pituitary. The increased ACTH stimulates growth of the remaining gland, which can then compensate by secreting more cortisol.** (This situation is very similar to the induction of thyroid goiter by a diet lacking in iodine, as the text explains. In the case of the adrenal gland, the enlargement is not visible, but goiters can be very noticeable and unsightly.)

43. (a) **↑ secretion of TRH and ↑ sensitivity of pituitary cells to TRH**
    (b) **↑ secretion of TSH**
    (c) **↑ secretion of TH**
    (d) **↑ total TH in blood**

    The key to answering this sort of question is that cells sensitive to TH levels and part of the negative feedback loop can interact only with free (unbound) hormone. Thus, an increase in the percentage of bound TH results in a decrease in the concentration of free hormones, which is the same from the cells' point of view as an absolute decline in levels. As long as the liver keeps secreting more binding proteins, the thyroid will keep secreting more TH, accounting for Sally's elevated total TH during her pregnancy (Question 14).

44. (a) **Correct.**
    (b) **Incorrect. This is also long-loop feedback.**
    (c) **Correct.**

(d) **Correct.** At this point you have to take this sort of assumption on faith, but in a normal person — in this case, one who does not have an endocrine disorder — adjustments in hormone levels have physiological utility. You will learn in Chapter 17 why GH levels change in these ways in response to these stimuli. If you answered "incorrect" because all the negative feedback loops made you think the levels of these hormones would be lowered to basal (or below) — don't feel bad. This example can seem overwhelmingly complicated. However, the negative feedback loops serve to keep hormone levels from rising too far over basal. Presumably, they will be maintained at or near an optimal level for the particular condition or stimulus.

45. (a) **Incorrect. It is melatonin.**

(b) **Incorrect. It is present in high concentrations in the blood at night.**

(c) **Incorrect. Its function is unknown in humans, "in contrast to several other animals."** For example, melatonin is known to be very important for synchronizing reproductive functions in animals that are seasonal breeders — that is, animals that breed once a year, either when days are growing longer (ground squirrels) or when they are growing shorter (deer). Melatonin probably does have effects on human body rhythms, but these effects are considerably more subtle than those seen in other animals.

46. Two candidate hormones from the anterior pituitary are **ß-lipotropin** and **ß-endorphin**.

47. (a) **Diagnosis: Primary hyposecretion of TH due to a dietary iodine deficiency. Treatment: Prescribe iodine supplements, such as iodized salt.** It is of course possible that the thyroid defect was caused not by lack of iodine but rather by some other mechanism in the gland that prevented normal secretion of TH in spite of elevated levels of TSH (indicated by the enlargement of the gland). In that case, a prescription for TH would be the proper treatment.

(b) **The easiest test would be to administer (inject) TRH to the patient. If the blood levels of TSH and TH rise in response, then the defect is secondary hyposecretion of TH at the level of the hypothalamus.** (That is, the hypothalamus is not secreting enough TRH.) **If TSH levels increase but TH levels do not, two defects are indicated — one at the level of the hypothalamus and the other at the level of the thyroid. If neither TSH nor TH levels rise, then one can be sure there is a defect at the level of the pituitary.** (In this case the thyroid may be defective as well, which could be demonstrated by observing no rise in blood TH after injecting TSH. The hypothalamus might also by defective, but there would be no easy way to diagnose this. Recall that measurement of hypothalamic hormones in the peripheral blood is not easily done.) In any case, the best treatment would probably be a prescription for TH. Even if the defect is in the hypothalamus or pituitary, TH (in fact, thyroxine) would alleviate the symptoms of hypothyroidism, and thyroxine is active when taken orally, while TSH and TRH are not.

(c) **Patient C is apparently a victim of hyporesponsiveness to TH. His problem may be too few TH receptors or defective receptors.** (Unfortunately, we have not yet developed methods for treating people with receptor defects. Perhaps genetic engineering will make treatments possible in the future.)

PRACTICE

True or false (correct the false statements) :

P1. In general, the rate of excretion of many hormones or their metabolites is comparable to their secretion rate.

P2. In general, steroid hormones bind to receptors on cell membranes whereas peptide hormones bind to receptors inside of cells.

P3. After thyroid hormones bind to their receptor, they activate mRNA transcription on segments of DNA.

P4. Thyroid hormones exert a permissive effect on the actions of epinephrine.

P5. Most molecules of thyroxine and thyrotropin in blood are bound to proteins.

P6. Stimuli for the secretion of hormones include changes in plasma concentrations of ions, nutrients, and other hormones.

P7. The name of the chemical class of hormones that are not soluble in water is "amines."

P8. A precursor of cortisol is cholesterol.

P9. The adrenal cortex is part of the autonomic nervous system.

P10. Receptors for estradiol, T_3, and vasopressin are located in the nuclei of target cells.

P11. Hormones of the posterior lobe of the pituitary are synthesized in neuronal cell bodies of the hypothalamus and transported to the pituitary in axonal fluid.

P12. The pituitary portal circulation refers to the blood vessels linking the anterior pituitary with the target glands of anterior pituitary hormones.

P13. The liver is a target gland of growth hormone.

Multiple choice (correct each incorrect choice):

P14-16. The hormones secreted by the posterior pituitary include

P14. vasopressin

P15. epinephrine

P16. oxytocin

P17-19. Stimuli for the secretion of growth hormone include

P17. waking from sleep.

P18. exercising.

P19. increased levels of IGF-I in blood.

Fill in the blanks:

P20-24. Indicate how altering the secretion rate of the hormones listed below affects the secretion of hormones from the anterior pituitary:

P20. Increasing TRH ____________________

P21. Increasing somatostatin ____________________

P22. Decreasing dopamine ____________________

P23. Increasing cortisol ____________________

P24. Increasing GnRH ____________________

P25-26. Elevated TH levels in blood ____________________ (increase/decrease) the sensitivity of pituitary cells to TRH. This is an example of ____________ (short/long) loop negative feedback.

P27-29. Elevated prolactin levels in blood will cause a(n) ________________ (increase/decrease) in PIH secretion from the __________________________________. This is an example of ________________- loop negative feedback.

P30. Depressed levels of TH in blood resulting from a defect in the cells that secrete TSH are an example of ________________________ (primary/secondary) hyposecretion of TH.

11: MUSCLE

Serious diseases caused by genetic defects (such as cystic fibrosis, discussed in the Guide's introduction to Chapter 6) are especially tragic because they so often rob children of their chances to have a normal childhood, not to mention their hopes to grow to adulthood. No disease illustrates this tragedy better than muscular dystrophy (MD), a crippler that strikes 1 out of every 4000 boys born in this country.

Muscular dystrophy is a disease of gradual muscle deterioration, weakness, paralysis — and finally death from failure of the heart or the muscles that control breathing. Until very recently, there was no cure or even a therapy on the horizon, but new breakthroughs in molecular biology have allowed scientists to identify the gene that causes the problem and to determine that a protein, *dystrophin*, is not synthesized in the muscle cells of MD victims.

The identification of dystrophin has opened one door of hope for MD victims. Another very promising development was the discovery of a strain of mouse having the same genetic defect, so that different types of therapy can be tested in the animal model. Very recent research has shown that transplanted immature muscle cells from genetically related, but normal, mice can divide and replace the cells that lack dystrophin in the defective mice. Clinical trials of the same sort of procedure in human MD victims are planned for the near future.

In this chapter, you will learn how muscles — the remarkable tissues that allow us to walk, run, talk, write answers to Study Guide questions, and so much more (and which most of us take completely for granted) — work.

Instructions for answering questions in this Study Guide

1. True or false: Correct each false statement. Whenever you must correct a false statement, there will almost always be more than one way to do so. In the Guide, the answer requiring the least correction will generally be given, with the understanding that other correct answers are possible. It is usually not sufficient, in terms of demonstrating understanding, to simply insert a "no" or "not," however.

2. Multiple choice: Any, all, or none of the responses may be correct. Explain why each incorrect response is incorrect.

3. Fill in the blanks: Choose the best word or words to complete each statement.

4. Directions will be given for other types of questions as they appear.

❐ Introduction (text pages 283-284)

1. It is only by controlling the activity of muscles that the human mind ultimately expresses itself. Explain.

____

____

____

____

2. Indicate whether each of the following is descriptive of skeletal (Sk), cardiac (C), and/or smooth (Sm) muscle, or none (N) of them.

_______ under voluntary control

_______ controlled by autonomic nervous system

_______ controlled by motor neurons

_______ can contract spontaneously

_______ controlled by hormones

_______ heart muscle

_______ muscle of uterus

_______ attached to eyes

_______ found in eyes

_______ attached to bone

_______ found in blood vessels

❐ Structure of Skeletal Muscles and Muscle Fibers (text pages 284-288)

3. Skeletal-muscle cells
   (a) are striped.
   (b) are muscle fibers.
   (c) are spherical.
   (d) have one nucleus each.
   (e) are among the longest cells in the body.

4. True or false: Skeletal muscles are made up of bundles of muscle fibers held together by sheaths called tendons.

5. True or false: Skeletal muscle can be distinguished from smooth and cardiac muscle under the microscope by its characteristic striations.

6. A myofibril is
   (a) a skeletal-muscle cell.
   (b) composed of filaments.
   (c) composed of proteins.

7. True or false: The name given to the unit of repeating pattern in a myofibril is "sarcomere."

8. Match the single correct name from the column on the right with its description on the left:

________	boundary of one sarcomere	A band
________	region of myofibril containing no thick filaments	H zone
________	region of myofibril containing no thin filaments	I band
________	region occupied by thick filaments in a myofibril	M line
________	anchor for thin filaments	Z line
________	anchor for thick filaments	

9. All of the lines, bands, and zones listed in Question 8 are located within a single sarcomere except the ________________.

10. Thick filaments, which are composed of the contractile protein ____________, are about ________________ (how much?) as thick as thin filaments. Thin filaments are composed of three proteins, including the contractile protein ______________. Cross bridges between the two kinds of filaments are portions of ______________ molecules.

❒ Molecular Mechanisms of Contraction (text pages 288-299)

11. True or false: The force-generating sites in a myofibril are the Z lines.

12. According to the sliding filament theory of skeletal-muscle contraction, during contraction the
    (a) thick filaments stay the same size but the thin filaments shorten.
    (b) sarcomeres shorten.
    (c) A bands shorten.
    (d) I bands shorten.

13. True or false: The cross-bridge cycle refers to the sequence of events between the time an individual cross bridge binds to a thin filament at the start of a muscle contraction and the time the cross bridge releases the filament at the end of the contraction.

14. True or false: The energy for muscle contraction is provided by the hydrolysis of ATP.

15. Arrange the events of the cross-bridge cycle in proper order beginning with the first event that occurs following an increase in intracellular calcium levels:
    (a) energy from the hydrolysis of ATP is stored in myosin.
    (b) movement of the myosin head slides the thin filament with the bound actin slightly toward the center of the sarcomere.
    (c) myosin catalyses the hydrolysis of ATP.
    (d) binding of myosin to actin discharges the energy stored in myosin, causing the myosin head, still bound to actin, to swivel in the direction of its tail.
    (e) energized myosin binds to actin.
    (f) binding of ATP to myosin causes the actin-myosin bond to be broken.

16. True or false: Actin is an ATPase.

17. True or false: The binding of ATP to myosin causes an allosteric change in myosin's actin-binding site such that the affinity of myosin for actin is decreased.

18. True or false: The phenomenon of rigor mortis demonstrates that myosin can bind to actin in the absence of ATP, but the bond cannot then be broken.

19. The protein in thin filaments that has myosin-binding sites is __________________. When a muscle is at rest, these sites are partially covered by molecules of ________________________, which are held in position by globular proteins called ________________________. Binding of __________________ to these globular proteins causes a change in their conformation such that they move the __________________ molecules away from __________________________.

20. True or false: Binding of myosin to actin cannot take place in the absence of calcium ion.

21. True or false: Excitation-contraction coupling refers to the binding of energized myosin to actin.

22. True or false: The site of calcium-ion storage in muscle cells is the lateral sacs of the transverse tubules.

23. Action potentials that originate in the plasma membrane of muscle fibers are propagated into the interior of the fibers along the membrane of the ____________________. Action-potential activity near the sarcoplasmic reticulum membrane triggers ____________________ by as yet unknown mechanisms. When action-potential activity ceases in the fiber, the ____________________ proteins in the membrane of the sarcoplasmic reticulum pump calcium ion into ____________________.

24. A single action potential in a skeletal muscle fiber lasts about 2 milliseconds, but the contractile activity stimulated in the muscle fiber by that action potential lasts about fifty times longer. Explain this discrepancy.

____

____

____

____

25. Motor neurons
   (a) innervate skeletal muscle.
   (b) innervate smooth muscle.
   (c) are unmyelinated.
   (d) have cell bodies in the peripheral nervous system.
   (e) propagate action potentials more quickly than any other neurons.

26. A motor unit consists of a
   (a) skeletal muscle and all the motor neurons that innervate it.
   (b) motor neuron and all the muscle fibers it innervates.

27. All the muscle fibers in a given motor unit
   (a) are located in the same muscle.
   (b) lie adjacent to each other.
   (c) contract simultaneously when stimulated by the motor neuron.

28. A neuromuscular junction is *not* a synapse, but there are some similarities between them. The motor axon terminal is analogous to the ____________________ of a presynaptic neuron, and the motor end plate corresponds roughly to the ____________________ of the postsynaptic cell.

29. Which of these statements regarding the properties of synapses and neuromuscular junctions is/are correct?
    (a) In both, the stimulus for exocytosis of vesicles containing neurotransmitter is influx of calcium ions into the axon terminal.
    (b) Unlike synaptic transmission, there is a one-to-one relationship between action potentials arriving in the axon terminal of a neuromuscular junction and action potentials in the motor end plate.
    (c) EPPs are always depolarizing, but EPSPs can be depolarizing or hyperpolarizing.
    (d) Action potentials generated in the motor end plate are propagated in both directions down the muscle membrane, whereas action potentials in post-synaptic cells travel in only one direction down the axon.
    (e) Motor end-plate membranes contain acetylcholinesterase, whereas the sub-synaptic membrane of neurons in cholinergic synapses does not.

30. True or false: The drug curare binds to ACh receptors on the motor end plate but does not activate them. Curare is therefore an ACh agonist.

31. True or false: Organophosphate "nerve gases" induce paralysis by blocking neural stimulation of the neuromuscular junctions, while botulinus toxin produces the same effect by inhibiting acetylcholinesterase.

❒ Mechanics of Single-Fiber Contraction (text pages 299-303)

32. Identify the type(s) of contraction that would occur under these load/tension conditions:
    (a) tension equals load ______________________________
    (b) tension exceeds load ______________________________
    (c) load exceeds tension ______________________________

33. Now identify the type of contraction that would occur during the following activities (consider only the muscles involved in the activity):
    (a) walking on a level surface ______________________
    (b) holding arm out in front of your face ______________________
    (c) walking down stairs ______________________
    (d) pushing on a brick wall ______________________
    (e) losing a wrist-wrestling match ______________________
    (f) lifting a cup to your mouth ______________________

34. True or false: The chemical changes the contractile proteins in skeletal-muscle fibers undergo are the same during the three different kinds of possible contractions.

35. True or false: The term "twitch" refers to the mechanical response of a muscle fiber during one cross-bridge cycle.

36. During an isometric twitch of a muscle fiber,
    (a) the tension generated is measured against time.
    (b) the length of time that tension can be detected is designated the contraction time.
    (c) there is a latent period corresponding to the events of excitation-contraction coupling.

37. If you compare the properties of an isotonic twitch with those of an isometric twitch measured in the same muscle fiber, you will see that
    (a) the latent period is longer for the isotonic twitch.
    (b) the measurable duration of contraction is longer for the isotonic twitch.

38. An isotonic contraction cannot be considered purely isotonic. Explain.

____

____

____

39. True or false: Sustained contraction and tension in a skeletal muscle is called tetanus.

40. A single action potential in a muscle fiber stimulates the release of sufficient Ca^{2+} to saturate all the calcium-binding sites on troponin. Nevertheless, during tetanus the fiber develops tension levels greater than twitch tension. Explain.

____

____

____

____

____

____

____

41. The optimal length of a skeletal muscle is
    (a) the length at which the muscle can generate its maximal tetanic tension.
    (b) the shortest length the muscle can achieve while attached to bone, because the amount of overlap between thick and thin filaments is maximal then.
    (c) approximately the same as its resting length.

42. True or false: The shortening velocity of a skeletal muscle fiber increases with decreasing load because the rate at which the cross-bridge cycle can occur increases with decreasing load.

❐ **Skeletal-Muscle Energy Metabolism** (text pages 303-306)

43. The ATP required for continuing contractile activity in a muscle is supplied by
    (a) preformed ATP present in the muscle for the first few minutes of contraction.
    (b) the phosphorylation of ADP by creatine phosphate for the first few minutes of contraction.
    (c) the breakdown of stored muscle glycogen to glucose followed by glycolysis and oxidative phosphorylation for the first few minutes of contraction.
    (d) the catabolism of blood-borne glucose and fatty acids after the first few minutes of contraction.

44. Although anaerobic glycolysis and substrate phosphorylation are relatively inefficient sources of ATP, these pathways are used preferentially (instead of oxidative phosphorylation) to supply ATP during very intense exercise. Explain.

    ____

    ____

    ____

    ____

45. After heavy exercise, breathing remains labored for some time, indicating a continuing need for increased oxygen. In muscle cells recovering from exercise,
    (a) oxygen is used for aerobic glycolysis to generate ATP.
    (b) oxygen is used for oxidation of lactic acid metabolites to generate ATP.
    (c) ATP is used to synthesize glycogen from glucose.
    (d) ATP is used to phosphorylate creatine.

46. Muscle fatigue
    (a) is caused by depletion of muscle ATP stores.
    (b) after short-duration, high-intensity exercise is caused by depletion of muscle glycogen stores.
    (c) after long-duration, low-intensity exercise is caused by excessive hydrogen ion concentrations that interfere with contractile protein activity.

47. Muscle fatigue is thought to have adaptive value. Explain.

____

____

____

❒ **Types of Skeletal-Muscle Fibers** (text pages 306-308)

48. Two properties that distinguish different types of muscle fibers from one another are
    (a)____
    (b)____

49. True or false: Fast fibers can be distinguished from slow fibers by the rate at which their myosin-ATPases split ATP.

50. True or false: Characteristics of oxidative fibers include numerous mitochondria, abundant myoglobin, and large stores of glycogen.

51. True or false: The difference in color between white muscle and red muscle is accounted for by the greater abundance of glycogen in white muscle.

52. In slow-oxidative fibers,
    (a) there are many more surrounding capillaries than in fast-glycolytic fibers.
    (b) the fiber diameters are narrower than in fast-glycolytic fibers.
    (c) the fibers are red.

53. True or false: The fiber type intermediate between the two extremes of fast-glycolytic and slow-oxidative is slow-glycolytic.

❐ **Whole-Muscle Contraction** (text pages 308-313)

54. True or false: Muscles used for delicate, finely controlled movements have smaller motor units than more coarsely controlled muscles.

55. True or false: In general, activating a slow-oxidative motor unit in a muscle will generate more tension than activating a fast-glycolytic motor unit.

56. True or false: Recruitment of different types of motor units in a muscle is the primary means of varying the amount of tension generated in that muscle.

57. In a muscle composed of different types of motor units,
    (a) motor units with the smallest motor neurons are the first to be recruited.
    (b) slow-oxidative motor units are the first to be recruited.
    (c) motor units that generate the least amount of tension are the first to be recruited.

58. True or false: The decrease in mass and strength of muscle as a result of damage to the nerves innervating the muscle is called disuse atrophy.

59. True or false: Long-duration, low-intensity exercise preferentially stimulates hypertrophy of the slow-oxidative fibers of the exercising muscles.

60. Some of the finest athletes in the world compete in the Olympic Games. Picture in your mind the running of the hundred-meter dash, where the winner gets the title of "world's fastest human." Next, picture the running of the marathon, the Olympics' greatest endurance race. Now, describe the differences you can see, and the unseen differences you would predict to be present, in the legs of the two kinds of athletes.

    ____
    ____
    ____
    ____
    ____
    ____
    ____

61. True or false: A flexor muscle pulls on bone to decrease the angle at a joint, whereas an extensor pushes the bone to increase the angle.

62. Examine text Figure 11-35 and the description of it on text page 312.

(a) Substitute a 5-kg ball for the load and calculate the new isometric tension exerted by the biceps muscle to hold the load steady.

(b) Now substitute a weighted 10-kg arm band and slide it 15 cm closer to the elbow. What is the isometric tension in this case?

63. Myesthenia gravis is an autoimmune disease in which the body's immune system attacks the ACh receptors on the motor end plate. Review the neuromuscular junction inhibitors described on text pages 297 to 299 and in Questions 30 and 31. Which one of these inhibitors (curare, organophosphates, or botulinus toxin) might be useful as a treatment for myesthenia gravis? Why?

____

____

____

____

❐ Smooth Muscle (text pages 314-321)

64. Unlike skeletal muscle, smooth muscle

(a) is not striated.

(b) does not have sarcomeres.

(c) does not have thick and thin filaments.

(d) has thin filaments attached to dense bodies or to the plasma membrane.

(e) does not use troponin-tropomyosin to regulate cross-bridge activity.

(f) does not use changes in cytosolic calcium concentrations to regulate cross-bridge activity.

65. True or false: The calcium-binding protein in smooth muscle is calmodulin.

66. True or false: The myosin in smooth muscle cells differs from that in skeletal muscle in that smooth-muscle myosin requires phosphorylation before it can bind to ATP.

67. Again comparing smooth-muscle contractile activity with that in skeletal muscle:
    (a) The velocity of smooth-muscle contraction is slower than that of skeletal muscle.
    (b) Smooth muscles are more resistant to fatigue than are skeletal muscles.
    (c) Unlike skeletal muscles, smooth muscles do not have a sarcoplasmic reticulum to store calcium ions.
    (d) Unlike skeletal muscles, smooth muscles can use extracellular calcium to regulate cross-bridge activity.
    (e) Smooth-muscle twitches are of longer duration than skeletal-muscle twitches.
    (f) Smooth-muscle cells can generate tension over a narrower range of lengths that can skeletal-muscle cells.

68. A low level of contractile activity in smooth muscle in the absence of external stimulation is called smooth-muscle ____________________.

69. True or false: In smooth muscle cells that can produce action potentials, cell membrane depolarization opens voltage-dependent sodium channels.

70. A pacemaker potential in a smooth muscle is the
    (a) depolarization generated by interactions of neurotransmitter molecules on the membrane of the cell.
    (b) spontaneous depolarization of smooth-muscle membrane toward action potential threshold.

71. Regarding the innervation of smooth muscles,
    (a) smooth-muscle cells do not have specialized areas of plasma membrane that correspond to the motor end plate of skeletal muscle.
    (b) a single smooth-muscle cell may be innervated by an autonomic neuron and a somatic neuron.
    (c) a single smooth-muscle cell may be innervated by a parasympathetic neuron and a sympathetic neuron.
    (d) smooth-muscle cells may receive both depolarizing and hyperpolarizing neural input.
    (e) a given neurotransmitter may be depolarizing for one type of smooth muscle and hyperpolarizing for another.

(f) in some smooth-muscle cells a neurotransmitter may have excitatory effects on tension without causing a change in membrane polarization.

72. List six factors other than nerves and hormones that can alter smooth-muscle cell tension.

_______________, _______________,

_______________, _______________,

_______________, and _______________.

73. True or false: Smooth muscles are classified by the number of properties they share with either single-unit or multiunit smooth muscles.

74. Single-unit smooth-muscle cells
 (a) have individual innervation of each cell.
 (b) have many gap junctions between cells.
 (c) may have pacemaker activity.
 (d) respond to stretch by contracting.
 (e) have tone.

75. Multiunit smooth-muscle cells
 (a) have individual innervation of each cell.
 (b) have many gap junctions between cells.
 (c) may have pacemaker activity.
 (d) respond to stretch by contracting.
 (e) have tone.

ANSWERS

Boldface type indicates the answers you should have given. Words in medium-face (ordinary) type explain or expand upon the answer.

1. **Our only means of communicating with anyone other than ourselves is through movement of our muscles — in speaking, writing, using sign language, or even making facial expressions. A totally paralyzed person could communicate nothing to anyone.)**

2. **Sk** under voluntary control
   C, Sm controlled by autonomic nervous system
   Sk controlled by motor neurons
   C, Sm can contract spontaneously
   C, Sm controlled by hormones
   C heart muscle
   Sm muscle of uterus
   Sk attached to eyes
   Sm found in eyes
   Sk attached to bone
   Sm found in blood vessels

3. (a) **Correct.**
   (b) **Correct.** ("Muscle cell" and "muscle fiber" are synonymous.)
   (c) **Incorrect. They are cylindrical.**
   (d) **Incorrect. They are multinucleated.**
   (e) **Correct.**

4. **False.** Skeletal muscles are made up of bundles of muscle fibers held together by sheaths ~~called tendons~~ **of connective tissue**.

5. **False.** Skeletal **and cardiac** muscle can be distinguished from smooth ~~and cardiac~~ muscle under the microscope by ~~its~~ **their** characteristic striations.

6. (a) **Incorrect. Muscle cells** (or fibers) **contain myofibrils**.
   (b) **Correct.**
   (c) **Correct.** (The filaments are composed of proteins.)

7. **True.**
   Note: The prefix "myo-" means "muscle." The prefix "sarco-" means "flesh" (which is another term for muscle). Whenever you see a word that begins with "sarco-" or "myo-," that word refers in some way to muscle.

8. **Z line** boundary of one sarcomere
   I band region of myofibril containing no thick filaments
   H zone region of myofibril containing no thin filaments
   A band region occupied by thick filaments in a myofibril
   Z line anchor for thin filaments
   M line anchor for thick filaments

9. All of the lines, bands, and zones listed in Question 8 are located within a single sarcomere except the **I band**

10. Thick filaments, which are composed of the contractile protein **myosin**, are about **twice** as thick as thin filaments. Thin filaments are composed of three proteins, including the contractile protein **actin**. Cross bridges between the two kinds of filaments are portions of **myosin** molecules.

11. **False.** The force-generating sites in a myofibril are the ~~Z lines~~ **cross bridges**.

12. (a) **Incorrect. The thin filaments slide past the thick filaments, but neither kind of filament shortens.**
 (b) **Correct.**
 (c) **Incorrect. The A band is the area occupied by the thick filaments in each sarcomere. They do not shorten, and neither does the A band.**
 (d) **Correct.** (Recall that the I band is the area occupied by thin filaments that are not overlapping with thick filaments. As the thin filaments slide past the thick filaments, propelled by the swiveling of the cross bridges, the length of thin filaments not overlapping thick filaments decreases, shortening the I band.)

 Note: Some students are puzzled by the fact that the I band, which spans two sarcomeres, shortens, although emphasis is usually placed on the shortening of the sarcomere itself. One source of confusion is the tendency to think of the thin filaments in adjoining sarcomeres as being one long chain. They are not; they are connected to the Z line and not to each other. As you can see from examining text Figure 11-8, during contraction both of the diagrammed sarcomeres shorten simultaneously with the thin filaments in each being pulled toward their respective M lines. This shortening is made possible by the anchorage of the thin filaments to the Z lines.

13. **False.** The cross-bridge cycle refers to the sequence of events between the time an individual cross bridge binds to a thin filament ~~at the start of~~ **during** a muscle contraction and the time the cross bridge ~~releases the filament at the end of the contraction~~ **binds to a different site on the thin filament**. (Many cross-bridge cycles are repeated in the course of a single contraction.)

14. **True.**

15. **d, b, f, c, a, e**

16. **False.** ~~Actin~~ **Myosin** is an ATPase.

17. **True.**

18. **True.**

19. The protein in thin filaments that has myosin-binding sites is **actin**. When a muscle is at rest, these sites are partially covered by molecules of **tropomyosin**, which are held in position by globular proteins called **troponin**. Binding of **calcium ions** to these globular proteins causes a change in their conformation such that they move the **tropomyosin** molecules away from **myosin-binding sites on actin**.
 Memorization hint: You can easily remember which regulatory molecule does what by keeping in mind that tropomyosin covers the myosin-binding sites. Another way to remember is that troponin is the smaller molecule, and also the shorter name.

20. **True.**

21. **False.** Excitation-contraction coupling refers to the ~~binding of energized myosin to actin~~ **sequence of events by which an action potential in the plasma membrane of a muscle fiber leads to cross-bridge activity by increasing cytosolic calcium concentration**. (Note that increased cytosolic calcium is the "coupler" here of the excitation — an action potential in muscle plasma membrane — and the contraction — cross-bridge activity. Recall from Chapter 10 that when a stimulus is received by endocrine cells to secrete peptide hormones stored in secretion granules, the "coupler" between stimulus and secretion is also increased cytosolic calcium.)

22. **False.** The site of calcium-ion storage in muscle cells is the lateral sacs of the ~~transverse tubules~~ **sarcoplasmic reticulum**. (*Sarco*plasmic reticulum means the endoplasmic reticulum of muscle. Recall from Chapter 7 that, in many cells, the second messenger inositol triphosphate causes release into the cytosol of calcium stored in the *endo*plasmic reticulum. Muscle cells have an enlarged version of the "standard" endoplasmic reticulum, which is specialized for calcium storage.)

23. Action potentials that originate in the plasma membrane of muscle fibers are propagated into the interior of the fibers along the membrane of the **transverse (t) tubule**. Action-potential activity near the sarcoplasmic reticulum membrane triggers **release of stored calcium** by as yet unknown mechanisms. When action-potential activity ceases in the fiber, the **Ca-ATPase (active transport)** proteins in the membrane of the sarcoplasmic reticulum pump calcium ion into **the lumen of the lateral sacs**.

24. **Action potentials in t-tubule membranes near the sarcoplasmic reticulum lead to immediate release of Ca^{2+}, which quickly binds to troponin and allows binding of myosin to actin. Once the depolarization is over, Ca-ATPase pumps begin sequestering the Ca^{2+}, but this pumping mechanism is much slower than the release event, leaving enough calcium ion in the cytoplasm to allow cross-bridge activity to continue for about 100 milliseconds.** (Keep in mind that, in the presence of ATP, the amount of calcium ion in the cytosol is the regulator of contractile activity. Action potentials themselves are not required. In other words, there is no direct cause-and-effect relationship between electrical activity in the muscle membrane and the contractile activity of the filaments.

25. (a) **Correct.**
    (b) **Incorrect. Autonomic neurons innervate smooth muscle.**
    (c) **Incorrect. They are heavily myelinated.**
    (d) **Incorrect. Their cell bodies are in the spinal cord or brainstem.**
    (e) **Correct.** (Because their axons are heavily myelinated and have the largest diameter of any neurons.)

26. (a) **Incorrect. (b) is correct.** (There is coordination among the different motor units in a single muscle, of course.)
    (b) **Correct.**

27. (a) **Correct.**
    (b) **Incorrect. They are interspersed among fibers belonging to other motor units.**
    (c) **Correct.**

28. A neuromuscular junction is *not* a synapse, but there are some similarities between them. The motor axon terminal is analogous to the **axon terminal** of a presynaptic neuron, and the motor end plate corresponds roughly to the **subsynaptic membrane** of the postsynaptic cell.

29. (a) **Correct.** (This is sometimes termed "excitation-secretion coupling.")
    (b) **Correct.**
    (c) **Incorrect. EPSPs are also always depolarizing.** A correct statement would be that EPPs are always excitatory, but postsynaptic potentials may be either excitatory (EPSPs) or inhibitory (IPSPs).
    (d) **Correct.** (This difference is simply one of anatomy. If an action potential is generated electrically in the middle of an axon, it will be propagated in both directions as well. There are no important differences in the membranes of muscle fibers and axons that make propagation direction different.)
    (e) **Incorrect. Pre- and post-synaptic membranes in cholinergic synapses also contain this enzyme.**

30. **False.** The drug curare binds to ACh receptors on the motor end plate but does not activate them. Curare is therefore an ACh ~~agonist~~ **antagonist**.

31. **False.** ~~Organophosphate "nerve gases" induce~~ **Botulinus toxin induces** paralysis by blocking neural stimulation of the neuromuscular junctions, while ~~botulinus toxin produces~~ **organo-phosphate "nerve gases" produce** the same effect by inhibiting acetylcholinesterase. (Therefore, the term "nerve gas" is a misnomer because it does not affect the nerves. As we shall see later in this chapter, there are therapeutic uses for some organophosphates. Botulinus toxin is simply a toxin — a very dangerous one.)

32. (a) **isometric**
    (b) **isotonic**
    (c) **lengthening or isometric, depending on the conditions**. (For example, if a heavy barbell you are trying to raise over your head forces your arm down, that would cause lengthening contractions in some of your arm muscles. Pulling on a barbell that is too heavy for you to budge would be an example of isometric contractions in other muscles.)

33. (a) **isotonic**
    (b) **isometric**
    (c) **lengthening**
    (d) **isometric**
    (e) **lengthening**
    (f) **isometric**

34. **True.**

35. **False.** The term "twitch" refers to the mechanical response of a muscle fiber ~~during one cross-bridge cycle~~ **to a single action potential.**

36. (a) **Correct.**
    (b) **Incorrect. Contraction time is the time interval between the first detectable tension and the maximal tension achieved.**
    (c) **Correct.**

37. (a) **Correct.**
    (b) **Incorrect. The measurable duration of contraction for an isotonic twitch is the duration of shortening, which is dependent on the weight of the load being moved but is always shorter than the measurable tension for an isotonic twitch.**

38. **During the time it is developing tension sufficient to move the load, the fiber contracts isometrically, as it does at the end of the contraction when tension is decreasing.**

39. **True.**

40. **After a single action potential in a skeletal muscle, Ca-ATPase pumps begin to sequester the calcium ions immediately. The length of time during which cytosolic calcium levels are high enough to saturate all the troponin and free all the binding sites on actin is therefore very short. There simply isn't time, in the first few milliseconds of a single twitch, for all the possible cross bridges to form. Therefore, maximal tension cannot be developed before unbound troponin begins to move tropomyosin to block the cross-bridge-binding sites on actin. Action potentials closely following one another, however, can keep the cytosolic calcium sufficiently elevated to allow maximal cross-bridge binding.**

41. (a) **Correct.**
    (b) **Incorrect. When fibers are very short, the overlap of thin filaments sliding past each other begins to interfere with cross-bridge binding.**
    (c) **Correct.**

42. **True.** (This is another way of saying that the velocity decreases with increasing load.)

43. (a) **Incorrect. There is enough preformed ATP available in muscle cells for only a few hundred milliseconds of activity.**

    (b) **Incorrect. There is enough creatine phosphate present in muscle cells to supply the phosphorylation of ADP for only a few seconds of activity.**

    (c) **Correct.**

    (d) **Correct.**

44. **The glycolytic reactions are very rapid compared to the reactions of the Krebs cycle and oxidative phosphorylation. The glycolytic reactions are so much faster, in fact, that muscle cells can obtain 38 molecules of ATP from the anaerobic breakdown of 19 molecules of glucose to lactic acid considerably faster than they can from the complete oxidation of one molecule of glucose to CO_2 and H_2O.** (This kind of "wasteful" usage of fuel cannot go on indefinitely, of course.)

45. (a) **Incorrect. After exercise is over, muscle cells stop breaking down glucose.**

    (b) **Correct.** (In the presence of oxygen, lactic acid is converted to pyruvic acid, which enters the Krebs cycle to be oxidized to CO_2 and H_2O.)

    (c) **Correct.**

    (d) **Correct.**

46. (a) **Incorrect. ATP concentrations in fatigued muscle are only slightly lower than in resting muscle.**

    (b) **Incorrect. It is thought to be caused by a failure of excitation-contraction coupling.**

    (c) **Incorrect. It is thought to be related to depletion of glycogen stores.** (During low-intensity, long-duration exercise, very little acid builds up in cells because the cells rely on oxidative pathways for ATP generation.)

47. **Muscle fatigue sets in before ATP is completely depleted in muscle cells, preventing the rigor that would occur if ATP levels fell too low. Rigor is damaging to the contractile proteins.**

48. (a) **Differences in shortening velocity (fast versus slow)**

    (b) **Differences in ATP-forming pathways (oxidative versus glycolytic)**

49. **True.**

50. **False.** Characteristics of oxidative fibers include numerous mitochondria, abundant myoglobin, and ~~large~~ **small** stores of glycogen.

51. **False.** The difference in color between white muscle and red muscle is accounted for by the greater abundance of ~~glycogen~~ **myoglobin** in ~~white~~ **red** muscle.

52. (a) **Correct.**
    (b) **Correct.**
    (c) **Correct.** (Myoglobin, like its sister oxygen-binding protein, hemoglobin, is a red pigment. Hemoglobin gives blood its red color.)

53. **False.** The fiber type intermediate between the two extremes of fast-glycolytic and slow-oxidative is ~~slow-glycolytic~~ **fast-oxidative**.

54. **True.**

55. **False.** In general, activating a slow-oxidative motor unit in a muscle will generate ~~more~~ **less** tension than activating a fast-glycolytic motor unit.

56. **True.**

57. (a) **Correct.** (Don't be confused by the seeming contradiction between this statement and the fact that neurons having the largest-diameter axons propagate action potentials the fastest. These neurons require greater amounts of excitatory synaptic input, in the form of temporal and spatial summation, to trigger action potentials. But once these largest motor neurons are depolarized sufficiently, then the action potentials are propagated fastest down their axons.)
    (b) **Correct.**
    (c) **Correct.**

58. **False.** The decrease in mass and strength of muscle as a result of damage to the nerves innervating the muscle is called ~~disuse~~ **denervation** atrophy. (The disuse atrophy associated with immobilization is probably also a result of diminished nervous stimulation of the muscle.)

59. **True.**

60. **Sprinter: Has heavily muscled legs of large diameter. Microscopic inspection of leg muscles would show hypertrophied (large diameter, increased number of myofibrils) fast-glycolytic muscle fibers.**

    Marathoner has slender legs. Microscopic inspection would reveal changes in the slow- and fast-oxidative fibers, including increased numbers of mitochondria and extensive blood supply. (These fibers would actually be more slender than their counterparts in the sprinter's leg muscles, and so use of the term "hypertrophy" is not quite correct.)

61. **False.** A flexor muscle pulls on bone to decrease the angle at a joint, whereas an extensor ~~pushes~~ **pulls on** the bone **from the opposite direction** to increase the angle.

62. (a) 5 kg x 35 cm = 5 cm x X

$$X = \frac{(5)(35)}{5} = \mathbf{35\ kg}$$

(b) 10 kg x 20 cm = 5 cm x X

$$X = \frac{(10)(20)}{5} = \mathbf{40\ kg}$$

(This example helps illustrate why it is easier to carry weights close to the body than to carry them extended at arms' length.)

63. **Organophosphate derivatives, which inhibit the breakdown of acetylcholinesterase, are helpful in treating the disease because they make the remaining ACh receptors more effective. (The ACh remains on the receptor longer, causing more depolarization than before.)**

64. (a) **Correct.**

(b) **Correct.** (This lack of sarcomeres accounts for the lack of striation.)

(c) **Incorrect. It has thick filaments containing myosin and thin filaments containing actin.**

(d) **Correct.**

(e) **Correct.**

(f) **Incorrect. Changes in cytosolic calcium in smooth muscle regulate enzymatic phosphorylation of myosin.**

65. **True.**

66. **False.** The myosin in smooth muscle cells differs from that in skeletal muscle in that smooth-muscle myosin requires phosphorylation before it can bind to ~~ATP~~ **actin**. (The ATP is used to phosphorylate the myosin.)

67. (a) **Correct.**

(b) **Correct.**

(c) **Incorrect. Smooth muscles have a sarcoplasmic reticulum that stores calcium, but it is considerably smaller than the sarcoplasmic reticulum of skeletal muscle.**

(d) **Correct.**

(e) **Correct.** (Because the rate of calcium removal from the cytosol is slower for smooth muscle.)

(f) **Incorrect. Smooth-muscle cells generate tension over a wider range of lengths than do skeletal-muscle cells.**

68. A low level of contractile activity in smooth muscle in the absence of external stimulation is called smooth-muscle **tone**.

69. **False.** In smooth muscle cells that can produce action potentials, cell membrane depolarization opens voltage-dependent ~~sodium~~ **calcium** channels.

70. (a) **Incorrect. (Answer b is correct.)**
    (b) **Correct.**

71. (a) **Correct.**
    (b) **Incorrect. Somatic (motor) neurons innervate only skeletal muscle.**
    (c) **Correct.**
    (d) **Correct.**
    (e) **Correct.**
    (f) **Correct.** (These neurotransmitters or hormones probably use inositol triphosphate as a second messenger.)

72. **paracrines**, **acidity**, **oxygen concentration**, **osmolarity**, **ionic composition of extracellular fluid**, and **stretch**.

73. **True.**

74. (a) **Incorrect. Only a few cells are innervated.**
    (b) **Correct.**
    (c) **Correct.**
    (d) **Correct.**
    (e) **Correct.**

75. (a) **Correct.**
    (b) **Incorrect. They have few gap junctions.**
    (c) **Incorrect. They do not have spontaneous activity.**
    (d) **Incorrect. They do not respond to stretch.**
    (e) **Incorrect. They do not have tone.**

PRACTICE

True or false (correct the false statements):

P1. During skeletal-muscle contraction, the distance between the Z lines of a sarcomere decreases because of shortening of the filaments.

P2. During skeletal-muscle contraction, the I band and H zone shorten but the A band stays the same.

P3. Curare blocks neuromuscular transmission by preventing the release of neurotransmitter from the motor neuron.

P4. During isometric contraction of a skeletal-muscle fiber, tension increases but the fiber length stays the same.

P5. The latent period of an isotonic twitch is shorter than the latent period of an isometric twitch.

P6. Muscles in the back have a higher proportion of fast-glycolytic fibers than do muscles in the arms.

P7. Muscles in the hands have larger motor units than muscles in the back.

P8. The biceps muscle is an extensor, and its antagonist, the triceps, is a flexor.

P9. A skeletal muscle generates its greatest twitch tension when it is stretched to twice its resting length.

P10. Endurance exercise, such as long-distance swimming, causes a preferential increase in glycolytic enzymes and increased mass of the exercising muscles.

P11. The larger the diameter of a skeletal-muscle fiber, the greater the tension it can generate.

P12. Myesthenia gravis is caused by a genetic defect in the acetylcholinesterase pathway.

P13. Nervous stimulation of skeletal muscle is always excitatory, whereas nervous stimulation of smooth muscle may be excitatory or inhibitory.

P14. Smooth-muscle cells may contract in the absence of nervous stimulation.

P15. Multiunit smooth muscle is characterized by many gap junctions between cells.

Multiple choice (correct each incorrect choice):

P16-18. Skeletal-muscle tension can be increased by increasing the

P16. number of motor units stimulated.

P17. rate of stimulation to each motor unit.

P18. number of muscle fibers in a muscle through exercise.

Matching:

P19-30. For each description related to the stimulation of skeletal-muscle contraction, on the left, match the single best molecule/structure/event listed on the right. An answer may be used more than once.

P19.___ neurotransmitter	(a) calcium ions
P20.___ generate(s) action potential in muscle	(b) ATP
P21.___ store(s) neurotransmitter	(c) tropomyosin
P22.___ terminate(s) action of neurotransmitter	(d) end-plate potential
P23.___ bind(s) to troponin	(e) transverse tubules
P24.___ needed for cross-bridge movement	(f) vesicles
P25.___ binding with neurotransmitter causes end-plate depolarization	(g) acetylcholine receptor
	(h) acetylcholinesterase
P26.___ develop(s) muscle tension	(i) sarcoplasmic reticulum
P27.___ store(s) calcium ions	(j) acetylcholine
P28.___ inhibit(s) binding of actin to myosin	(k) myosin cross bridge
P29.___ conduct(s) action potentials into muscle-fiber interior	
P30.___ required for release of actin from myosin	

12: CONTROL OF BODY MOVEMENT

Just as muscular dystrophy is a crippling disease primarily affecting the young, Parkinson's disease also cripples, but its victims are older adults — numbering more than 1,000,000 in this country alone. Frequently these are people who have lived full, active lives but are gradually losing control of their muscles and their movements. They feel that the disease is robbing them of their lives. Actor/comedian Terry Thomas, who suffers from Parkinson's, said in an interview for television that the best thing about him now was his earlobes — the disease had "messed up everything else."

Parkinson's is not a disease of the muscles themselves, nor of the motor neurons controlling them. The cause of the disease is not known, but its symptoms, described later in this chapter, are a result of the degeneration of a discrete nucleus in the brainstem called the *substantia nigra*. In a normal brain, neurons in this nucleus release the neurotransmitter dopamine in a set of subcortical nuclei, the basal ganglia, which are important for establishing the program of step-by-step instructions for body movement that is ultimately relayed to the motor neurons. In the absence of dopamine, the basal ganglia cannot function properly and planned movements become more and more difficult to make.

Until recently, the only treatment for Parkinson's disease was the drug L-dopa, a precursor of dopamine that, unlike the neurotransmitter itself, crosses the blood-brain barrier. L-dopa is effective only temporarily, however, and it has unpleasant side effects. More promising treatment may soon be at hand, however; scientists searching for more effective therapies for Parkinson's have found a drug that, when combined with L-dopa, seems to slow the onset of the disease by blocking the destruction of substantia nigra cells. And a more radical treatment that has reached clinical trials after more than ten years of careful animal testing is the transplantation of dopamine-synthesizing tissue into the basal ganglia of the patient. Tiny bits of adrenal medulla from the patient's own body have been tried, as well as the more controversial transplant of brain tissue from aborted human fetuses. Although more years of research are in store before we can say we have cured Parkinson's disease, the results of these experiments offer hope for the future.

In this chapter, you will learn some of the fundamentals of the immensely complex tasks performed by our brains, nerves, and muscles every time we walk across the street or pick up a pen to write our names or offer a hand in greeting.

Instructions for answering questions in this Study Guide

1. True or false: Correct each false statement. Whenever you must correct a false statement, there will almost always be more than one way to do so. In the Guide, the answer requiring the least correction will generally be given, with the understanding that other correct answers are possible. It is usually not sufficient, in terms of demonstrating understanding, to simply insert a "no" or "not," however.

2. Multiple choice: Any, all, or none of the responses may be correct. Explain why each incorrect response is incorrect.

3. Fill in the blanks: Choose the best word or words to complete each statement.

4. Directions will be given for other types of questions as they appear.

❐ Introduction (text pages 325-328)

1. True or false: The motor neuron pool of a skeletal muscle comprises the innervation of all the motor units of that muscle.

2. True or false: The fewer motor neurons of a given motor neuron pool that are activated, the finer is the control of muscle movement.

3. True or false: Inhibitory input to motor neurons is as important for normal muscle function as excitatory input.

4. According to the hierarchy of motor control described in the text,
   (a) the highest level is composed entirely of areas of cerebral cortex.
   (b) the middle-level structures include part of the cerebral cortex, the cerebellum, parts of the basal ganglia, and some brainstem nuclei.
   (c) the task of the middle level is, in brief, to reduce complex plans of movement to discrete programs and subprograms of instruction for movement of individual muscles.
   (d) structures in the middle level receive information from both the highest level and the lowest level of the hierarchy.
   (e) the lowest level controls individual motor neuron pools.
   (f) motor neurons are part of the lowest level in the hierarchy.

5. True or false: Once a planned movement is under way, it is usually carried out with no further modifications until it is finished.

6. True or false: Practicing a movement allows for "fine tuning" the original program so that the movement can be executed with fewer corrections.

❐ Voluntary and Involuntary Actions (text page 328)

7. Indicate whether the following actions are primarily or usually voluntary (v), or primarily or usually involuntary (i). (In answering this question, do not consider the muscular activity that supports the action, such as the postural muscles in the example of threading a needle.)

   ___walking ___putting down an object ___breathing ___swallowing

   ___blinking ___picking up an object ___typing ___slam-dunking a basketball

❒ Local Control of Motor Neurons (text pages 328-333)

8. True or false: Afferent information about body movement is integrated at the level of the interneurons controlling the firing of motor neurons.

9. Local interneurons controlling the firing of motor neurons
   (a) receive input from receptors in the muscle that is innervated by the motor neurons.
   (b) receive information from muscles that are antagonistic to the one innervated by the motor neurons.
   (c) receive information from the joints and tendons associated with the muscle that is innervated by the motor neurons.
   (d) may be inhibitory.

10. The receptors that monitor changes in muscle length are called ________________ receptors. Their afferent nerve endings are wrapped around special muscle fibers called ________________ and are enclosed in a structure called the ________________. The receptors respond to ________________ of the muscle fibers by increasing their rate of firing action potentials in the ________________ neurons.

11. True or false: Afferent fibers from muscle-spindle stretch receptors in a muscle make excitatory synaptic contact with motor neurons that innervate the skeletomotor fibers of the same muscle.

12. The stretch reflex
    (a) is monosynaptic.
    (b) is accompanied by polysynaptic reflexes.
    (c) occurs only in the muscles controlling the knee.

13. True or false: Information from the muscle-spindle stretch receptors is not conveyed above the level of the spinal cord and thus does not reach consciousness.

14. Spindle fibers
    (a) are not true muscle fibers because they cannot contract.
    (b) are innervated by alpha motor neurons.
    (c) are innervated by gamma motor neurons.
    (d) function to maintain tension on spindle receptors.

15. True or false: During motor activity, alpha-gamma coactivation provides important information regarding muscle length to the higher motor control centers.

16. The receptors that monitor tension in muscles during contraction are called ______________________________; these receptors are located in ______________ attached to muscles. When the associated muscle contracts, the tension on the ______________ is detected by the receptor, which then conveys the information in the form of ______________ to the spinal cord.

17. True or false: Stimulation of a Golgi tendon organ causes a reflex contraction of the muscle whose tension the receptor is monitoring.

18. True or false: The muscle spindle receptors and Golgi tendon organs associated with the same muscle are activated at the same time.

19. The flexion reflex
    (a) allows for reflex withdrawal of a limb from a painful stimulus.
    (b) stimulates contraction of the ipsilateral extensor and inhibits contraction of the ipsilateral flexor.
    (c) is a monosynaptic reflex.

20. The crossed-extensor reflex in a limb
    (a) is initiated by a painful stimulus in the ipsilateral limb.
    (b) stimulates contraction of the contralateral extensor and inhibits contraction of the contralateral flexor.
    (c) accompanies the stretch reflex.

21. Assume you touch a hot stove with your right hand. The stimulus will activate a ______________ (what kind?) reflex that will result in ______________ of your right biceps muscle and ______________ of your right triceps muscle.

❒ Descending Pathways and the Brain Centers That Control Them

(text pages 333-341)

22. The sensorimotor cortex, which occupies part of the ________________ level of the motor control hierarchy, includes the ____________________ cortex and ______________ area in the frontal lobe and the __________________ cortex and part of the ______________________cortex next to it in the ____________________ lobe.

23. True or false: The skeletal muscles of the body are represented in the motor cortex proportionately to their size.

24. True or false: Neurons in the primary motor cortex are involved with discrete movements of individual muscles and muscle groups, whereas neurons in the premotor area direct more complicated, integrated aspects of movement.

25. True or false: Sensory information relayed from the parietal lobe association areas to the premotor area is important for planning strategies of movement.

26. The highest hierarchical level of motor control
    (a) includes the premotor area.
    (b) includes the supplementary motor area.
    (c) includes association cortex in the frontal and parietal lobes.
    (d) includes limbic-system structures.
    (e) directs "what to do" and "when to do it."

27. The readiness potential
    (a) is recorded from individual neurons in the brain.
    (b) arises in the primary motor cortex.
    (c) precedes movement by about 55 milliseconds.

28. The basal ganglia
    (a) are part of the middle level of the motor-control hierarchy.
    (b) receive excitatory input from dopaminergic neurons of the substantia nigra.
    (c) receive excitatory input from the cerebral cortex.
    (d) form part of a feedback loop with the motor cortex and the hypothalamus.

(e) selectively inhibit output from the thalamus to the motor cortex.
(f) are important for changing programs of movement.

29. In Parkinson's disease
(a) input from the substantia nigra to the basal ganglia is impaired.
(b) ability to initiate voluntary movement is not impaired.
(c) ability to suppress unwanted movement is impaired.

30. The cerebellum
(a) initiates body movements.
(b) helps to coordinate body movements.
(c) is important in maintaining posture.
(d) receives no input from sensory pathways.

31. True or false: One of the primary roles of the cerebellum is to compare a given program for movement with the actual movement as it occurs, and to make adjustments in the movement and in the program.

32. Symptoms of cerebellar damage include
(a) paralysis.
(b) tremor at rest.
(c) tremor while reaching for an object.
(d) difficulty maintaining balance.
(e) inability to make smooth, quick movements.

33. The corticospinal pathway
(a) is a descending motor pathway.
(b) begins in the cerebellar cortex.
(c) ends in the white matter of the spinal cord.
(d) crosses from right to left and vice versa in the spinal cord.
(e) consists of many neurons linked in a chain by synapses.

34. True or false: The motor pathway controlling voluntary movements of the head and neck is called the pyramidal tract.

35. Corticospinal fibers
    (a) may synapse directly on alpha motor neurons.
    (b) may synapse directly on gamma motor neurons.
    (c) may synapse directly on interneurons.
    (d) may form inhibitory synapses on motor neurons.
    (e) generally affect large groups of muscles.
    (f) may synapse with afferent neurons in ascending pathways.

36. In the brainstem and spinal cord, descending fiber tracts that have their cell bodies in nuclei of the brainstem are collectively called the ________________ or ________________ pathways. The cell bodies of these fibers receive information from the ________________, ________________, ________________, ________________, and ________________ in a series of complicated loops. The fibers ultimately synapse on ________________ or, more commonly, on ________________ in the brainstem and spinal cord. In general, these pathways are involved with ________________ ________________ ________________.

❐ Muscle Tone (text page 341)

37. Muscle tone
    (a) is defined as the active resistance to stretching a muscle.
    (b) is absent in a normal relaxed limb.
    (c) increases as a relaxed person becomes alert.
    (d) is tested clinically to assess the state of health of various parts of the motor system.

38. In general, damage to the descending motor pathways results in abnormally high muscle tone, known as ________________, whereas damage to the alpha motor neurons leads to ________________. Therefore, damage to the descending pathways selectively impairs ________________(inhibitory/excitatory) input to the alpha motor neurons.

39. Two kinds of hypertonia that can be distinguished clinically are ______________, which is accompanied by increased responsiveness to motor reflexes, and ______________, which is not.

❒ Maintenance of Upright Posture and Balance (text pages 341-342)

40. Receptors of the postural reflexes include
    (a) rods and cones.
    (b) hair cells.
    (c) muscle spindles and other proprioceptors.

41. True or false: Crossed-extensor reflexes but not stretch reflexes are important postural reflexes.

❒ Walking (text pages 342-344)

42. The cyclic pattern of movement that results in walking is
    (a) "wired in" at the level of the brainstem and spinal cord.
    (b) controlled by central pattern generators.
    (c) independent of control from higher brain areas.

ANSWERS

Boldface type indicates the answers you should have given. Words in medium-face (ordinary) type explain or expand upon the answer.

1. **True.**

2. **True.**

3. **True.**

4. (a) **Incorrect. The highest level is composed of areas of cerebral cortex and the limbic system, including non-cortical areas such as the thalamus and hypothalamus.** (The limbic system structures are involved in emotion and memory. Review the description of the limbic system on text page 177 and text Figure 8-12.)

   (b) **Correct.**

   (c) **Incorrect. The subprograms determine movements of individual joints, not muscles.** (Remember that at least two muscles are required to move a joint.)

   (d) **Incorrect. Middle-level structures receive "command" information from the highest level of the motor control hierarchy and afferent information from receptors.** (The lowest-level nuclei don't "talk back" directly to the "programmers," the middle level of the hierarchy.)

   Note: You should be aware that there is also much "cross-talk" between the middle-level structures.

   (e) **Correct.** (This is the level of control of individual muscles.)

   (f) **Incorrect. The motor neurons are the "final output stage" of the control system, not part of the hierarchy of control.**

5. **False.** Once a planned movement is under way, it is usually ~~carried out with no further modifications until it is finished~~ **continuously adjusted so that the actual movement is the same as the intended movement.**

6. **True.**

7. **v** walking **v** putting down an object **i** breathing **i** swallowing
   i blinking **v** picking up an object **v** typing **v** slam-dunking a basketball

   Note that all the involuntary actions can be performed voluntarily and that all the voluntary actions can become automatic, sliding down the scale toward involuntary. In some cases, such as the act of dropping a hot potato, a voluntary action becomes strictly, or almost strictly, involuntary — that is, a reflex.

8. **True.** (Afferent information is also integrated at higher levels, of course, as described in the introduction.) Note, too, that the descending neurons that send the program down to the level of the interneurons in the vicinity of the motor neurons are themselves technically interneurons.

9. (a) **Correct.**

   (b) **Correct.**

   (c) **Correct.**

   (d) **Correct.**

10. The receptors that monitor changes in muscle length are called **stretch** receptors. Their afferent nerve endings are wrapped around special muscle fibers called **spindle fibers** and are enclosed in a structure called the **muscle spindle**. The receptors respond to **stretch** of the muscle fibers by increasing their rate of firing action potentials in the **afferent** neurons.

11. **True.**

12. (a) **Correct.**
    (b) **Correct.** (The inhibition of contraction of the antagonist muscle is polysynaptic and an integral part of the stretch reflex.)
    (c) **Incorrect. It occurs in all skeletal muscles.** (Although this is not stated explicitly in the text, the ability to detect stretch and to stimulate contraction opposing the stretch is important for virtually all the skeletal muscles.)

13. **False.** Information from the muscle-spindle stretch receptors is ~~not~~ conveyed above the level of the spinal cord and ~~thus does not reach~~ **contributes to** consciousness **of perception of limb or joint position.**

14. (a) **Incorrect. Spindle fibers are specialized fibers that contract at each end, keeping tension on the noncontractile middle portion in contact with the receptor endings.**
    (b) **Incorrect. Alpha motor neurons innervate skeletomotor fibers.**
    (c) **Correct.**
    (d) **Correct.**

15. **True.**

16. The receptors that monitor tension in muscles during contraction are called **Golgi tendon organs**; these receptors are located in **tendons** attached to muscles. When the associated muscle contracts, the tension on the **tendon** is detected by the receptor, which then conveys the information in the form of **action potentials** to the spinal cord.

17. **False.** Stimulation of a Golgi tendon organ causes a reflex **inhibition of** contraction of the muscle whose tension the receptor is monitoring.

18. **True.** (because of alpha-gamma coactivation)

19. (a) **Correct.**

    (b) **Incorrect. It stimulates contraction of the ipsilateral flexor and inhibits contraction of the ipsilateral extensor.**

    (c) **Incorrect. It is polysynaptic.**

    (Examine text Figure 12-9 carefully if you had trouble with this question.)

20. (a) **Incorrect. The stimulus is in the contralateral limb.**

    (b) **Correct.**

    (c) **Incorrect. It accompanies** (is part of) **the flexor reflex.**

21. Assume you touch a hot stove with your right hand. The stimulus will activate a **flexor** reflex that will result in **contraction** of your right biceps muscle and **relaxation** of your right triceps muscle.

22. The sensorimotor cortex, which occupies part of the **middle** level of the motor control hierarchy, includes the **primary motor** cortex and **premotor** area in the frontal lobe and the **somatosensory** cortex and part of the **association** cortex next to it in the **parietal** lobe.

23. **False.** The skeletal muscles of the body are represented in the motor cortex proportionately to ~~their size~~ **the skill of the movements they can perform.** Text Figure 12-12 shows what is called the "homunculus" — a representation of a "little man" in the brain. The "mapping" was done on patients undergoing brain surgery after the anesthesia was allowed to wear off (the brain has no pain receptors). This figure should remind you of one that shares similarities with it — the representation of the body in the somatosensory cortex (text Figure 9-14). Note, in Figure 12-12, that the left hemisphere controls the right side of the body and vice versa.

24. **True.**

25. **True.**

26. (a) **Incorrect. The premotor area is in the middle level.**

    (b) **Correct.**

    (c) **Correct.**

    (d) **Correct.**

    (e) **Correct.**

27. (a) **Incorrect. It is recorded from electrodes placed on the scalp.**
    (b) **Incorrect. It arises in the supplementary motor area.**
    (c) **Incorrect. It precedes movement by 800 milliseconds.** (The *motor* potential, which arises in the motor cortex, precedes movement by 55 milliseconds.)

    As the text mentions, control of body movement is imperfectly understood, particularly with respect to exactly where the "idea" to move comes from. But new techniques have given us some interesting information. For example, there are methods of scanning changes in blood flow in the brain in unanesthetized humans. When a person moves, say, a finger, researchers can detect a surge of blood flow in the area of the motor cortex that corresponds to that finger. If the person taps out something in Morse code with that finger, blood flow increases in the motor cortex as before, but also in the supplementary motor area. Finally, when the person mentally "rehearses" the tapping without moving the finger, blood flow increases in the supplementary cortex as before but not in the motor cortex. This method, like the measurement of readiness potentials, offers a way to view "thinking" that is intriguing. (However, as the text explains, the supplementary motor area is not thought to be the ultimate initiator of movement.)

28. (a) **Correct.**
    (b) **Incorrect. Dopamine is inhibitory to neurons in the basal ganglia.**
    (c) **Correct.**
    (d) **Incorrect. They form part of a feedback loop with the motor cortex and the *thalamus*.**
    (e) **Correct.**
    (f) **Correct.**

29. (a) **Correct.**
    (b) **Incorrect. Ability to initiate voluntary movement *is* impaired.**
    (c) **Correct.**

    Observation of the defects that result from damage to a particular area of the brain can aid our understanding of the normal functioning of that area. Parkinson's disease provides one example to help us understand the role of the basal ganglia in motor control. Another example is the destruction of one of the major nuclei in the basal ganglia. The result of such destruction is a condition called *chorea*, characterized by excessive, uncontrolled, and abnormal body movements. Thus, a major overall effect of that nucleus, and the basal ganglia as a whole, is to suppress such movements by selective inhibition at the level of the thalamus.

    The defects in Parkinson's disease, to reiterate, result from loss of inhibitory input from the substantia nigra. This loss of inhibition leaves the basal ganglia in a state of hyperfunctioning, responding mainly to excitatory stimuli from the cerebral cortex. In turn, input to the thalamus from the basal ganglia is unbalanced and results in the two kinds of impairments described in Question 29-b and -c. The ability to "start up" a movement is gradually lost (even taking a single step is excruciatingly difficult for someone with severe Parkinson's), and tremor and rigid paralysis interfere even with the patient's rest. As we shall see, the rigidity reflects an abnormally heightened muscle tone.

30. (a) **Incorrect. It does not initiate movement.** (Stimulation of neurons in the cerebellum does not result in movement of muscle groups, as would stimulation of the sensorimotor cortex.)
    (b) **Correct.**
    (c) **Correct.**
    (d) **Incorrect. It receives information from all receptors affected by movement.**

31. **True.**

32. (a) **Incorrect. Cerebellar damage does not cause paralysis.**
    (b) **Incorrect. Resting tremor is a symptom of Parkinson's disease, not cerebellar damage.**
    (c) **Correct.** ("Intention tremor" is a hallmark of cerebellar damage, and it illustrates the importance of the cerebellum in helping coordinate *intention* (to reach an object) with the precise, exact movements required to reach it.)
    (d) **Correct.**
    (e) **Correct.**

33. (a) **Correct.**
    (b) **Incorrect. It begins in the sensorimotor** (cerebral) **cortex.**
    (c) **Incorrect. It ends in the gray matter of the spinal cord.**
    (d) **Incorrect. It crosses in the brainstem medulla.**
    (e) **Incorrect. It has no intervening synapses between the cell bodies in the sensorimotor cortex, and the motor neurons and local interneurons in the cord.**

It is not uncommon for students to become confused regarding the "simple" nature of the corticospinal pathway after so much emphasis about multiple loops and integration of hierarchical signals involving a host of brain areas. However, the signal to proceed — that is, input to the motor neurons (and local interneurons around them) — comes directly from the motor area of the cortex via this pathway. (Information is integrated by the cortical neurons and by neurons presynaptic to them.)

34. **False.** The motor pathway controlling voluntary movements of the head and neck is called the ~~pyramidal~~ **corticobulbar** tract. (As the text mentions, the pyramidal tract is another name for the corticospinal pathway. The plural term "corticospinal pathways" includes both the long pathway and the shorter corticobulbar pathway.)

35. (a) **Correct.**
    (b) **Correct.**
    (c) **Correct.**
    (d) **Incorrect. They may excite inhibitory interneurons and so effectively inhibit motor neurons, but their synaptic input is excitatory** (that is, depolarizing).
    (e) **Incorrect. They generally affect the activity of only one muscle** (one motor neuron pool).
    (f) **Correct.**

36. In the brainstem and spinal cord, descending fiber tracts that have their cell bodies in nuclei of the brainstem are collectively called the **multineuronal** or **extrapyramidal** pathways. The cell bodies of these fibers receive information from the **sensorimotor cortex, the basal ganglia, the thalamus, the cerebellum,** and **the brainstem** in a series of complicated loops. The fibers ultimately synapse on **motor neurons** or, more commonly, on **interneurons** in the brainstem and spinal cord. In general, these pathways are involved with **coordination of large muscle groups used in the maintenance of upright posture, locomotion, and turning to a specific stimulus.**

37. (a) **Incorrect. It is defined as the resistance of muscle to continuous passive stretch.**
    (b) **Incorrect. It is low but present in relaxed muscle.**
    (c) **Correct.**
    (d) **Correct.**

38. In general, damage to the descending motor pathways results in abnormally high muscle tone, known as **hypertonia**, whereas damage to the alpha motor neurons leads to **hypotonia**. Therefore, damage to the descending pathways selectively impairs **inhibitory** input to the alpha motor neurons.

39. Two kinds of hypertonia that can be distinguished clinically are **spasticity**, which is accompanied by increased responsiveness to motor reflexes, and **rigidity**, which is not. (Rigidity, which is one of the symptoms of Parkinson's disease, is a result of simultaneous contraction of antagonistic muscles.)

40. (a) **Correct.** (These are the photoreceptors.)
    (b) **Correct.** (These are the receptors in the vestibular system.)
    (c) **Correct.**

41. **False.** Crossed-extensor reflexes ~~but not~~ **and** stretch reflexes are **both** important postural reflexes.

42. (a) **Correct.**
    (b) **Correct.**
    (c) **Incorrect. All motor activity, even spinal reflexes, is dependent on descending motor pathways to some extent.** (We have already seen that people with Parkinson's disease or cerebellar damage have difficulty walking. However, you should appreciate the adaptive value of having walking motions be essentially "instinctive.")

PRACTICE

True or false (correct the false statements):

P1. The motor neuron pool refers to all the motor neurons innervating muscles in a limb.

P2. The activity of alpha motor neurons innervating an extensor muscle can be inhibited by activation of the Golgi tendon organs near the muscle.

P3. When a person steps on a tack with her left foot, flexor muscles on the right leg and extensor muscles on the left leg will be stimulated to contract.

P4. In the primary motor cortex, the hands are represented by a larger strip than the legs.

P5. Electrodes placed on the scalp over the primary motor cortex can record electrical activity known as the readiness potential almost 1 second before movement occurs.

P6. The thalamus is an important relay station for feedback of information to the basal ganglia from the motor cortex.

P7. In the motor cortex, representation of the left half of the body is on the right side of the brain.

P8. In general, the multineuronal pathways have greater influence over motor neurons controlling muscles involved in fine movements, and the corticospinal pathways are more involved in coordination of large muscle groups.

Multiple choice (correct each incorrect choice):

P9-12. The knee-jerk reflex

- P9. is a withdrawal reflex.
- P10. contributes to maintenance of posture.
- P11. can be modified by alpha-gamma coactivation.
- P12. is monosynaptic.

P13-15. The cerebellum

- P13. contributes fibers to the descending motor pathways.
- P14. has direct input to the sensorimotor cortex.
- P15. is essential for smoothly coordinated movements.

P16-19. Regarding the postural reflexes,

- P16. the afferent pathways include information from the eyes.
- P17. the efferent pathways are the skeletal muscles.
- P18. the integrating centers are in the cerebral cortex.
- P19. they are absent in blind people.

P20-24. Indicate whether the following are symptoms of Parkinson's disease (P), cerebellar damage (C), or neither (N):

P20. ____ resting tremor

P21. ____ intention tremor

P22. ____ shuffling gait

P23. ____ rigidity

P24. ____ wide, rolling gait

P25. ____ total paralysis

13: CIRCULATION

If Claude Bernard is considered "the father of physiology" (Chapter 1), then perhaps William Harvey, the seventeenth-century physician to English kings, deserves the title of physiology's "grandfather." It was Harvey who discovered that the blood in the body is propelled in a circle (it circulates) by the heart, a muscular pump.

The prevailing theory in Harvey's time, which had been formulated centuries earlier by Greek scholars, was that blood was produced in the liver and that it "ebbed and flowed," first in one direction and then in another, in any given blood vessel. The heart was thought to be a chamber that "dilated" when it received blood from veins, and this dilation then allowed the "natural spirits" in blood to mix with substances from the air to produce "vital spirits." According to this theory, blood had to pass from the right side of the heart to the left through invisible pores, and it did not travel to the lungs.

Harvey questioned many aspects of the classical theory, and set about to use "direct ocular observations, dissections, and experimentation" to understand the functions of the heart and blood vessels. He wrote:

> *When I first tried animal experimentation for the purpose of discovering the motions and functions of the heart by actual inspection and not by other people's books, I found it so truly difficult that I almost believed Fracastorius, that the motion of the heart was to be understood by God alone...*
>
> *Finally, using greater care every day, with very frequent experimentation, observing a variety of animals, and comparing many observations, I felt my way out of this labyrinth, and gained accurate information, which I desired, of the motions and functions of the heart and arteries.*
>
> Chapter I, de Motu Cordis

In this chapter, you will learn what Harvey discovered and much more about the physiology of the circulatory system, our "river of life."

Instructions for answering questions in this Study Guide

1. True or false: Correct each false statement. Whenever you must correct a false statement, there will almost always be more than one way to do so. In the Guide, the answer requiring the least correction will generally be given, with the understanding that other correct answers are possible. It is usually not sufficient, in terms of demonstrating understanding, to simply insert a "no" or "not," however.

2. Multiple choice: Any, all, or none of the responses may be correct. Explain why each incorrect response is incorrect.

3. Fill in the blanks: Choose the best word or words to complete each statement.

4. Directions will be given for other types of questions as they appear.

SECTION A. BLOOD

❒ Introduction (text page 350)

1. Blood consists of cells suspended in a liquid called ____________. The most numerous cells, the ______________, occupy about ________ percent of the total blood volume in a normal man. This percentage is known as the ______________. The other cell types, the ______________ and ____________, make up less than ________ percent of the total volume.

2. The normal hematocrit in a woman is about 42%. Assuming a blood volume of 5.3 L, what would her plasma volume be?

❒ Plasma (text pages 350-351)

3. Most of the weight of plasma is ________________, which accounts for ____________ percent of it. The only other category of molecules that adds measurably to the weight of plasma is the ________________. Of these, the ________________ and ________________ are the most abundant. The difference between plasma and ________________ is that clotting proteins have been removed from the latter.

4. True or false: The concentration of albumins (in millimoles per liter) in plasma is greater than the concentration of sodium ion.

❒ The Blood Cells (text pages 351-357)

5. Erythrocytes
   (a) have a high surface-to-volume ratio.
   (b) carry oxygen and carbon dioxide in blood.
   (c) are red.
   (d) contain myoglobin.

6. Hemoglobin
   (a) is a protein.
   (b) contains four globin molecules attached to a single heme group.
   (c) contains four iron atoms.

7. True or false: The function of iron in hemoglobin is to impart a red color to the pigment.

8. Erythrocytes are produced in the ____________________ and are destroyed in the ________________ after a life span of about _______ months. Mature erythrocytes contain no __________________ or __________________. Immature erythrocytes, called ______________________, contain some ribosomes. These immature blood cells __________(are/are not) found in large numbers in blood under normal conditions.

9. Match the compound from the list on the right with its correct description on the left:

___ storage protein for iron	(a) folic acid
___ stimulates erythrocyte production	(b) ferritin
___ necessary for DNA synthesis	(c) vitamin B_{12}
___ necessary for folic acid function	(d) vitamin A
___ transport protein for iron	(e) erythropoietin
	(f) myoglobin
	(g) transferrin

10. True or false: The primary stimulus for erythropoietin secretion is decreased iron concentrations in the liver.

11. Anemia
    (a) can be caused by too little iron in the diet.
    (b) can be caused by too little vitamin B_{12} in the diet.
    (c) can be caused by kidney failure.
    (d) can be caused by abnormal hemoglobin.
    (e) refers to reduced oxygen-carrying capacity of the blood.

12. You make the diagnosis: Three patients come to see you complaining of weakness and looking pale. Blood samples allow you to determine quickly the hematocrit, average cell size, and hemoglobin concentration [Hb] per cell in each patient. Match the lab results and other information given with the most likely diagnosis. Explain each answer.

    Diagnosis choices: internal bleeding; iron deficiency; vitamin B_{12} deficiency; folic acid deficiency

    (a) Symptoms: low hematocrit; microcytosis; low [Hb]
    Diagnosis: ____________________

____

____

____

(b) Symptoms: low hematocrit; normocytosis; normal [Hb]

Diagnosis: ____________________

____

____

____

(c) Symptoms: low hematocrit; macrocytosis; strict vegetarian

Diagnosis: ____________________

____

____

____

13. There are five classes of leukocytes. Three are grouped together under the name ____________________ because they have ____________________ nuclei and abundant ____________________. The granules of the three cell types have different affinities for dyes: the ____________________ stain preferentially with the red dye ____________________; the ____________________ with blue (basic) dyes; and ____________________ with both.

The other two types of leukocytes are the ____________________ and the ____________________. All five of these cell types have functions in the defense systems of the body. The most abundant of them are the ____________________, followed by the ____________________. Another group of cells, the ____________________, are really fragments of larger cells called ____________________ and function in blood clotting.

14. All blood cell types are

(a) produced in the bone marrow.

(b) descendants of myeloid stem cells.

(c) regulated by hormones called colony-stimulating factors.

SECTION B. OVERALL DESIGN OF THE CARDIOVASCULAR SYSTEM

❒ Introduction (text pages 357-360)

15. The two basic functions of the cardiovascular system are to ________________ ________________________________ to cells and to ________________________ ________________________ from cells. These functions are carried out in the ____________________________, which at any given moment contain about ______ percent of the total blood volume.

16. The two circulations discovered by William Harvey are the ____________________ circulation, in which blood is pumped by the ______________________________ of the heart to the lungs, and the ____________________ circulation, in which blood pumped by the ___________________________ of the heart reaches all the tissues of the body except the lungs.

17. Beginning at the right atrium, trace the 18-step path an erythrocyte takes on one complete circuit through the circulation. Use the text description and text Figures 13-8 and 13-9.

18. True or false: Blood vessels that carry well-oxygenated blood are called arteries.

19. True or false: At rest, more blood flows per minute to the kidneys and abdominal organs than to the muscles and skin.

20. True or false: At rest, more blood flows per minute to the skin and abdominal organs than to the lungs.

❐ Pressure, Flow, and Resistance (text pages 360-362)

21. Write an equation that describes how the flow of a fluid in a tube is related to the pressure exerted by the fluid and to the resistance to flow exerted by both the tube and the fluid.

22. The equation in Question 21 states that
    (a) the flow of fluid in a tube depends upon the absolute pressure at the beginning of the tube.
    (b) the flow of fluid from point A to point B in a tube depends upon the difference in pressure between A and B.
    (c) the flow of fluid in a tube is directly proportional to the pressure difference in the tube.
    (d) the greater the resistance to flow in a tube, the greater the flow for any given pressure difference.

23. (a) List the three variables that affect the resistance to flow of a fluid in a tube: ______________________, ______________________, and ______________________

    (b) Which variable plays the major role in determining blood flow in the circulatory system? ______________________

24. Fluid is flowing in a straight tube (such as in text Figure 13-11).

    (a) P_1 = 100 mm Hg, P_2 = 90 mm Hg, R = 1 mm Hg/mL/min.
    What is the value of F?

    (b) The values for P_1 and P_2 are the same as in 24-a, but the radius of the tube is increased threefold. All other parameters are the same. What is the value of F in this case?

25. From text Table 13-3: The components of the cardiovascular system
    (a) that produce the pressures that drive blood through the pulmonary and systemic systems are ______________________________.
    (b) that are the major sites of resistance to flow are the ____________________.
    (c) that are the site of nutrient and waste product exchange are ______________.
    (d) that act as pressure reservoirs for maintaining blood flow are ____________.
    (e) whose diameters are regulated are ________________, ________________, and ________________________.

SECTION C. THE HEART

❐ Anatomy (text pages 362-365)

26. The heart is composed primarily of __________________________, which are collectively termed the ________________________. This structure is surrounded by a fibrous sac, the __________________________. The heart chambers, the ________________________ and ________________________, are lined by ______________________.

27. The atrioventricular valves prevent the flow of blood
    (a) between the two atria and between the two ventricles.
    (b) from the atria to the ventricles.
    (c) from the pulmonary trunk into the right ventricle and from the aorta into the left ventricle.

28. True or false: The function of the papillary muscles is to keep the atrioventricular valves from becoming pushed back into the atria.

29. Cardiac-muscle cells
    (a) are striated.
    (b) are connected by intercalated disks.
    (c) are connected by gap junctions.
    (d) have cholinergic and adrenergic receptors.

30. True or false: The conducting system of the heart is a branch of the vagus nerve.

31. True or false: The myocardial cells receive their blood supply from the cardiac arteries that branch off from the aorta.

❐ **Heartbeat Coordination** (text pages 365-370)

32. Compare the membrane potential recording from a ventricular cell (text Figure 13-16) with the same sort of recording from a neuron (text Figure 8-30, page 195). Compare, too, the relative membrane permeabilities of the two kinds of cells. Note particularly the difference in units on the time axes (seconds versus milliseconds). Which of the following statements is/are correct?

    (a) Like neurons, ventricular muscle cells at rest have a higher permeability to potassium ions than to sodium ions.
    (b) Like neurons, ventricular muscle cells have a resting membrane potential closer to the sodium equilibrium potential than to the potassium equilibrium potential.
    (c) In both neurons and ventricular muscle cells, the permeability to potassium increases just as the voltage-gated sodium channels begin to close.
    (d) In both kinds of cells, increased potassium permeability contributes to repolarization after depolarization.
    (e) The shape of the action potential in ventricular cells resembles closely the shape of the calcium permeability curve.
    (f) The duration of the ventricular muscle action potential is almost one hundred times that of the nerve cell action potential.
    (g) During the depolarization and plateau phases of the myocardial action potential, potassium ions do not leave the cell.
    (h) During the initial depolarization phase of the myocardial action potential, both sodium and calcium ions enter through voltage-gated channels.

33. True or false: A major difference between contractile cardiac cells and myocardial cells in the conducting system is that the contractile cells do not undergo spontaneous depolarization.

34. True or false: The atrioventricular node is normally the heart's pacemaker because it has the fastest spontaneous rate of depolarization.

35. Arrange in proper order the following events in the sequence of heart excitation:
    (a) depolarization of the atrioventricular node
    (b) depolarization of the atria
    (c) depolarization of the ventricular contractile cells
    (d) depolarization of the sinoatrial node
    (e) depolarization of the bundle of His
    (f) depolarization of the Purkinje fibers

36. True or false: The wave of contraction that spreads through the ventricles starts at their tips and flows upward.

37. The electrocardiogram is a
    (a) recording of the electrical activity of individual cardiac cells.
    (b) recording of currents generated in the extracellular fluid by the electrical events of the heart.
    (c) useful diagnostic tool for detecting defects in the electrical conduction mechanisms of the heart.

38. In the ECG, the P wave corresponds to ______________________________ ______________________________, the QRS complex to ______________________________ ______________________________, and the T wave to ______________________________ ______________________________. The electrical event that is not seen is ______________________________.

39. Comparing excitation-contraction coupling in cardiac muscle with that in skeletal muscle:
    (a) Extracellular calcium plays a major role in cardiac but not skeletal muscle.
    (b) The stimulus for calcium release from the sarcoplasmic reticulum is the same in both muscles.
    (c) Troponin sites are always saturated immediately after calcium release in both muscles.
    (d) There is no net change in total intracellular calcium concentration in either muscle.

40. The heart cannot undergo tetanus. Explain.______________________________
____
____
____

41. True or false: An ectopic focus is an area of the myocardial conducting system other than the atrioventricular node that initiates cardiac depolarization.

❐ Mechanical Events of the Cardiac Cycle (text pages 371-375)

42. During the cardiac cycle, the period of time when the ventricles are contracting and blood is being forced into the pulmonary trunk and aorta is called ____________________. The period of ventricular relaxation and filling with blood is called ________________________. In between these events are brief periods in which ventricular volume does not change, the ____________________________ ventricular contraction and relaxation. Normally the period of ____________________ lasts the longest.

43. True or false: The determinant for whether blood ejection occurs during ventricular contraction or whether filling occurs during ventricular relaxation is the pressure difference across the heart valves.

44. True or false: The only time during the cardiac cycle that the AV valves are open is systole.

45. True or false: During systole of the left heart, left ventricular pressure exceeds aorta pressure, allowing blood ejection.

46. True or false: At all times in the cardiac cycle other than systole, aortic pressure exceeds left ventricular pressure.

47. The difference between the end-diastolic volume and the end-systolic volume in the ventricle is the ____________________________ — the volume of blood ejected during ________________________.

48. True or false: The left ventricle ejects a larger volume of blood during systole than does the right ventricle because the left ventricular pressure is greater than the right.

49. True or false: Normal heart sounds are produced by the slamming shut of (a) the AV valves at the end of systole and (b) the aortic and pulmonary valves at the end of diastole.

50. Abnormal heart sounds called ______________________ may indicate heart disease, particularly problems with the valves. For example, aortic stenosis, which causes a murmur during ______________________, is a result of an abnormal ______________________. Left AV insufficiency, which also causes a murmur during ______________________, is caused by ______________________.

❐ The Cardiac Output (text pages 375-381)

51. Cardiac output is the
    (a) volume of blood pumped per minute by both ventricles.
    (b) volume of blood flowing through the systemic circulation each minute.
    (c) product of the number of heart beats per minute and the volume pumped per beat.

52. An average adult at rest has a cardiac output of about 5 L/min, while in a trained athlete cardiac output can reach 35 L/min. Could this increase be accounted for by increased heart rate alone?

53. True or false: The inherent pacemaker discharge rate of the sinoatrial node is 72 beats per minute.

54. The discharge rate of the SA node is
    (a) increased by sympathetic stimulation.
    (b) decreased by parasympathetic stimulation.
    (c) increased by acetylcholine.
    (d) decreased by epinephrine.
    (e) decreased by increased activity of the vagus nerve.

55. True or false: Acetylcholine reduces the slope of the pacemaker potential by increasing permeability to potassium, whereas norepinephrine increases the slope of the potential by increasing permeability to sodium and calcium.

56. True or false: During stimulation of the parasympathetic nerves, plasma levels of epinephrine increase.

57. True or false: Sympathetic stimulation of the heart increases conduction through the AV node.

58. Starling's law of the heart states that
    (a) increasing end-diastolic volume increases stroke volume because there is more blood to pump.
    (b) increasing end-diastolic volume increases the strength of contraction of heart muscle and thus increases stroke volume.
    (c) the force of contraction of heart muscle increases as the muscle is stretched over a finite range.

59. True or false: The Starling curve (a plot of stroke volume versus ventricular end-diastolic volume) is a length-tension curve.

60. In a heart-lung preparation (text Figure 13-31), the end-diastolic volume can be increased by raising the reservoir and thus increasing the ______________________. In an intact person or animal, end-diastolic volume is enhanced by increasing ______________________.

61. Ventricular contractility is
    (a) the strength of ventricular contraction.
    (b) the strength of ventricular contraction at any given end-diastolic volume.
    (c) increased by parasympathetic stimulation.
    (d) increased by sympathetic stimulation.

62. The effects of epinephrine and norepinephrine on cardiac contractility are mediated
    (a) through alpha-adrenergic receptors.
    (b) by the second messenger cyclic GMP.
    (c) by activated protein kinase.

63. Effects of epinephrine and norepinephrine on cardiac cells include
    (a) increased calcium permeability during depolarization.
    (b) decreased pumping of calcium back into the sarcoplasmic reticulum.
    (c) increased binding of calcium to troponin.
    (d) increased velocity of the cross-bridge cycle.

64. A person with a heart transplant has no innervation of the heart. What is the heart rate at rest in such a person? Can cardiac output be regulated at all? Explain.

____

____

____

____

____

SECTION D. THE VASCULAR SYSTEM

❒ Introduction (text page 381)

65. Endothelial cells
   (a) line every branch of the circulatory vessels and the heart.
   (b) are a passive layer of cells with no specialized functions.

❒ Arteries (text pages 381-384)

66. The arteries
   (a) are thick-walled, high-resistance tubes.
   (b) contain large amounts of elastic tissue.
   (c) have large diameters.
   (d) "store" pressure during diastole to drive blood flow through the rest of the blood vessels during systole.

67. True or false: Systolic pressure is the peak pressure in systemic arteries and occurs at the beginning of systole.

68. Pulse pressure
   (a) can be felt as a throb in arteries of the wrist and neck.
   (b) is the difference between systolic and diastolic pressures.
   (c) decreases with increasing stroke volume.
   (d) is an indication of the health of the cardiovascular system.

69. Your blood pressure is measured with a sphygmomanometer; the systolic pressure (SP) is 120 mm Hg, and the diastolic pressure (DP) is 80 mm Hg. What is
    (a) your pulse pressure?

    (b) your mean arterial pressure?

70. Regarding the measurement of blood pressure:
    (a) The device used to measure it takes advantage of sounds made by blood flowing through unconstricted arteries.
    (b) When the first sounds are heard through the stethoscope, systolic pressure is just slightly greater than the pressure on the manometer.
    (c) When the last sounds are heard through the stethoscope, diastolic pressure is just slightly less than the pressure on the manometer.

❒ Arterioles (text pages 384-389)

71. True or false: The flow rate of blood through an organ is determined by the relative dilation or constriction of the walls of the organ's small arteries.

72. True or false: The smooth muscle in arteriole walls can be stimulated either to contract or to relax, and thus either decrease or increase the amount of blood flowing through arteriolar beds.

73. The increase in blood flow through an organ that is actively metabolizing is termed ______________________. This change in blood flow is a form of ______________________ that is independent of nerves and hormones. The chemical stimuli that produce the ______________________ (vasodilation/ vasoconstriction) include decreased ______________________, increased levels of ______________________, ______________________, ______________________, ______________________, ______________________, ______________________, and increased ______________________. These same chemical signals also play a role in regulating blood flow in the face of increased or decreased blood pressure, termed ______________________. The phenomenon of reactive hyperemia is an example of an extreme form of ______________________.

74. Regarding the innervation of arteriolar smooth muscle:
   (a) Most arterioles receive extensive parasympathetic innervation.
   (b) Most arterioles receive extensive sympathetic innervation.
   (c) Epinephrine from sympathetic nerves interacts with alpha-adrenergic receptors on the arteriolar smooth muscle to induce vasoconstriction.

75. True or false: Because sympathetic stimulation of the arterioles causes vasoconstriction, the sympathetic nerves play no role in reflex vasodilation.

76. Endothelium-derived relaxing factors
   (a) are secreted by the arteriolar endothelium.
   (b) are secreted by the arterial endothelium.
   (c) may mediate the vasodilator effects of many chemicals.

77. Examine text Table 13-6 on page 390.
   (a) Which two organs utilize pressure autoregulation as a major source of vascular control?____
   (b) Which organ uses mainly local metabolites for vascular control? ____________________
   (c) Which two organs are controlled mainly by local metabolic factors during specific activities? ________________________________

❒ Capillaries (text pages 389-396)

78. True or false: At any one time, approximately 5 percent of a person's total blood volume is in the set of blood vessels with the longest combined length.

79. True or false: Angiogenesis is the growth of new capillary beds.

80. True or false: Because capillaries have no smooth muscle, flow of blood through capillary beds cannot be controlled.

81. In the systemic circulation, the greatest resistance to flow is found in the ____________________, with the second greatest in the ____________________. The greatest cross-sectional area is in the ____________________. There is a/an ____________________ (direct/inverse) correlation between flow rate and cross-

sectional area, such that the highest flow rate is in the ____________________ and the lowest flow rate is in the ____________________.

82. With regard to diffusion of molecules across the capillary wall,
   (a) polar molecules diffuse through the endothelial plasma membranes.
   (b) water-soluble molecules diffuse across intercellular clefts or other water-filled channels in the endothelial cells.
   (c) capillary "leakiness" varies from organ to organ, with liver capillaries among the least leaky.
   (d) carrier-mediated transport is especially important for providing nutrients to the brain.

83. True or false: The concentrations of molecules in the intracellular fluid are the same as those in blood plasma.

84. True or false: The hydrostatic pressure difference between the plasma inside the capillary beds and the interstitial fluid surrounding the cells favors filtration of fluid.

85. True or false: The osmotic pressure difference between the plasma inside the capillary beds and the interstitial fluid surrounding the cells favors filtration of fluid.

86. Crystalloids in plasma and interstitial fluid are ____________________ (penetrating/nonpenetrating) solutes, and colloids are ____________________ (penetrating/nonpenetrating) solutes. During filtration, the concentration of ____________________ (crystalloids/colloids) stays the same in both fluid compartments.

87. In the average capillary bed,
   (a) the net filtration pressure at the alveolar end favors filtration.
   (b) the net filtration pressure at the venous end favors filtration.
   (c) the overall result is usually little net loss of fluid from the capillaries.

88. True or false: The amount of nutrients and waste products moved across the capillary wall by bulk flow is greater than the amount that diffuses across under normal conditions.

89. True or false: Strong vasoconstriction of arterioles leading to a capillary bed will probably result in loss of plasma by filtration in the capillary bed.

90. True or false: In the normal lung, capillary osmotic pressure always exceeds filtration pressure and so absorption of fluid is the rule.

❐ **Veins** (text pages 396-398)

91. True or false: Unlike arterioles, venules have no smooth muscle in their walls.

92. Veins
    (a) are large-volume, low-pressure conduits.
    (b) carry most of the total blood volume at any given time.
    (c) are innervated by parasympathetic neurons.

93. Venous pressure is
    (a) increased by sympathetic stimulation.
    (b) increased by skeletal muscle contraction.
    (c) increased by breathing deeply.
    (d) an important determinant for stroke volume.

94. True or false: The purpose of valves in veins is to keep blood flowing in only one direction — back to the heart.

❐ **The Lymphatic System** (text pages 398-400)

95. The lymphatic system
    (a) is part of the cardiovascular system.
    (b) consists of small organs called nodes that pump the lymph through the lymphatic vessels.
    (c) begins with lymphatic capillaries.
    (d) carries filtered fluid and protein back to the circulatory system.
    (e) is important for preventing edema.

SECTION E. INTEGRATION OF CARDIOVASCULAR FUNCTION: REGULATION OF SYSTEMIC ARTERIAL PRESSURE

❒ Introduction (text pages 400-403)

96. True or false: The regulated variable of the cardiovascular system is the mean systemic arterial blood pressure.

97. Write the flow/pressure/resistance equation so that pressure depends on flow and resistance. What do these three terms mean with respect to the cardiovascular system as a whole?

98. One way to look at the homeostatic regulation of systemic blood pressure is to assume that the one thing the cardiovascular system must do, no matter what, is deliver adequate amounts of blood to the brain. Explain.

____

____

____

____

____

❒ Baroreceptor Reflexes (text pages 403-407)

99. True or false: Receptors for arterial blood pressure are located in the walls of arteries that supply the brain with blood.

100. Receptors sensitive to blood pressure in the walls of arteries are special types of mechanoreceptors called ______________________. They are located in the ______________________ and the ______________________. Together these mechanoreceptors are known as the ______________________. Analogous receptors are also found in the ________________, the ________________, and the ________________.

101. True or false: Increasing the mean arterial pressure or the pulse pressure in the carotid artery will cause decreased firing of action potentials in the afferent fibers associated with the baroreceptors.

102. The integrating center that receives information about blood pressure and volume and then coordinates the responses of effector tissues is the ______________________________ in the brainstem. When arterial or other baroreceptors increase their firing rate to this center, it ______________ (increases/decreases) the activity of sympathetic nerves to the ______________, ______________, and ______________ and ______________ (increases/decreases) the activity of the parasympathetic nerves to the ______________. These effects combine to ______________ (increase/decrease) cardiac output, ______________ (increase/decrease) total peripheral resistance, and, as a result, ______________ (increase/decrease) mean arterial blood pressure.

❒ Blood Volume and Long-Term Regulation of Arterial Pressure (text page 407)

103. True or false: Blood-volume changes are an important determinant of long-term regulation of blood pressure.

❒ Other Cardiovascular Reflexes and Responses (text pages 407-409)

104. True or false: The medullary cardiovascular center is the only part of the brain involved with the regulation of blood pressure.

SECTION F. CARDIOVASCULAR PATTERNS IN HEALTH AND DISEASE

❒ Hemorrhage and Other Causes of Hypotension (text pages 410-412)

105. Hypotension
   (a) refers to blood loss.
   (b) can be caused by sweating, vomiting, or emotional responses.
   (c) stimulates increased firing of the arterial baroreceptors.

106. Indicate whether the listed variables will increase (I), decrease (D), or remain unchanged (N) following hemorrhage, compared to pre-hemorrhage values.

(a) ____ cardiac output

(b) ____ heart rate

(c) ____ total peripheral resistance

(d) ____ venous return

(e) ____ activity of the parasympathetic nerves to the heart

(f) ____ capillary filtration

(g) ____ firing of baroreceptors

(h) ____ activity of sympathetic nerves to the heart

(i) ____ stroke volume

(j) ____ percent of the blood volume in the veins

(k) ____ arteriolar vasoconstriction

(l) ____ blood flow to the brain

107. True or false: The condition of shock is an example of failure of homeostatic mechanisms to compensate when one or more parts of the reflex control system are damaged.

❐ The Upright Posture (text pages 412-413)

108. When a person is quietly standing erect, with his arms at his sides,

(a) blood pressure in the capillary beds of the feet is about fourfold higher than that of capillary beds in the neck.

(b) blood pressure in the capillary beds of the hands is approximately half that of the feet.

(c) fainting may occur unless he alternately contracts and relaxes the muscles of his legs.

❐ Exercise (text pages 413-416)

109. During exercise, blood flow

(a) increases to the skeletal muscles and the heart.

(b) decreases to the kidneys, abdominal organs, and skin.

(c) remains the same to the brain.

110. True or false: During exercise, sympathetic stimulation of the heart and arteriolar smooth muscle is enhanced because of reflexes triggered by elevated mean arterial pressure and elevated pulse pressure.

111. True or false: During exercise, "exercise centers" in the brain override baroreceptor reflexes to allow increased cardiac output and vasoconstriction in internal organs in the face of increased pulse pressure.

112. During exercise,
   (a) the arterioles in skeletal muscles dilate because of decreased sympathetic activity.
   (b) blood flow to skin increases because of active hyperemia.
   (c) blood flow to the brain increases because of the increased arterial pressure.

113. Physical endurance training
   (a) enhances a person's maximal work load.
   (b) enhances a person's maximal cardiac output.
   (c) enhances a person's maximal stroke volume.
   (d) enhances a person's maximal heart rate.
   (e) increases a person's resting heart rate.

❒ **Hypertension** (text pages 416-417)

114. Hypertension
   (a) refers to a chronic state of elevated blood pressure.
   (b) is defined as a systolic pressure greater than 120 mm Hg and a diastolic pressure greater than 80 mm Hg.
   (c) is a result of chronically elevated cardiac output.
   (d) is usually caused by kidney disease.
   (e) can sometimes be treated with drugs that increase excretion of water in the urine.

115. Which of the following types of drugs might be useful for treating hypertension? Explain your answers.
   (a) A drug that increases heart rate and contractility

   ____

   ____

(b) A drug that mimics the effects of decreased O_2 in arterioles

____

____

❒ **Heart Failure** (text page 418)

116. True or false: A failing heart does not respond to increased end-diastolic volume by increasing stroke volume.

117. True or false: A serious consequence of failure of the left ventricle is edema in the legs and feet.

❒ **"Heart Attacks" and Atherosclerosis** (text pages 418-420)

118. The medical term for heart attack is ______________________________ ______________________________. The immediate cause of heart attack is insufficient blood flow to the ______________________________ and subsequent damage to the myocardium. Sudden death from heart attack is usually a result of ______________________________, in which contractions of ventricular cells are not coordinated. Death can be averted if CPR — which stands for ______________________________ — is begun right away.

119. Atherosclerosis
   (a) is the major cause of heart attacks.
   (b) is a thinning in the walls of arteries.
   (c) is associated with low blood levels of cholesterol, obesity, and cigarette smoking.
   (d) may result in occlusion of the vessel.
   (e) may result in coronary occlusion (stroke).

120. The occasional pain experienced during exertion or anxiety by someone with atherosclerosis of a coronary artery is called ______________________________. One new treatment for this problem is ______________________________, the enlargement of a vessel by threading a balloon on a catheter into the occluded vessel.

ANSWERS

Boldface type indicates the answers you should have given. Words in medium-face (ordinary) type explain or expand upon the answer.

1. Blood consists of cells suspended in a liquid called **plasma**. The most numerous cells, the **erythrocytes**, occupy about **45** percent of the total blood volume in a normal man. This percentage is known as the **hematocrit**. The other cell types, the **leukocytes** and **platelets**, make up less than **one** percent of the total volume.

2. 0.42 x 5.3 L = 2.226 L (≈ 2.2 L)
   5.3 L - 2.2 L = **3.1 L**

3. Most of the weight of plasma is **water**, which accounts for **93** percent of it. The only other category of molecules that adds measurably to the weight of plasma is the **plasma proteins**. Of these, the **albumins** and **globulins** are the most abundant. The difference between plasma and **serum** is that clotting proteins have been removed from the latter.

4. **False.** The concentration of albumins (in millimoles per liter) in plasma is ~~greater~~ **less** than the concentration of sodium ion.

5. (a) **Correct.**
   (b) **Correct.**
   (c) **Correct.**
   (d) **Incorrect. They contain hemoglobin.**

6. (a) **Correct.**
   (b) **Incorrect. It contains four heme groups attached to a globin molecule.**
   (c) **Correct.**

7. **False.** The function of iron in hemoglobin is to ~~impart a red color to the pigment~~ **bind oxygen**. (It imparts a red color, true, but that is not its function.)

8. Erythrocytes are produced in the **bone marrow** and are destroyed in the **spleen** after a life span of about **four** months. Mature erythrocytes contain no **nuclei** or **organelles**. Immature erythrocytes, called **reticulocytes**, contain some ribosomes. These immature blood cells **are not** found in large numbers in blood under normal conditions.

9. **b** storage protein for iron
   e stimulates erythrocyte production
   a necessary for DNA synthesis

c necessary for folic acid function
g transport protein for iron

10. **False.** The primary stimulus for erythropoietin secretion is decreased ~~iron concentrations in the liver~~ **oxygen delivery to the kidneys.**

11. (a) **Correct.**
(b) **Correct.**
(c) **Correct.**
(d) **Correct.**
(e) **Correct.** (This is what all the conditions leading to anemia — too few erythrocytes or too little hemoglobin per erythrocyte or both — have in common.)

Note: The ending "-emia" on a word always refers to a condition of the blood. For example, leukemia is a type of cancer characterized by overproduction of leukocytes and great increases in the numbers of these cells in the blood.

12. (a) Diagnosis: **iron deficiency. Microcytosis is the most telling clue. In patients with iron deficiency, erythropoietin levels are high because the blood is carrying insufficient O_2 to the kidneys. Under the strong stimulus to increase production of erythrocytes, the production machinery in the marrow turns out as many cells as possible, with minimal amounts of Hb. Thus, the cells are small.**

(b) Diagnosis: **internal bleeding. Again, cell size is important, because it means the erythrocyte production mechanisms are functioning normally. Therefore abnormal blood loss is the most likely diagnosis.** (The diagnosis specifies *internal* bleeding because a person who has lost blood from a cut or other external wound would presumably seek treatment for the wound.)

Note: A low hematocrit means that the plasma volume lost during bleeding is being restored faster than erythropoiesis can keep up. You will learn later on in this chapter how blood volume is regulated.

(c) Diagnosis: **vitamin B_{12} deficiency. The lab finding of macrocytosis does not rule out folic acid deficiency, but folic acid is found in vegetables and vitamin B_{12} is not.** Strict vegetarians (no animal products in the diet) are at risk of vitamin B_{12} deficiency.

13. There are five classes of leukocytes. Three are grouped together under the name **polymorphonuclear granulocytes** because they have **multilobed** nuclei and abundant **membrane-bound granules**. The granules of the three cell types have different affinities for dyes: the **eosinophils** stain preferentially with the red dye **eosin**; the **basophils** with blue (basic) dyes; and **neutrophils** with both.

The other two types of leukocytes are the **monocytes** and the **lymphocytes**. All five of these cell types have functions in the defense systems of the body. The most abundant are the **neutrophils**, followed by the **lymphocytes**. Another group of cells,

the **platelets**, are really fragments of larger cells called **megakaryocytes** and function in blood clotting.
Note the suffix "-phil" for the three types of polymorphonuclear granulocytes. Recall that it means "to love," as "hydrophilic" means "loving water."

14. (a) **Correct.**
    (b) **Incorrect. They are descendants of pluripotent stem cells**. (Myeloid stem cells are precursors for all cells but the lymphocytes.)
    (c) **Incorrect. All but the erythrocytes are regulated by CSFs.**

15. The two basic functions of the cardiovascular system are to **carry nutrients** (and hormones) to cells and to **remove metabolic waste products** from cells. These functions are carried out in the **capillaries**, which at any given moment contain about **five** percent of the total blood volume.

16. The two circulations discovered by William Harvey are the **pulmonary** circulation, in which blood is pumped by the **right ventricle** of the heart to the lungs, and the **systemic** circulation, in which blood pumped by the **left ventricle** of the heart reaches all the tissues of the body except the lungs.

17. **right atrium** → **right ventricle** → **pulmonary trunk** → **pulmonary artery (right or left)** → **pulmonary arteriole** → **pulmonary capillary**
↓
systemic artery ← **aorta** ← **left ventricle** ← **left atrium** ← **pulmonary vein** ← **pulmonary venule**
↓
systemic arteriole → **systemic capillary** → **systemic venule** → **systemic vein** → **vena cava (superior or inferior)** → **right atrium**

18. **False.** Blood vessels that carry ~~well oxygenated~~ blood **away from the heart** are called arteries.

19. **True.**

20. **False.** At rest (or at any time), ~~more blood flows per minute to the skin and abdominal organs than~~ **blood flow to the lungs equals blood flow to the rest of the body.**

21. $\mathbf{F} = \dfrac{\Delta \mathbf{P}}{\mathbf{R}}$

22. (a) **Incorrect. It depends on the difference in pressure, as in 22-b and -c.**
    (b) **Correct.**
    (c) **Correct.** (This is a less specific way of stating 22-b.)

(d) **Incorrect. The less the resistance, the greater the flow.** (This is the meaning of "inversely proportional.")

23. (a) **fluid viscosity**, **tube length**, **tube radius**
(b) **tube radius**

24. (a) $F = \frac{\Delta P}{R} = \frac{100 \text{ mm Hg} - 90 \text{ mm Hg}}{1 \text{ mm Hg/mL/min}}$
$= \mathbf{10\ mL/min.}$

(b) We must determine the new resistance, R_b. Since the only change from part (a) is in the radius of the tube, we need to concern ourselves only with the effect of radius (r) on resistance (R). We know that

$$R \approx \frac{1}{r^4}$$

Since r_b (the radius of the second tube) is 3 x r_a (the radius of the first tube), then

$r_b = 3r_a$ and $R_b = R_a/3^4$ = (1 mm Hg/mL/min)/(3 x 3 x 3 x 3) = (1/81) mm Hg/mL/min

Therefore,

$$F = \frac{(100 - 90) \text{ mm Hg}}{(1/81) \text{ mm Hg/mL/min}} = 10 \times 81 = \mathbf{810\ mL/min}$$

25. (a) **the ventricles of the heart.**
(b) **the arterioles.**
(c) **the capillaries.**
(d) **the arteries.**
(e) **the arterioles**, **venules**, and **veins.**

This information is not given explicitly in text Table 13-3, but you should understand from the discussion on resistance that the only regulated (by homeostatic mechanisms) variable in the resistance equation is tube diameter (diameter = 2 x radius). Therefore, since you are told (a) that arterioles are the major sites of resistance to flow and (b) that they participate in the regulation of arterial blood pressure, you may have correctly surmised that their diameter is regulated. Similarly, you are told that venules participate in the regulation of capillary blood pressure, and again you may have guessed that this is done by altering their diameter. You are also told that the capacity of the veins for blood is adjusted, which implies regulation.

Do not feel concerned if the relationships between flow, resistance, and pressure are not crystal clear to you at this point. We will be revisiting them again when you have more knowledge about the workings of the cardiovascular system.

26. The heart is composed primarily of **cardiac-muscle cells**, which are collectively termed the **myocardium** ("myo-" = muscle and "cardium" = heart). This structure is

surrounded by a fibrous sac, the **pericardium** ("peri-" = around). The heart chambers, the **atria** and **ventricles**, are lined by **endothelium** (which is a kind of epithelial tissue; endothelium is characteristic of the cardiovascular system).

27. (a) **Incorrect. The interatrial septum and the interventricular septum do this.** (See text Figure 13-13.)
    (b) **Incorrect. They prevent flow from the ventricles back to the atria.**
    (c) **Incorrect. The pulmonary valve does the former, and the aortic valve does the latter.**

28. **True.**

29. (a) **Correct.**
    (b) **Correct.**
    (c) **Correct.**
    (d) **Correct.**

30. **False.** The conducting system of the heart is a ~~branch of the vagus nerve~~ **network of specialized cardiac-muscle fibers.**

31. **False.** The myocardial cells receive their blood supply from the ~~cardiac~~ **coronary** arteries that branch off from the aorta. (The blood supply of the heart is named "coronary" after the Latin word for *crown*, or *something that encircles*. The coronary circulation encircles the heart.)

32. (a) **Correct.**
    (b) **Incorrect. In both, the resting membrane potential is closer to the potassium equilibrium potential.**
    (c) **Incorrect. In the cardiac cells, potassium permeability *decreases* almost tenfold as the sodium channels close.**
    (d) **Correct.**
    (e) **Correct.**
    (f) **Correct.**
    (g) **Incorrect. They are propelled out of the cell along both their concentration gradient and the electrical gradient.** (Permeability to potassium ion does not fall to zero.)
    (h) **Incorrect. During the initial depolarization, calcium ions cannot enter because of the delay in opening of their slow channels.**

33. **True.**

34. **False**. The ~~atrioventricular~~ **sinoatrial** node is normally the heart's pacemaker because it has the fastest spontaneous rate of depolarization..

35. **d, b, a, e, f, c**

36. **True.** (Compare text Figures 13-19 and 13-13. The blood exits the ventricles upward, into the pulmonary trunk and the aorta.)

37. (a) **Incorrect. See answer (b).**
    (b) **Correct.**
    (c) **Correct.**

38. In the ECG, the P wave corresponds to **atrial depolarization**, the QRS complex to **ventricular depolarization**, and the T wave to **ventricular repolarization**. The electrical event that is not seen is **atrial repolarization.** (Because it normally occurs at about the same time as ventricular depolarization, atrial repolarization is lost in the QRS complex.)
    Note: Do not be confused by the fact that the T wave, which signals ventricular repolarization, is an *upward* deflection on the ECG. (Again, the ECG is not a recording of potential differences between the inside and outside of cells, but rather an indirect measure of the flow of cardiac electrical activity.)

39. (a) **Correct.**
    (b) **Incorrect. In cardiac muscle, the stimulus is the increase in intracellular calcium through slow channels, not an action potential in T tubules.**
    (c) **Incorrect. In cardiac cells, there is ordinarily not enough calcium released to bind all the troponin.** (This fact offers a means of regulating the strength of cardiac muscle contraction.)
    (d) **Correct.**

40. **Tetanus in a skeletal muscle is a prolonged contraction made possible when an action potential stimulates the muscle to contract while it is still generating tension as a result of a previous contraction. The tensions add together to create a summation — tetanus. In cardiac muscle, the absolute refractory period lasts almost as long as the (first) contraction, and so summation of tensions is almost impossible.**

41. **False.** An ectopic focus is an area of the myocardial conducting system other than the ~~atrioventricular~~ **sinoatrial** node that initiates cardiac depolarization.

42. During the cardiac cycle, the period of time when the ventricles are contracting and blood is being forced into the pulmonary trunk and aorta is called **systole**. The period of ventricular relaxation and filling with blood is called **diastole**. In between these events are brief periods in which ventricular volume does not change, the **isovolumetric** ventricular contraction and relaxation. Normally the period of **diastole** lasts the longest.

43. **True.**

44. **False.** The only time during the cardiac cycle that the AV valves are open is ~~systole~~ **diastole**.

45. **True.**

46. **True.** (Note that, in the early printings of your text, there is an error in Figure 13-25. In the top left heart, the one labeled "ISOMETRIC VENTRICULAR CONTRACTION," the arrow pointing toward the aortic valve should point in the opposite direction:

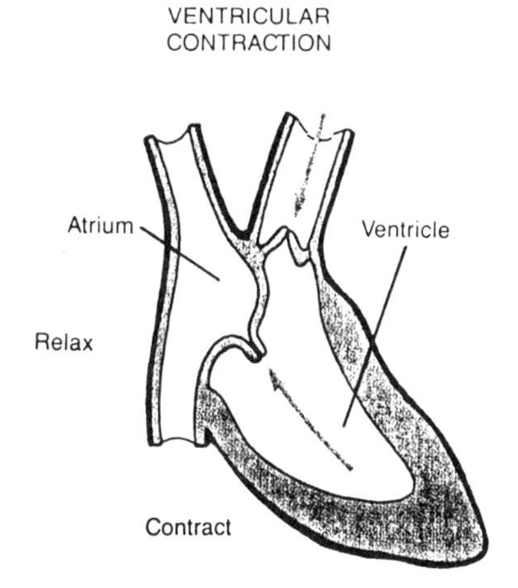

AV valve:	Closed	Correct drawing for isovolumetric ventricular contraction in text Figure 13-25.
Aortic and pulmonary valves:	Closed	

You should correct your text if yours is one of the early, incorrect printings.)

47. The difference between the end-diastolic volume and the end-systolic volume in the ventricle is the **stroke volume** — the volume of blood ejected during **systole**.

48. **False.** The left ventricle ejects ~~a larger~~ **the same** volume of blood during systole ~~than~~ **as** does the right ventricle ~~because~~ **even though** the left ventricular pressure is greater than the right. (This is a case of the *flow* of blood in two different parts of the circulatory system being the same despite disparities in pressure differences. Therefore, you may correctly surmise that there must be differences in the *resistance* of the two systems.)

49. **False.** Normal heart sounds are produced by the slamming shut of (a) the AV valves at the end of ~~systole~~ **diastole** and (b) the aortic and pulmonary valves at the end of ~~diastole~~ **systole**.

50. Abnormal heart sounds called **murmurs** may indicate heart disease, particularly problems with the valves. For example, aortic stenosis, which causes a murmur during **systole**, is a result of an abnormal **narrowing of the aortic valve.** Left AV

insufficiency, which also causes a murmur during **systole**, is caused by **backward flow of blood from ventricle to atrium because of a leaky valve**.

51. (a) **Incorrect. It is the volume pumped by each ventricle, not both together.**
    (b) **Correct.** (or through the pulmonary circulation)
    (c) **Correct.** (CO = HR x SV)

52. Average adult at rest: 72 beats/min x 0.07 L/beat = 5.0 L/min (from text page 375)

    Athlete's output relative to average adult's: (35 L/min)/(5 L/min) = 7-fold increase
    7 x 72 = 504 beats/min
    Refractory period of ventricular muscle = 250 ms = 0.25 s
    Maximum number of beats/s = 4 (since each one requires 0.25 s)
    Maximum number of beats/min = 4 x 60 = 240

    Clearly, stroke volume must increase as well as heart rate, since a heart rate of 504 beats/min is impossible. (In addition, at very high heart rates there is too little time to fill the ventricles and the pump becomes very inefficient.)

53. **False.** The inherent pacemaker discharge rate of the sinoatrial node is ~~72~~ **100** beats per minute.

54. (a) **Correct.**
    (b) **Correct.**
    (c) **Incorrect. It is decreased by acetylcholine** (the neurotransmitter of the parasympathetic postganglionic fibers).
    (d) **Incorrect. It is increased by epinephrine.**
    (e) **Correct.**

55. **True.**

56. **False.** During stimulation of the ~~para~~**sympathetic** nerves, plasma levels of epinephrine increase.

57. **True.**

58. (a) **Incorrect. See (b) and (c).**
    (b) **Correct.**
    (c) **Correct.**

59. **True.**

60. In a heart-lung preparation (text Figure 13-31), the end-diastolic volume can be increased by raising the reservoir and thus increasing the **ventricular filling pressure**. In an intact person or animal, end-diastolic volume is enhanced by increasing **stroke volume**.

61. (a) **Incorrect. See (b).**
 (b) **Correct.**
 (c) **Incorrect. It is decreased** (slightly) **by parasympathetic stimulation.**
 (d) **Correct.**

62. (a) **Incorrect. They are mediated through beta-adrenergic receptors.**
 (b) **Incorrect. They are mediated by cyclic AMP.**
 (c) **Correct.**

63. (a) **Correct.**
 (b) **Incorrect. Active transport of calcium into the reticulum is increased.**
 (c) **Correct.** (Since more calcium is available to bind.)
 (d) **Correct.**

64. **Resting heart rate would be about 100 beats/min (the spontaneous rate of the SA node). Yes, the person would still respond to circulating epinephrine with increased heart rate and increased contractility, and so cardiac output could be increased.** (Venous return, and thus end-diastolic volume, would also vary with different conditions to affect cardiac output, as we shall see.)

65. (a) **Correct.**
 (b) **Incorrect. They have many specialized functions, as can be seen in text Table 13-5.**

66. (a) **Incorrect. They are thick-walled, low-resistance tubes.**
 (b) **Correct.**
 (c) **Correct.**
 (d) **Incorrect. They "store" pressure during systole to drive blood flow through the rest of the blood vessels during diastole.**

67. **False.** Systolic pressure is the peak pressure in systemic arteries and occurs ~~at the beginning~~ **in the middle** of systole.

68. (a) **Correct.**
 (b) **Correct.**
 (c) **Incorrect. It increases with increasing stroke volume.**
 (d) **Correct.** (A very high pulse pressure indicates decreased compliance of the arterial walls. A very low pulse pressure might indicate cardiac failure. When a doctor takes your pulse, she or he checks both its rate and its strength.)

69. (a) pulse pressure (PP) = SP - DP
= 120 mm Hg - 80 mm Hg
= **40 mm Hg**

(b) mean arterial pressure (MAP) = DP + PP/3
= 80 mm Hg + (40/3) mm Hg
= 80 mm Hg + 13.3 mm Hg
= **93 mm Hg** (approximately)

70. (a) **Incorrect. The device takes advantage of sounds made by turbulent flow of blood through constricted arteries.**

(b) **Correct.**

(c) **Incorrect. Diastolic pressure is just slightly greater than the pressure on the manometer.**

71. **False.** The flow rate of blood through an organ is determined by the relative dilation or constriction of the walls of the organ's ~~small arteries~~ **arterioles.**

72. **True.**

73. The increase in blood flow through an organ that is actively metabolizing is termed **active hyperemia**. This change in blood flow is a form of **autoregulation** that is independent of nerves and hormones. The chemical stimuli that produce the **vasodilation** include decreased **oxygen concentrations**, increased levels of **carbon dioxide**, **hydrogen ion**, **metabolites**, **eicosanoids**, **extracellular potassium**, **bradykinin**, and increased **osmolarity**. These same chemical signals also play a role in regulating blood flow in the face of increased or decreased blood pressure, termed **pressure autoregulation**. The phenomenon of reactive hyperemia is an example of an extreme form of **pressure autoregulation**. (Note the ending "-emia" again. "Hyperemia" means "increased flow of blood.")

74. (a) **Incorrect. Very few arteriolar beds receive parasympathetic innervation.**

(b) **Correct.**

(c) **Incorrect. Norepinephrine from sympathetic nerves does this.** Note that circulating epinephrine does interact with alpha-adrenergic receptors to induce vasoconstriction (*any* chemical that activates those receptors would cause vasoconstriction) but it can also act on specific ß-receptors on the same cells to stimulate vasodilation. As the text mentions, these ß-receptors are of little importance except in the arteriolar beds supplying blood to skeletal muscle.

75. **False.** Because sympathetic stimulation of the arterioles causes vasoconstriction, **inhibition of** the sympathetic nerves play**s** ~~no~~ **an important** role in reflex vasodilation.

76. (a) **Correct.**
    (b) **Correct.**
    (c) **Correct.**

77. (a) **the brain and the kidneys** (Sympathetic control is also very important for the kidneys, but not for the brain.)
    (b) **the heart**
    (c) **skeletal muscle during exercise and the abdominal organs following ingestion of a meal**

78. **True.** (The capillaries also have the greatest total cross-sectional area, as you will see.)

79. **True.** ("Angio-" refers to blood vessels, and "genesis" to new beginnings. Capillary beds have to either grow or regress with changes in the body. Gaining weight in the form of stored fat or increasing muscle mass through exercise requires stimulating angiogenesis. Losing weight requires blood vessel regression, or shrinkage, the opposite of angiogenesis.)

80. **False.** ~~Because~~ **Even though** capillaries have no smooth muscle, flow of blood through capillary beds ~~cannot~~ **can** be controlled **by contraction or relaxation of precapillary sphincters**.

81. In the systemic circulation, the greatest resistance to flow is found in the **arterioles**, with the second greatest in the **capillaries**. The greatest cross-sectional area is in the **capillaries**. There is an **inverse** correlation between flow rate and cross-sectional area, such that the highest flow rate is in the **arteries** and the lowest flow rate is in the **capillaries**. (Which makes perfect sense because the blood is merely travelling in the arteries; it is "doing work" in the capillaries.)

82. (a) **Incorrect. Nonpolar** (lipid-soluble) **molecules diffuse through these membranes.**
    (b) **Correct.**
    (c) **Incorrect. Liver capillaries are among the leakiest.**
    (d) **Correct.**

83. **False.** The concentrations of molecules in the intracellular fluid are the same as those in blood plasma **except for the high concentration of proteins in the plasma**.

84. **True.**

85. **False.** The osmotic pressure difference between the plasma inside the capillary beds and the interstitial fluid surrounding the cells favors ~~filtration~~ **absorption** of fluid. (The terms "filtration" and "absorption" refer to capillary events — fluid is filtered *out of* and absorbed *into* the capillary.)

86. Crystalloids in plasma and interstitial fluid are **penetrating** solutes, and colloids are **nonpenetrating** solutes. During filtration (that is, movement of fluid out of the capillaries), the concentration of **crystalloids** stays the same in both fluid compartments.

87. (a) **Correct.** Net hydrostatic pressure exceeds net osmotic pressure due to proteins.
    (b) **Incorrect. Normally, the net filtration pressure favors absorption into the capillary bed at the venous end.**
    (c) **Correct.**

88. **False.** The amount of nutrients and waste products moved across the capillary wall by bulk flow is ~~greater~~ **much less** than the amount that diffuses across under normal conditions.

89. **False.** Strong vasoconstriction of arterioles leading to a capillary bed will probably result in ~~loss of plasma by filtration~~ **absorption of fluid into** the capillary bed.

90. **True.** (The net osmotic force due to proteins in the plasma and interstitial fluid is about the same as in other beds, that is, about 25 mm Hg favoring absorption of liquid into the capillaries. As we shall see in the next chapter, lung tissues cannot function efficiently when there is fluid accumulated around them.)

91. **False.** ~~Unl~~**Like** arterioles, venules have ~~no~~ smooth muscle in their walls. (But they have considerably less than do the arterioles.)

92. (a) **Correct.**
    (b) **Correct.**
    (c) **Incorrect. They are innervated by the sympathetic nerves.**

93. (a) **Correct.**
    (b) **Correct.**
    (c) **Correct.**
    (d) **Correct.**

94. **True.**

95. (a) **Incorrect. It returns filtered fluid back to the cardiovascular system, but it is not part of the system.**
    (b) **Incorrect. The nodes are not pumps.** (Lymph is pumped by the skeletal muscle and respiratory pumps much as blood in the veins is pumped. In addition, smooth muscle in the lymphatic vessels contracts to help pump the lymph.)
    (c) **Correct.**
    (d) **Correct.**
    (e) **Correct.**

96. **True.** (This is one of *the* take-home messages of this chapter. When in doubt, ask yourself how arterial blood pressure might be affected by whatever change in the cardiovascular system you are being asked to analyze, and then — most important — ask yourself how the body's homeostatic reflexes can address the problem.)

97. **$\Delta P = FR$**
ΔP = mean aortic pressure
F = cardiac output
R = total peripheral resistance
(Memorize this answer. It is very important for your understanding of how the cardiovascular system can compensate for different physiological conditions — that is, how it can achieve homeostasis.)

98. **The brain must receive adequate oxygen and nutrients or else it will stop functioning, and the whole organism will die. Thus, a matter of paramount importance is to keep the driving force of the blood (the mean arterial pressure) high enough to ensure adequate blood flow to the brain (against gravity).** (Of course, it is possible for blood pressure to be too high and cause blood vessels in the brain to rupture. Thus, the pressure must be maintained within rather narrow limits.)

99. **True.** (This fact makes very good sense in light of the discussion of Question 98.)

100. Receptors sensitive to blood pressure in the walls of arteries are special types of mechanoreceptors called **baroreceptors**. They are located in the **carotid sinus** and the **aortic arch**. Together these mechanoreceptors are known as the **arterial baroreceptors**. Analogous receptors are also found in the **walls of the heart**, the **large veins**, and the **pulmonary vessels**.

101. **False.** Increasing the mean arterial pressure or the pulse pressure in the carotid artery will cause ~~decreased~~ **increased** firing of action potentials in the afferent fibers associated with the baroreceptors.

102. The integrating center that receives information about blood pressure and volume and then coordinates the responses of effector tissues is **the medullary cardiovascular center** in the brainstem. When arterial or other baroreceptors increase their firing rate to this center, it **decreases** the activity of sympathetic nerves to the **heart**, **arterioles**, and **veins** and **increases** the activity of the parasympathetic nerves to the **heart**. These effects combine to **decrease** cardiac output, **decrease** total peripheral resistance, and, as a result, **decrease** mean arterial blood pressure.

103. **True.** (Text Figure 13-63 shows the relationship between arterial pressure and blood volume. We shall see in Chapter 15 how the kidneys and the nerves and hormones that act on them fit into the arterial pressure/blood volume feedback loops — in other words, the details of the kidneys' blue boxes in Figure 13-63.

104. **False. Higher brain centers** (including the hypothalamus), **as well as** the medullary cardiovascular center, ~~is the only part of the brain~~ **are** involved with the regulation of blood pressure.

105. (a) **Incorrect. It refers to low blood pressure.**
(b) **Correct.**
(c) **Incorrect. It causes decreased firing of the baroreceptors.**

106. (a) **D** (g) **D**
(b) **I** (h) **I**
(c) **I** (i) **D**
(d) **D** (j) **D**
(e) **D** (k) **I**
(f) **D** (l) **NC** (or small D)

Some of these answers may not seem intuitively obvious, but instead require careful thought. Keep in mind that the reflexes tend to restore mean arterial pressure (and plasma volume) toward pre-hemorrhage levels but do not fully succeed for many hours. Thus, in (d), venous return will be decreased following hemorrhage even though sympathetic constriction of the veins will drive a higher percentage of the blood in the veins toward the heart (j), and decreased filtration pressure (f) in the capillary beds (a result of the decreased mean arterial pressure and the increased arteriolar vasoconstriction) will increase the absorption of fluid there. Eventually these mechanisms will compensate completely, but during compensation venous return will remain lower than normal.

The answer to (l) reflects the fact that brain arterioles are not constricted by increased sympathetic activity, and blood flow to the brain will not be impaired unless the hemorrhage is severe. (Recall that the brain uses pressure autoregulation as a means of regulating its own blood flow. When arterial blood pressure falls and brain blood flow decreases momentarily, oxygen levels will fall and concentrations of carbon dioxide and metabolites will increase, causing vasodilation of the brain arterioles and compensating for moderate declines in arterial pressure.)

107. **True.** (As the text explains, in cases of prolonged shock the homeostatic negative feedback responses become ever more damaging positive feedback cycles.)

108. (a) **Correct.**
(b) **Correct.** (The column of blood in the arms is about half the length of the column in the legs.)
(c) **Correct.**

109. (a) **Correct.**
(b) **Incorrect. Blood flow to the skin increases.**
(c) **Correct.**

110. **False.** During exercise, sympathetic stimulation of the heart and arteriolar smooth muscle is enhanced ~~because of reflexes triggered by elevated~~ **even though** mean arterial pressure and pulse pressure **are elevated**.

111. **True.**

112. (a) **Incorrect. They dilate because of active hyperemia.**
   (b) **Incorrect. Blood flow increases because increased body temperature inhibits sympathetic nerves supplying the arterioles in the skin.**
   (c) **Incorrect. The blood flow to the brain does not change because autoregulation causes compensatory vasoconstriction of brain arterioles.**

113. (a) **Correct.**
   (b) **Correct.**
   (c) **Correct.**
   (d) **Incorrect. The maximal heart rate is unchanged by exercise.**
   (e) **Incorrect. It decreases one's resting heart rate.**

114. (a) **Correct.**
   (b) **Incorrect. It is defined as systolic greater than 140 mm Hg and diastolic greater than 90 mm Hg.** (The diastolic value is usually considered the more important.)
   (c) **Incorrect. It is a result of chronically increased arteriolar vasoconstriction.**
   (d) **Incorrect. Kidney disease is the cause of less than 5 percent of cases. The cause of most hypertension is unknown.**
   (e) **Correct.** (Diuretics cause increased excretion of water and sodium, which effectively decreases blood volume, and thus pressure.)

115. (a) **No, this drug would not be useful. Increasing cardiac output would worsen the elevated blood pressure.**
   (b) **Yes, this drug would be useful. Decreased O_2 is one signal for local vasodilation of arterioles, and a drug that mimicked this effect would cause a decrease in peripheral resistance.** (ß-blockers are thought to act this way. ß-blockers also inhibit contractility of the heart to some extent.)

116. **False.** A failing heart does not respond **as strongly as a normal heart** to increased end-diastolic volume by increasing stroke volume. (The failing heart has a lower Starling curve, as shown in text Figure 13-71.)

117. **False.** A serious consequence of failure of the left ventricle is **pulmonary** edema ~~in the legs and feet.~~

118. The medical term for heart attack is "**myocardial infarction**." The immediate cause of heart attack is insufficient blood flow to the **coronary circulation** and subsequent

damage to the myocardium. Sudden death from heart attack is usually a result of **ventricular fibrillation**, in which contractions of ventricular cells are not coordinated. Death can be averted if CPR — which stands for **cardiopulmonary resuscitation** — is begun right away.

119. (a) **Correct.**
    (b) **Incorrect. It is a thickening in the walls of arteries.**
    (c) **Incorrect. It is associated with *high* blood levels of cholesterol, obesity, and cigarette smoking.**
    (d) **Correct.**
    (e) **Correct.**

120. The occasional pain experienced during exertion or anxiety by someone with atherosclerosis of a coronary artery is called **angina pectoris**. One new treatment for this problem is **coronary angioplasty**, the enlargement of a vessel by threading a balloon on a catheter into the occluded vessel.

PRACTICE

True or false (correct the false statements):

P1. A person with a low hematocrit and macrocytosis is probably suffering from iron deficiency.

P2. An anemic person is likely to have higher than normal plasma levels of erythropoietin.

P3. The sinoatrial node is normally the pacemaker of the heart.

P4. The QRS wave of the ECG corresponds to depolarization of the atria.

P5. An electrocardiogram would be useful for diagnosing diseases involving the valves of the heart.

P6. Cardiac muscle cannot undergo tetanus because its absolute refractory period lasts almost as long as the muscle twitch.

P7. The left ventricle has to pump more blood than the right ventricle because the left has to pump blood through the whole body, not just the pulmonary system.

P8. The left ventricle has a thicker wall than the right ventricle because the left has to pump blood against a higher pressure.

P9. Stimulation of the vagus nerve lowers the heart rate.

P10. Starling's law of the heart states that an increased venous return will normally result in a higher heart rate.

P11. The stroke volume of the heart can be increased by recruiting more cardiac muscle fibers into activity.

P12. Mean arterial pressure can be estimated by determining the average of the systolic and diastolic pressures.

P13. Valves are important for promoting one-way flow of blood through the heart and veins.

P14. Total peripheral resistance in the circulatory system is primarily determined by the degree of vasoconstriction in the veins.

P15. An athlete who has just run 2 kilometers would be expected to have both an increased mean arterial blood pressure and increased total peripheral resistance.

P16. Hypertension is often treated with drugs that increase total peripheral resistance.

P17. Left ventricular failure often results in increased net pulmonary capillary filtration into the interstitial fluid of the lungs and a resulting decrease in the diffusional exchange of O_2 and CO_2 between lung and blood.

P18. Increased concentrations of plasma proteins, increased venous pressure, and lymphatic obstruction all tend to cause tissue edema.

Multiple choice (correct each incorrect choice):

P19-21. Consider these measurements for two individuals:

	A	B
heart rate (beats/min)	70	100
stroke volume (mL/beat)	60	90
mean arterial pressure (mm Hg)	100	100

P19. Cardiac output is greater in A.

P20. Total peripheral resistance is greater in A.

P21. Coronary blood flow is greater in B.

P22-26. Following a mild hemorrhage, there will be a decrease in blood volume and in arterial pressure. Physiological compensation for this would include:

P22. increased frequency of action potentials from baroreceptors in the aorta and carotid sinus.

P23. decreased frequency of action potentials in the sympathetic nerves to the heart.

P24. increased heart rate and decreased peripheral resistance.

P25. increased contractility of cardiac muscle due to decreased stimulation of parasympathetic nerves.

P26. increased vasoconstriction of peripheral arterioles.

Fill in the blanks:

P27-29. In tissue capillaries, most net movement of nutrients from the blood occurs by the process of ______________________________. Fluid movement occurs in response to differences in ______________________________ pressure and ______________________________ pressure between the capillaries and interstitial fluid.

P30-35. Indicate whether blood flow during exercise increases (I), decreases (D), or stays the same (NC):

P30.____ brain P31.____ heart P32.____ skeletal muscles

P33.____ kidneys P34.____ stomach P35.____ skin

14: RESPIRATION

Breathing. We breathe in and out, on the average, about 720 times an hour — 17,280 times a day — without giving the process any thought. Only when something goes wrong — an asthma attack, a very bad chest cold — do we pay much attention to our lungs and the movements of our chest walls that permit breathing. We don't even worry about breathing when we go to sleep, knowing that somehow we will continue to breathe all through the night.

At least, most of us do not worry about breathing while we sleep. About one out of every hundred men (this is another problem affecting men disproportionately) experiences sleep *apnea* — the cessation of breathing while sleeping. Every minute or two, the apnea patient must wake up briefly in order to breathe, disturbing his rest. For some, the problem means going through days in a sleepy stupor; for others, the condition can be fatal because of cardiac arrhythmias that sometimes accompany the apnea.

Sleep apnea is thought to strike the newborn as well. It is the best candidate for causing "sudden-infant-death syndrome" — SIDS — where a baby dies in its sleep for no apparent reason. SIDS kills some seven thousand infants each year in the United States and is the single biggest killer of children aged one month to one year.

Today, researchers are exploring the relationship between sleep and breathing, and are knocking on the door of the understanding that is the first step for treating and (we hope) curing this dreadful malady.

In this chapter you will be introduced to the respiratory system and gain some understanding about what it is that happens when we breathe.

Instructions for answering questions in this Study Guide

1. True or false: Correct each false statement. Whenever you must correct a false statement, there will almost always be more than one way to do so. In the Guide, the answer requiring the least correction will generally be given, with the understanding that other correct answers are possible. It is usually not sufficient, in terms of demonstrating understanding, to simply insert a "no" or "not," however.

2. Multiple choice: Any, all, or none of the responses may be correct. Explain why each incorrect response is incorrect.

3. Fill in the blanks: Choose the best word or words to complete each statement.

4. Directions will be given for other types of questions as they appear.

❐ Introduction (text pages 427-428)

1. Functions of the respiratory system include
   (a) exchange of carbon dioxide from the air with oxygen from the blood.
   (b) regulation of H^+ concentration.
   (c) secretion of hormones.
   (d) phonation.

❐ Organization of the Respiratory System (text pages 428-433)

2. The respiratory cycle
   (a) consists of inspiration and expiration.
   (b) consists of inhalation and exhalation.
   (c) is timed such that inspiration coincides with cardiac diastole and expiration coincides with cardiac systole.

3. The sites for gas exchange with the blood are the ____________________. Most of these structures form clusters of ________________________ that are found at the ends of airway passages called ________________________. The larger airway passages from which the smaller ones branch are called ___________________ and are supported by the _________________ in their walls. These passages in turn branch from the tube beyond the larynx called the ___________________. The airways are divided into two functional zones, the _________________________ zone, across which gas exchange occurs, and the ________________________ zone, where no gas exchange occurs.

4. True or false: The blood vessels supplying the lungs are the pleural vessels.

5. Cells that contribute to the defense mechanisms of the lungs include
   (a) ciliated epithelial cells.
   (b) mucus-secreting cells.
   (c) phagocytes.
   (d) bacteria.

6. True or false: The upper airways of the respiratory system include the nose, mouth, pharynx, and esophagus.

7. In the alveoli
   (a) air mixes with blood.
   (b) O_2 and CO_2 are exchanged by bulk flow.
   (c) there are two kinds of epithelial cells.

8. True or false: The diaphragm is a large sheet of smooth muscle that separates the thorax from the abdomen.

9. Regarding the relationship of the lungs and the pleura,
   (a) each lung is covered by a sheet of pleural sac.
   (b) the outer layer of each pleural sac is attached to the chest wall and the diaphragm.
   (c) intrapleural fluid lies between the lungs and the pleura.

10. True or false: If the parietal pleura is cut during surgery, the lung on the side of the cut will expand and the chest wall will be compressed.

11. At the end of a forced expiration
   (a) the pressure of the intrapleural fluid is the same as atmospheric pressure.
   (b) the alveolar pressure is less than atmospheric pressure.
   (c) the transpulmonary pressure stretches the lungs.
   (d) the tendency for the lungs to recoil to an unstretched state is balanced by the tendency of the chest wall to expand.

❒ **Ventilation and Lung Mechanics** (text pages 433-442)

12. True or false: The process of ventilation includes both the transport of gases by bulk flow and the diffusion of gases into and out of the blood.

13. True or false: According to Boyle's law, a gas will expand to fill the volume available to it.

14. Write the bulk flow equation for ventilation.

15. The volume of air flowing into or out of the alveoli is increased when there is an increase in
    (a) airway resistance.
    (b) the pressure gradient from the atmosphere to the alveoli.
    (c) the diameter of the airways.

16. Air will flow into the lungs
    (a) when P_{alv} is greater than P_{atm}.
    (b) when the lungs expand.
    (c) when transpleural pressure becomes greater than 4 mm Hg.

17. True or false: The first step in inspiration is the enlargement of the thoracic cage.

18. True or false: During inspiration, the total alveolar volume decreases.

19. True or false: Expiration of air from the lungs requires contraction of the expiratory intercostal muscles.

20. A person with abnormally low lung compliance
    (a) must work harder than a normal person to inspire the same amount of air.
    (b) must have a greater than normal intrapleural pressure to inspire the same amount of air.
    (c) may have thicker lung tissues than a normal person.
    (d) may have greater alveolar surface tension than a normal person.
    (e) may have a defect in type II alveolar cells.

21. Low lung compliance can become part of a vicious circle of positive feedback that makes the condition worse. Use information from text Table 14-3 and the description of breathing in someone with low lung compliance to explain this statement.

    ____

    ____

    ____

    ____

22. True or false: Premature infants frequently develop respiratory distress syndrome of the newborn because they have too few alveoli to exchange O_2 and CO_2 efficiently.

23. Babies born with congenital defects of their adrenal glands often have lower than normal lung compliance. Explain.

____

____

____

____

24. True or false: The diameter of the airways in normal lungs is generally great enough that little resistance is offered to air flow.

25. Smooth muscle cells in the walls of airways leading to the alveoli contract in response to
    (a) epinephrine.
    (b) parasympathetic stimulation.
    (c) noxious stimuli.
    (d) increased CO_2 in the alveoli.
    (e) histamine.

26. True or false: Cigarette smoking is commonly associated with increased incidence of asthma.

27. During an asthma attack breathing is labored, with expiration relatively harder to perform than inspiration. Explain.

____

____

____

____

28. True or false: During the Heimlich maneuver, sudden force against the diaphragm causes a sudden decrease in alveolar pressure, forceful expiration, and expulsion of the object blocking the airways.

29. True or false: The total lung capacity in a normal person is usually roughly the same as the person's total blood volume.

30. Match the lung volume terms on the right with the correct description on the left.

____ volume inspired at rest
____ maximal volume that can be inspired and expired
____ vital capacity + residual volume
____ maximal expiration after resting expiration
____ volume remaining in lungs after maximal expiration
____ maximal volume that can be inspired
____ volume remaining in lung after resting expiration

(a) vital capacity
(b) residual volume
(c) inspiratory reserve volume
(d) resting tidal volume
(e) expiratory reserve volume
(f) total lung capacity
(g) inspiratory capacity
(h) forced vital capacity
(i) functional residual capacity

31. True or false: A clinical sign of restrictive lung disease would be a reduced FEV_1/FVC ratio.

32. The respiratory value analogous to cardiac output is the ____________________. It is determined by multiplying the ____________________ by the ____________________.

33. The anatomic dead space is
(a) about 1 liter in a normal person.
(b) the volume of the conducting airways.
(c) called that because the tissues surrounding it are no longer living.

34. A healthy young man has a total lung capacity of 6 L, a residual volume of 1 L, and an anatomic dead space of 150 mL. Which of the following statements is/are correct?
(a) With a tidal volume of 500 mL and a breathing rate of 10 breaths/minute, his minute volume would be 5 L/minute.
(b) Doubling the breathing rate of (a) would double his minute volume.
(c) With parameters as in (a), his alveolar ventilation would be 5 L/minute.
(d) Doubling the depth of respiration would double his alveolar ventilation.
(e) Following a maximal expiration, a maximal inspiration would bring 5850 mL of fresh air into the alveoli.

35. True or false: The total dead space in the lungs includes anatomic dead space and any portion of the alveoli obstructed from the air flow.

❐ Exchange of Gases in Alveoli and Tissues (text pages 442-448)

36. True or false: In a steady state, the rate of CO_2 production in the cells exceeds the rate of CO_2 expiration by the lungs.

37. True or false: In a person consuming a normal diet of carbohydrates, proteins, and fats, the production of CO_2 exceeds the consumption of O_2.

38. True or false: At sea level, P_{O_2} is about 760 mm Hg.

39. True or false: In the systemic circulation, the P_{O_2} of arterial blood exceeds the P_{O_2} of venous blood, while the reverse is true for the pulmonary circulation.

40. The concentration of a gas in a liquid is
    (a) directly proportional to the partial pressure of the gas in the air around the liquid.
    (b) directly proportional to the solubility of the gas in the liquid.
    (c) expressed as the partial pressure of the gas in the liquid.

41. Regarding the partial pressures of O_2 and CO_2, which of the following statements is/are true in a normal person at rest?
    (a) P_{O_2} of the air is greater than alveolar P_{O_2}.
    (b) P_{CO_2} of the air is greater than alveolar P_{CO_2}.
    (c) Alveolar P_{O_2} is greater than the P_{O_2} in the pulmonary capillaries.
    (d) P_{CO_2} in the systemic veins exceeds venous P_{O_2}.
    (e) P_{CO_2} in the systemic arteries is negligible.

42. Regarding the relationships between the rates of oxygen consumption/carbon dioxide production and alveolar ventilation,
    (a) increased ventilation without a similar increase in oxygen consumption is called hyperpnea.
    (b) in hypoventilation, alveolar P_{CO_2} increases above resting levels.
    (c) in hyperventilation, alveolar P_{O_2} can fall to zero.
    (d) in hyperpnea, resting values of alveolar P_{O_2} and P_{CO_2} are maintained.

43. True or false: During rest, many of the pulmonary capillaries are closed, and so they cannot participate in alveolar-blood gas exchange.

44. Ventilation-perfusion inequality
    (a) does not exist in normal lungs.
    (b) may result from increased alveolar dead space.
    (c) is associated with diseases such as emphysema.
    (d) is caused by too little or too much blood flow relative to ventilation.
    (e) is caused by too little or too much ventilation relative to blood flow.

45. True or false: Decreased alveolar P_{O_2} causes reflex vasodilation of the pulmonary arterioles.

❒ Transport of Oxygen in Blood (text pages 449-452)

46. True or false: Most of the O_2 carried in the blood is dissolved in plasma.

47. Hemoglobin (Hb) can exist in a form that has O_2 bound to it, called ____________________, and also in an unbound form, called ____________________. The percentage of the bound form to the total Hb is called the percent ____________________ of Hb. The most important factor determining this percentage is the ____________________.

48. According to the oxygen-hemoglobin dissociation curve,
    (a) the greater the P_{O_2} of the blood, the greater the dissociation of O_2 from hemoglobin.
    (b) at normal systemic arterial P_{O_2}, hemoglobin is almost 100% saturated.
    (c) at normal venous P_{O_2}, only about 25% of the hemoglobin is in the form of deoxyhemoglobin.
    (d) if lung disease results in an arterial P_{O_2} of only 80 mm Hg, the ability of the blood to carry O_2 to the tissues is severely compromised.

49. True or false: In the systemic capillaries, the P_{O_2} of the erythrocytes is greater than the P_{O_2} of the plasma, causing a shift from oxyhemoglobin to deoxyhemoglobin.

50. True or false: If a tissue increases its metabolism enough to drop its interstitial P_{O_2} to half the resting venous value, the amount of O_2 available from oxyhemoglobin will be more than twice that for the resting tissue.

51. Curve B represents the oxygen-hemoglobin dissociation curve for normal body temperature (37°), arterial hydrogen ion concentration, and 2,3-DPG concentration.

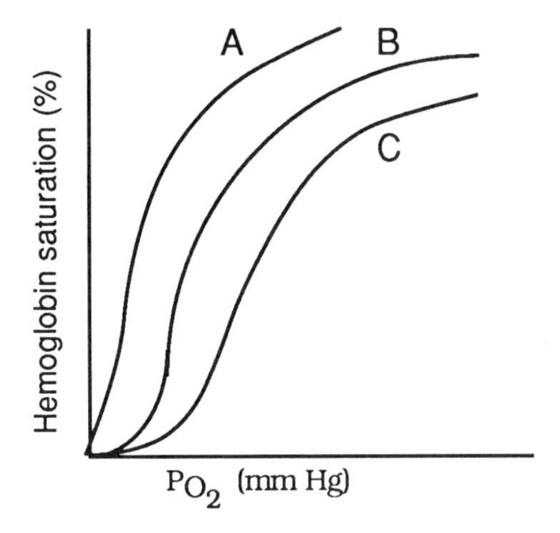

(a) Curve A may represent the dissociation at higher than normal body temperature.
(b) Curve A may represent the dissociation at very low 2,3- DPG levels.
(c) Curve C may represent the dissociation at lower than normal acidity.

52. True or false: The affinity of hemoglobin for O_2 is decreased in rapidly metabolizing tissues.

❒ **Transport of Carbon Dioxide** (text pages 452-454)

53. True or false: Unlike O_2, more of the CO_2 in blood is dissolved in plasma than is bound to hemoglobin.

54. Carbonic anhydrase
(a) catalyzes the conversion of carbonic acid to bicarbonate and hydrogen ions.
(b) is found in erythrocytes.
(c) is found in plasma.

55. Bicarbonate ion
    (a) carries most of the CO_2 in arterial and venous blood.
    (b) is dissolved in plasma.
    (c) is exchanged for chloride ion by erythrocytes in the lung capillaries.

56. True or false: Deoxyhemoglobin binds bicarbonate ion better than oxyhemoglobin does.

❐ **Transport of Hydrogen Ions Between Tissues and Lungs** (text page 454)

57. True or false: The majority of hydrogen ions generated in the formation of carbonic acid are carried dissolved in the plasma.

58. True or false: When a person hyperventilates, there are lower than normal levels of P_{CO_2} and hydrogen ion in arterial blood, a condition called respiratory acidosis.

❐ **Control of Respiration** (text pages 455-464)

59. True or false: The rhythmic contractions and relaxations of the inspiratory muscles are dependent upon spontaneous depolarizations of the diaphragm.

60. True or false: The neurons responsible for the cyclic nature of respiratory-muscle function are located in the brainstem.

61. True or false: The medullary inspiratory neurons receive inhibitory neural input from the pons and also from pulmonary stretch receptors.

62. The chemoreceptors important for controlling ventilation comprise two classes. The first, called ______________________ chemoreceptors, are sensitive to decreased ____________________ and increased ____________________ and ____________________ in blood. They are located in the ____ and ____. The other class, the

________________ chemoreceptors, are exquisitely sensitive to ________________ in the ________________. They are located in the ________________.

63. Will any of the following cause a significant increase in ventilation? Explain.

    (a) A decrease in inspired P_{O_2} such that arterial P_{O_2} is reduced to 80 mm Hg.

    ____

    ____

    (b) Iron deficiency that decreases hemoglobin concentrations by 20%.

    ____

    ____

    (c) Breathing a high concentration of carbon monoxide.

    ____

    ____

64. True or false: A slight fall in arterial P_{CO_2} is a much stronger stimulus for increasing ventilation than a comparable fall in arterial P_{O_2}.

65. True or false: The most important signal for regulating ventilation is hydrogen ion concentration in the blood.

66. True or false: Increased concentrations of lactic acid stimulate increased ventilation primarily via central chemoreceptors.

67. True or false: Increased arterial P_{CO_2} and hydrogen ion concentration and decreased arterial P_{O_2} are all readily demonstrated during moderate exercise, and thus are probably the signals for increased ventilation.

68. If a person hyperventilates for a minute or two before diving into water, she or he can swim under water for longer than normal but risks blacking out and drowning. Explain.

    ____

    ____

    ____

    ____

69. True or false: Protective ventilatory reflexes include both violent respiratory reflexes and reflex cessation of respiration.

70. True or false: Sleep apnea refers to the normal decrease in ventilation during sleep.

❒ Hypoxia (text pages 464-465)

71. Match the descriptions of hypoxias with their correct name:

____ blood flow too low	(a) histotoxic hypoxia
____ arterial P_{O_2} reduced	(b) ischemic hypoxia
____ cells unable to utilize O_2	(c) hypoxic hypoxia
____ total O_2 content of blood reduced but P_{O_2} normal	(d) anemic hypoxia

72. True or false: The most common cause of hypoxic hypoxia in disease is diffusion impairment of O_2 resulting from decreased alveolar surface area.

73. True or false: Exposure to high altitude is a form of ischemic hypoxia.

74. A hike up Mt. Whitney in the Sierra Nevada, elevation 4830 meters, is a popular outing among California's backpackers. Most people camp high up the mountain for the "assault push" the next day. Imagine you are there, waking up at about 3300 meters. Explain your answers to these questions:

(a) What is your ventilation at 3300 meters, compared to when you were driving up to the start of the trail at about sea level?

____

____

(b) How would your erythropoietin level at 3300 meters compare to what it was down below at the beginning of the trail?

____

____

(c) Assume you had a kit handy for testing blood DPG levels. How would your DPG level at 3300 meters compare to normal? What is the significance of this result?

❒ **Nonrespiratory Functions of the Lungs** (text page 465)

75. True or false: Nonrespiratory functions of the lungs include acting as "filters" for certain chemicals and blood clots.

ANSWERS

Boldface type indicates the answers you should have given. Words in medium-face (ordinary) type explain or expand upon the answer.

1. (a) **Incorrect. It exchanges oxygen from the air with carbon dioxide from the blood.**
   (b) **Correct.**
   (c) **Correct.**
   (d) **Correct.**

2. (a) **Correct.**
   (b) **Correct.**
   (c) **Incorrect. The respiratory cycle is not timed to coincide with the cardiac cycle.** (There are generally five or six cardiac cycles per respiratory cycle.)

3. The sites for gas exchange with the blood are the **alveoli**. Most of these structures form clusters of **alveolar sacs** that are found at the ends of airway passages called **bronchioles**. The larger airway passages from which the smaller ones branch are called **bronchi** and are supported by the **cartilage** in their walls. These passages in turn branch from the tube beyond the larynx called the **trachea**. The airways are divided into two functional zones, the **respiratory** zone, across which gas exchange occurs, and the **conducting** zone, where no gas exchange occurs.

4. **False.** The blood vessels supplying the lungs are the ~~pleural~~ **pulmonary** vessels.

5. (a) **Correct.**
   (b) **Correct.**
   (c) **Correct.**
   (d) **Incorrect. An important role of the other three kinds of cells is to defend *against* bacteria.**

6. **False.** The upper airways of the respiratory system include the nose, mouth, pharynx and ~~esophagus~~ **larynx**.

7. (a) **Incorrect. Air and blood are separated at least by the capillary endothelium and the alveolar epithelium.** (And usually by thin basement membrane and connective tissue as well.)
   (b) **Incorrect. O_2 and CO_2 are exchanged by diffusion.**
   (c) **Correct.** (the gas-exchanging cells and the surfactant-secreting cells)

8. **False.** The diaphragm is a large sheet of ~~smooth~~ **skeletal** muscle that separates the thorax from the abdomen.

9. (a) **Correct.**
   (b) **Correct.**
   (c) **Incorrect. The fluid lies in the pleural sac, between the visceral sheet and the parietal sheet.**

10. **False.** If the parietal pleura is cut during surgery, the ~~lung~~ **chest wall** on the side of the cut will expand and the ~~chest wall~~ **lung** will be compressed.

11. (a) **Incorrect. It is less than atmospheric pressure.**
    (b) **Incorrect. It is the same as atmospheric pressure.**
    (c) **Correct.**
    (d) **Correct.** (It is this balanced "pull" on the pleural sacs that creates the less-than-atmospheric pressure in the intrapleural fluid.)

12. **False.** The process of ~~ventilation~~ **respiration** includes both the transport of gases by bulk flow and the diffusion of gases into and out of the blood.
    Alternatively, the answer below is also correct:
    False. The process of ventilation includes ~~both~~ the transport of gases by bulk flow ~~and the the diffusion of gases~~ into and out of ~~blood~~ **the lungs.** (See text Figure 14-7.)

13. **False.** According to Boyle's law, ~~a gas will expand to fill the volume available to it~~ **there is an inverse correlation between the pressure of a fixed number of gas molecules in a container and the volume of the container**. (The initial [false] statement may be seen as a corollary to Boyle's law.)

14.

$$F = \frac{P_{atm} - P_{alv}}{R}$$

15. (a) **Incorrect. Air flow would be decreased with increased airway resistance.**
    (b) **Correct.**
    (c) **Correct.** (As with blood flow in the circulatory system, the diameter of the "tubing" is the most important component of airway resistance.)

16. (a) **Incorrect. The opposite is correct.**
    (b) **Correct.**
    (c) **Correct.**

17. **False.** The first step in inspiration is the ~~enlargement of the thoracic cage~~ **contraction of the diaphragm and inspiratory intercostal muscles.** (This contraction is the cause of expansion of the thoracic cage.)

    Note: Some students have difficulty reconciling the *expansion* of the chest with the *contraction* of muscles, because these words seem to be opposites. Although the inspiratory muscles do get shorter when they contract, in the process of shortening they allow the chest to expand. As the text mentions, the diaphragm is a dome-shaped muscle at rest. When it *con*tracts, it *re*tracts away from the thoracic cavity, enlarging the cavity. When the inspiratory intercostal muscles shorten, they lift the ribs up and out, enlarging the rib cage.

18. **False.** During inspiration, the total alveolar volume ~~decreases~~ **increases**.

19. **False.** Expiration of air from the lungs **does not** requires contraction of the expiratory intercostal muscles, **but only the cessation of contraction of the inspiratory muscles.**

20. (a) **Correct.**
    (b) **Incorrect. He/she must have a lesser** (more negative with respect to atmospheric pressure) **intrapleural pressure**.
    (c) **Correct.**
    (d) **Correct.**
    (e) **Correct.**

21. **A person with low lung compliance breathes shallowly (the lung volume is small). Pulmonary surfactant, which lowers surface tension on the alveolar cells and increases compliance, decreases in concentration when the lung volume is small and constant. Therefore, decreased compliance leads to decreased surfactant, which leads to decreased compliance.**

22. **False.** Premature infants frequently develop respiratory distress syndrome of the newborn because they have ~~too few alveoli to exchange O_2 and CO_2 efficiently~~ **immature type II alveolar cells and too little surfactant.**

23. **Defective adrenal glands would produce too little cortisol. Cortisol is necessary for maturation of the type II alveolar cells that secrete surfactant. Therefore, babies with adrenal defects would have low lung compliance because of low surfactant levels.**

24. **True.**

25. (a) **Incorrect. Epinephrine causes dilation of the airways.**
    (b) **Correct.**
    (c) **Correct.**
    (d) **Incorrect. This is a stimulus for dilation.**
    (e) **Correct.**

26. **False.** Cigarette smoking is commonly associated with increased incidence of ~~asthma~~ **chronic obstructive pulmonary diseases** (emphysema and chronic bronchitis). (Although smoking does not cause asthma, people who have asthma may find their condition worsened by cigarette smoke, either first- or second-hand.)

27. **During an asthma attack, airway smooth muscle contracts and resistance increases. During inspiration, the physical factors of increased transpulmonary pressure and the outward pull of the connective tissue fibers tend to open the airways wider than they are during expiration.** (An asthmatic person has to force expiration, which is both tiring and frightening.)

28. **False.** During the Heimlich maneuver, sudden force against the diaphragm causes a sudden ~~decrease~~ **increase** in alveolar pressure, forceful expiration, and expulsion of the object blocking the airways.

29. **False.** The total lung capacity in a normal person is usually ~~roughly the same as~~ **about a liter greater than** the person's total blood volume. (Six liters lung capacity versus five liters of blood in the "standard" 70 kg man.)

30. d volume inspired at rest
    a maximal volume that can be inspired and expired
    f vital capacity + residual volume
    e maximal expiration after resting expiration
    b volume remaining in lungs after maximal expiration
    g maximal volume that can be inspired
    i volume remaining in lung after resting expiration

31. **False.** A clinical sign of ~~restrictive~~ **obstructive** lung disease would be a reduced FEV_1/FVC ratio.
    Alternatively, the answer below is also correct:
    False. A clinical sign of restrictive lung disease would be a reduced **vital capacity with a normal** FEV_1/FVC ratio.

32. The respiratory value analogous to cardiac output is the **minute ventilation**. It is determined by multiplying the **tidal volume** by the **breathing rate**.

33. (a) **Incorrect. It is about 150 mL.**
    (b) **Correct.**
    (c) **Incorrect. It is called dead space because no gas is exchanged there. The tissues (bronchioles, bronchi, and so forth) are living.**

34. (a) **Correct.** (500 mL x 10/min = 5000 mL/min = 5 L/min)
    (b) **Correct.** (500 mL x 20/min = 10,000 mL/min = 10 L/min)
    (c) **Incorrect. Alveolar ventilation = minute ventilation - dead space ventilation = 5000 mL/min - (150 x 10 = 1500 mL/min) = 3500 mL/min**
    (d) **Incorrect. Alveolar ventilation = (1000 mL x 10/min) - (150 mL x 10/min) = 10,000 mL/min - 1500 mL/min = 8500 mL/min = 2.4 times (more than double)**
    (e) **Incorrect. A maximal expiration followed by a maximal inspiration is the vital capacity, or total lung capacity minus the residual volume, which is 6 L - 1 L = 5 L. Thus 5 L of air could be drawn into the lungs, but 150 mL is dead space air, leaving 4850 mL of fresh air.**

35. **False.** The total dead space in the lungs includes anatomic dead space and any portion of the alveoli ~~obstructed from the air flow~~ **that has little or no blood supply.** (The concept of "dead space" applies to *oxygenated* areas of the lungs that cannot, for whatever reason, exchange gases with the capillary blood. All healthy lungs have an anatomic dead space, but only unhealthy lungs have much alveolar dead space.)

36. **False.** In a steady state, the rate of CO_2 production in the cells ~~exceeds~~ **equals** the rate of CO_2 expiration by the lungs.

37. **False.** In a person consuming a normal diet of carbohydrates, proteins, and fats, the production of CO_2 ~~exceeds~~ **is less than** the consumption of O_2. (In a mixed diet, the RQ is about 0.8. Thus, as is diagrammed in text Figure 14-16, the lung cells expire 200 mL of CO_2 for every 250 mL of O_2 consumed. A diet very high in carbohydrate would yield an RQ closer to 1.0.)

38. **False.** At sea level, P_{O_2} is about ~~760~~ 0.21 x 760 = **160** mm Hg.

39. **True.**

40. (a) **Correct.**
    (b) **Correct.**
    (c) **Correct.**

41. (a) **Correct.**
    (b) **Incorrect. Alveolar P_{CO_2} is more than 100 times greater than the (negligible) P_{CO_2} in air.**
    (c) **Correct.**
    (d) **Correct.**
    (e) **Incorrect. P_{CO_2} in arterial blood is almost as high as it is in venous blood** (40 mm Hg versus 46 mm Hg).

42. (a) **Incorrect. It is called hyperventilation.**
    (b) **Correct.**
    (c) **Incorrect. This happens in extreme hypoventilation.**
    (d) **Correct.**

43. **True.**

44. (a) **Incorrect. It is present but minor in normal lungs.**
    (b) **Correct.**
    (c) **Correct.**
    (d) **Correct.**
    (e) **Correct.**

45. **False.** Decreased alveolar P_{O_2} causes reflex ~~vasodilation~~ **vasoconstriction** of the pulmonary arterioles. (As the text points out, this effect of low P_{O_2} on pulmonary vascular smooth muscle is opposite that of the same stimulus on systemic vascular smooth muscle. As is illustrated in text Figure 14-20, the increased contraction, vasoconstriction, and thus vascular resistance decrease the perfusion of lung areas not receiving adequate ventilation. As a result, the blood can be diverted to other

parts of the lung where ventilation is greater and blood-gas exchange can be more efficient.)

46. **False.** Most of the O_2 carried in the blood ~~is dissolved in plasma~~ **reversibly bound to hemoglobin**.

47. Hemoglobin (Hb) can exist in a form that has O_2 bound to it, called **oxyhemoglobin**, and also in an unbound form, called **deoxyhemoglobin**. The percentage of the bound form to the total Hb is called the percent **saturation** of Hb. The most important factor determining this percentage is the **blood P_{O_2}**.

48. (a) **Incorrect. The greater the blood P_{O_2}, the greater the association (or saturation) of hemoglobin with O_2.**
    (b) **Correct.**
    (c) **Correct.**
    (d) **Incorrect. At a P_{O_2} of 80 mm Hg, hemoglobin is still almost completely saturated, and so it can carry adequate O_2 to the tissues.**

49. **True.**

50. **True.** A decrease from 40% to 20% saturation is accompanied by an O_2 dissociation of from 25% (75% saturation of hemoglobin) to 65% (35% saturation). (If the O_2-hemoglobin dissociation curve did not have the properties of a steep portion at lower partial pressures and a plateau at higher ones, the O_2-carrying capacity of the blood would be reduced, as would be the efficiency of unloading O_2 to tissues.)

51. (a) **Incorrect. Curve A may represent low body temperature.**
    (b) **Correct.**
    (c) **Incorrect. Curve C may represent high acidity.**

52. **True.** (Because of the effects of increased P_{CO_2}, $[H^+]$, and temperature.)

53. **False.** ~~Unl~~**Like** O_2, more of the CO_2 in blood ~~is dissolved in plasma than is bound to hemoglobin~~ **bound to hemoglobin than is dissolved in plasma.**

54. (a) **Incorrect. It catalyzes the conversion of CO_2 and H_2O to carbonic acid, and the reverse.**
    (b) **Correct.**
    (c) **Incorrect. It is not found in plasma.**

55. (a) **Correct.**
(b) **Correct.**
(c) **Correct.** (This exchange is in the reverse direction of the exchange in the tissue capillaries; that is, the bicarbonate is taken back up into the erythrocytes so that it can be converted back to CO_2 and H_2O.)

56. **False.** Deoxyhemoglobin binds ~~bicarbonate ion~~ **CO_2** better than oxyhemoglobin does.

57. **False.** The majority of hydrogen ions generated in the formation of carbonic acid are carried ~~dissolved in the plasma~~ **bound to deoxyhemoglobin**.

58. **False.** When a person hyperventilates, there are lower than normal levels of P_{CO_2} and hydrogen ion in arterial blood, a condition called respiratory ~~acidosis~~ **alkalosis**. (Acidosis refers to a higher than normal acidity of the blood. The "respiratory" adjective is attached to differentiate this condition, which is caused by a problem with respiration, from another kind of acidosis/alkalosis, one caused by a problem with metabolism. We shall learn about this latter problem in the next section.)

59. **False.** The rhythmic contractions and relaxations of the inspiratory muscles are dependent upon ~~spontaneous depolarizations of the diaphragm~~ **cyclical respiratory-muscle excitation by the motor nerves innervating them.**

60. **True.** (The medulla is part of the brainstem.)

61. **True.**

62. The chemoreceptors important for controlling ventilation comprise two classes. The first, called **peripheral** chemoreceptors, are sensitive to decreased **P_{O_2}** and increased **hydrogen ion** and **P_{CO_2}** in blood. They are located in the **arch of the aorta** (aortic bodies) and **near the carotid sinus** (carotid bodies). The other class, the **central** chemoreceptors, are exquisitely sensitive to **hydrogen ion concentration** in the **brain interstitial fluid**. They are located in the **medulla**.

63. (a) **No. Arterial P_{O_2} must be decreased to below 60 mm Hg before a significant increase in ventilation will occur.**
(b) **No. The peripheral chemoreceptors are sensitive only to the amount of O_2 dissolved in blood, not to the amount bound to hemoglobin. The blood P_{O_2} is not changed in anemia.**
(c) **No. Carbon monoxide does not depress blood P_{O_2} levels.** (One reason carbon monoxide is so dangerous is that it is tasteless, odorless, and does not cause a reflex

increased ventilation. Thus, the only warning that one is being poisoned is the grogginess preceding collapse and unconsciousness.)

64. **False.** A slight ~~fall~~ **rise** in arterial P_{CO_2} is a much stronger stimulus for increasing ventilation than a comparable fall in arterial P_{O_2}.

65. **False.** The most important signal for regulating ventilation is hydrogen ion concentration in the ~~blood~~ **brain extracellular fluid.**

66. **False.** Increased concentrations of lactic acid stimulate increased ventilation primarily via ~~central~~ **peripheral** chemoreceptors. (Hydrogen ion cannot cross the blood-brain barrier very well, and so it cannot influence the central chemoreceptors much.)

67. **False.** Increased arterial P_{CO_2} and hydrogen ion concentration and decreased arterial P_{O_2} ~~are all~~ **cannot be** readily demonstrated during moderate exercise, and thus ~~are probably~~ the signals for increased ventilation **are not clear**.

68. **Hyperventilation blows off CO_2 and reduces arterial P_{CO_2} without increasing arterial P_{O_2}. Decreased arterial P_{CO_2} depresses the peripheral and central chemoreceptors, which in turn depresses the medullary inspiratory neurons. The exertion of swimming uses up O_2 and may lower arterial P_{O_2} to dangerous, black-out levels.**

69. **True.**

70. **False.** Sleep apnea refers to the ~~normal~~ **abnormally exaggerated** decrease in ventilation during sleep **that occurs in some persons.**

71. **b** blood flow too low
    c arterial P_{O_2} reduced
    a cells unable to utilize O_2
    d total O_2 content of blood reduced but P_{O_2} normal

72. **False.** The most common cause of hypoxic hypoxia in disease is ~~diffusion impairment of O_2 resulting from decreased alveolar surface area~~ **ventilation-perfusion inequality.**

73. **False.** Exposure to high altitude is a form of ~~ischemic~~ **hypoxic** hypoxia.

74. (a) **Elevated. At 3300 meters, P_{O_2} levels in air are reduced enough that arterial P_{O_2} will be significantly decreased.**

(b) **Elevated. Decreased arterial P_{O_2} leads to decreased O_2 delivery to the kidneys, signalling them to increase erythropoietin secretion.**

(c) **Elevated. Increased DPG shifts the O_2-hemoglobin curve to the right, which allows for increased O_2 unloading in the tissues.**

75. **True.**

PRACTICE

True or false (correct the false statements):

P1. One effect of cigarette smoke on the lungs is that the ciliated cells lining the airways are stimulated to have increased activity.

P2. During inspiration, the diaphragm contracts and the alveolar pressure is less than atmospheric pressure; during expiration, the diaphragm relaxes and the alveolar pressure is greater than atmospheric pressure.

P3. In a normal person at rest, the part of the respiratory cycle in which no air is flowing through the airways and the respiratory muscles are relaxed is at the end of inspiration, prior to expiration.

P4. During inspiration, intrapleural pressure is more negative than alveolar pressure.

P5. Emphysema is a disease characterized by low lung compliance, obstructed airways, and ventilation-perfusion inequality.

P6. Doubling the rate of breathing will cause a greater increase in alveolar ventilation than will doubling the depth of inspiration.

P7. A small amount of carbon monoxide in the air would alter neither the P_{O_2} of inspired air nor the P_{O_2} of arterial blood.

P8. A decrease in arterial P_{O_2} stimulates increased ventilation by causing increased firing of action potentials from the carotid and aortic body chemoreceptors.

P9. One cause of hypoxia is anemia.

P10. Acclimatization to high altitudes includes "resetting" the respiratory-center neurons so that ventilation does not increase in response to low arterial P_{O_2}.

Multiple choice (correct each incorrect choice):

P11-13. The airways leading to the alveoli contain

P11. smooth muscle cells that contract in response to irritating chemicals.

P12. smooth muscle cells that contract when CO_2 levels near them decrease.

P13. the anatomic dead space.

P14-16. Which of the following would result from breathing 5% CO_2 in air?

P14. The pH of arterial blood rises.

P15. Alveolar ventilation increases.

P16. The O_2-hemoglobin dissociation curve shifts to the left.

P17-19. Carbon dioxide

P17. levels in the blood are decreased by voluntary hypoventilation.

P18. accumulation in the blood is associated with an increase in acidity.

P19. will not stimulate ventilation when the nerves from the carotid and aortic bodies have been cut.

Fill in the blanks:

P20-22. Increasing concentrations of ____________ in the alveoli cause pulmonary arterioles to dilate, and increasing concentrations of __________ in the alveoli cause bronchioles to dilate. These changes help ensure efficient __________________________.

P23-26. During exercise, the O_2-hemoglobin curve shifts ________________ (up/down) and to the ______________ (left/right). Two reasons for this shift are ______________________ and ________________________.

P27-30. The efficient transport of CO_2 by the blood is possible because of the enzyme ____________________________, which is found in the ___________________________ and catalyzes the reaction _____________________________________. This reaction is responsible for the transport of __________ percent of the CO_2 in venous blood.

15: THE KIDNEYS AND REGULATION OF WATER AND INORGANIC IONS

In 1855, a physician named Thomas Addison noted distinctive symptoms revealed in the autopsies of several patients:"anemia, general languor and debility, remarkable feebleness of the heart's action, irritability of the stomach, and a peculiar change of color to the skin, occurring in connection with a diseased condition of the supra-renal capsules." Addison's observation began nearly a century of scientific investigation into the function of the adrenal glands. Following his report, physiologists removed the adrenals from laboratory animals and found that the animals died soon thereafter. The adrenal glands were necessary for life. But why?

Some scientists, noting the close proximity of the adrenals to the "filter organs" (the kidneys), maintained that the adrenals' life-maintaining role was the neutralization or elimination of toxic substances. Others argued that the variety of metabolic defects shown in the experimental animals prior to death, including markedly decreased blood levels of sodium and glucose, pointed to the lack of a regulator molecule such as a hormone.

Another question that stirred much controversy was whether it was the disruption of mineral balance or of carbohydrate balance that led to death. Scientists answered this question by showing that injections of salt water solutions prolonged life while injections of glucose, which increased blood glucose levels, were relatively ineffective.

Finally, in 1953, the life-maintaining hormone of the adrenal cortex was identified as the salt-retaining steroid *aldosterone*. In this chapter you will learn why maintenance of blood sodium levels is necessary for life, and how the kidneys achieve this maintenance, with the help of their close neighbors, the adrenal glands.

Instructions for answering questions in this Study Guide

1. True or false: Correct each false statement. Whenever you must correct a false statement, there will almost always be more than one way to do so. In the Guide, the answer requiring the least correction will generally be given, with the understanding that other correct answers are possible. It is usually not sufficient, in terms of demonstrating understanding, to simply insert a "no" or "not," however.

2. Multiple choice: Any, all, or none of the responses may be correct. Explain why each incorrect response is incorrect.

3. Fill in the blanks: Choose the best word or words to complete each statement.

4. Directions will be given for other types of questions as they appear.

❐ Introduction (text page 472)

1. The kidneys
   (a) are important for salt and water homeostasis.
   (b) function only to eliminate wastes in urine.
   (c) are the major regulators of the concentrations of all inorganic ions in the body.

SECTION A. BASIC PRINCIPLES OF RENAL PHYSIOLOGY

❐ Functions of the Kidneys (text pages 472-473)

2. True or false: The adjective that means "of the kidneys" is adrenal.

3. Metabolic waste products excreted by the kidneys include
   (a) urea.
   (b) uric acid.
   (c) sodium ion.
   (d) drugs.

4. The three hormones secreted by the kidneys are
   (a) ______________________
   (b) ______________________
   (c) ______________________

❐ Structure of the Kidneys and Urinary System (text pages 473-475)

5. The functional unit of the kidney is the ________________. Each unit is composed of a ______________________ that acts as a filter and a ____________________ that processes the filtrate. The filter component is in turn composed of a tuft of capillaries called the ____ and a capsule, called ____, that is continuous with the __________________________________. The capillary tuft is unusual in that it has ________________________________ on either side of it.

6. The nephron tubule can be divided into four distinct anatomic and functional divisions: the segment nearest the capsule is called the ______________________; the long, hairpin-shaped segment is called the ______________________________ and is further subdivided into ______________________ and ______________________ limbs; the ______________________; and the ______________________, which are joined by other nephrons.

7. True or false: The anatomic arrangement of nephrons in the kidneys is such that the glomerulus, proximal tubule, and distal tubule are in the renal cortex, while the loop of Henle and collecting ducts lie mainly in the renal medulla.

8. True or false: The distal tubules drain into the kidney pelvis, which in turn drains into the urethra, a tube that carries urine to the bladder.

9. Beginning with the renal artery and ending with the renal vein, trace the six-step course of an erythrocyte through the blood vessels of the kidney.

10. The juxtaglomerular apparatus is
 (a) close to the glomerulus.
 (b) composed of parts of the ascending loop of Henle and the efferent arteriole.
 (c) composed of granular cells and the macula densa.
 (d) the site of renin secretion.

❐ **Basic Renal Processes** (text pages 475-480)

11. The composition of the glomerular filtrate
 (a) is identical to that of blood plasma.
 (b) is identical to that of urine.
 (c) changes as the filtrate passes through the tubule.

12. True or false: The three basic processes of kidney function are filtration from the glomerulus to the capsule, secretion from the tubule to the peritubular capillaries, and reabsorption from the capillaries into the tubular lumen.

13. True or false: Different substances in the glomerular filtrate differ in the extent that they can be secreted and reabsorbed.

14. True or false: Large amounts of protein in a person's urine indicate that the person is eating a high-protein diet.

15. Filtration across the glomerular capillaries into Bowman's capsule occurs
    (a) by diffusion.
    (b) when the hydrostatic pressure in the capillaries exceeds the fluid pressure in the capsule.
    (c) when the hydrostatic pressure in the capillaries exceeds the sum of the fluid pressure in the capsule plus the osmotic force due to proteins in the plasma.

16. The glomerular filtration rate (GFR)
    (a) is the volume of fluid filtered into the Bowman's capsule of each nephron of both kidneys per unit time.
    (b) exceeds the cardiac output.
    (c) in a day is about 60 times the plasma volume.
    (d) is constant and unchanging.

17. True or false: The filtered load of water is 180 L/day.

18. True or false: Water, sodium, and glucose all undergo tubular reabsorption, but urea does not.

19. True or false: The kidneys regulate the plasma concentrations of water, sodium, and glucose.

20. True or false: When a substance is actively reabsorbed from the tubule into the peritubular capillary, the active transport systems are located in the the capillary wall.

21. Tubular reabsorption of
    (a) glucose is active.
    (b) glucose is saturable.
    (c) vitamin C has a transport maximum.
    (d) urea would not occur if water were not being reabsorbed simultaneously.

22. Tubular cells may
    (a) actively transport molecules across their basolateral membranes in order to secrete the molecules into the tubular lumen.
    (b) synthesize substances in order to secrete them into the tubular lumen.
    (c) secrete protein into the tubular lumen.

❐ **Micturition** (text pages 480-481)

23. True or false: The spinal reflex for micturition involves stretch receptors in the wall of the bladder that send messages about distension to sympathetic nerves in the spinal cord.

24. True or false: Voluntary control of micturition involves controlling motor input to the muscles of the pelvic diaphragm.

SECTION B. REGULATION OF SODIUM AND WATER BALANCE

❐ **Total-Body Balance and Internal Distribution of Sodium and Water** (text pages 481-482)

25. A person who is in balance for total body water
    (a) must ingest more water than he or she loses in the urine.
    (b) must ingest more water than she or he loses by all output pathways combined.
    (c) will have a balanced intake and output so that total body water will stay the same.

26. True or false: Total-body balance of water and sodium ion is largely maintained by regulating urinary loss of these molecules.

27. True or false: Water will move by diffusion from a compartment with high osmolarity to a compartment with low osmolarity.

28. True or false: Most of the body's water is located inside cells, whereas most of the body's sodium ion is in the interstitial fluid and bone.

29. True or false: The concentration of water in the extracellular fluid is dependent upon the concentration of sodium in the extracellular fluid.

❒ **Basic Renal Processes for Sodium and Water** (text pages 482-488)

30. Both water and sodium are
    (a) filtered at the glomerulus.
    (b) actively reabsorbed by the tubule.
    (c) secreted by the tubule.

31. The active transport of sodium
    (a) in all segments of the tubule is dependent upon Na,K-ATPase pumps in the luminal membrane.
    (b) in all segments of the tubule follows passive transport of sodium into the epithelial cells.
    (c) in the proximal tubule allows for secondary active cotransport of hydrogen ion and countertransport of glucose and amino acids.

32. True or false: Water movement passively follows the active transport of sodium into the interstitial fluid outside the tubule, where both kinds of molecules then diffuse into the peritubular capillaries.

33. The permeability to water of the tubular epithelium
    (a) is high along all segments of the tubule.
    (b) is always high in the proximal tubule.
    (c) may be high or low in the collecting ducts.

34. Antidiuretic hormone
    (a) is secreted by the anterior pituitary gland.
    (b) is a peptide.
    (c) uses cyclic AMP as a second messenger.
    (d) acts on cells of the proximal tubule.
    (e) increases the number of Na,K-ATPase pumps in the luminal membrane of the tubular epithelium.

35. True or false: A consequence of lack of ADH is excretion of sugar in the urine, diabetes mellitus.

36. True or false: A person lacking ADH would have to drink up to 180 L of water per day to make up for the water lost in the urine.

37. True or false: Unlike the renal cortical interstitial fluid, the interstitial fluid of the medulla is hyperosmotic.

38. The countercurrent multiplier system
    (a) requires the collecting ducts to be close to the loops of Henle.
    (b) requires the peritubular capillaries to be close to the collecting ducts.
    (c) requires active transport of sodium and chloride out of the descending limb of the loop of Henle.
    (d) would not function if the descending limb of the loop of Henle were permeable to water.
    (e) would not function if the ascending limb of the loop of Henle were permeable to water.

39. True or false: Movement of water out of the collecting ducts takes place by bulk flow.

40. True or false: In the absence of ADH, urine is isosmotic with plasma.

41. The kangaroo rat lives in the desert and can survive for long periods of time without water. This animal has nephrons with unusually long loops of Henle. Explain how this feature would be adaptive for an animal in an arid environment.

____

____

____

____

❐ **Renal Sodium Regulation** (text pages 488-492)

42. True or false: Increasing the GFR will increase the excretion rate of sodium.

43. True or false: The most important receptors in the reflexes regulating sodium homeostasis are the cardiovascular baroreceptors.

44. Aldosterone
    (a) increases sodium reabsorption by the proximal tubule.
    (b) stimulates the opening of sodium channels in its target cells.
    (c) is required for the reabsorption of most of the sodium filtered by the nephron.

45. The primary signal for the release of aldosterone from the ____________________ is the hormone ____________________, which is cleaved from a molecule called ____________________ in the presence of ____________________ enzyme. This transformation takes place primarily in the capillaries of the ____________. The initial prohormone for the reaction, ____________________, is produced by the liver and cleaved by the enzyme ____________________, which is secreted by the ____________________ cells of the afferent arterioles of the kidney. The rate-limiting factor in these reactions is the concentration of ____________________.

46. Stimuli that increase the concentration of renin in the blood include
    (a) increased mean arterial blood pressure.
    (b) decreased sodium and chloride concentrations in the tubular fluid.
    (c) decreased renal blood pressure.

47. True or false: Atrial distension is a stimulus for the secretion of atrial natriuretic factor, which stimulates the reabsorption of sodium by the kidneys.

48. Which of the following will decrease blood pressure?
    (a) A drug that interferes with aldosterone synthesis.
    (b) A drug that is an agonist of atrial natriuretic factor.
    (c) A drug that decreases sympathetic stimulation of renal arterioles.
    (d) A drug that enhances the activity of converting enzyme.
    (e) A drug that decreases liver production of angiotensinogen.

❐ **Renal Water Regulation** (text pages 492-494)

49. True or false: Following hemorrhage, ADH secretion increases because of increased firing of hypothalamic osmoreceptors.

50. True or false: Ingestion of a salt tablet triggers reflexes that bring about relatively greater excretion of sodium than water.

❐ **A Summary Example: The Response to Sweating** (text page 494 and Figure 15-22)

51. True or false: Because fluid lost as sweat is isosmotic, the baroreceptor reflexes are the most important in restoring fluid volume.

52. The response to sweating includes
    (a) reflexes that increase the GFR.
    (b) increased aldosterone secretion.
    (c) decreased firing of atrial baroreceptors.
    (d) decreased water permeability of the collecting ducts.
    (e) increased osmolarity of urine.

❐ Thirst and Salt Appetite (text pages 494-496)

53. True or false: Nuclei in the hypothalamus mediate the sensation of thirst.

54. True or false: Loss of 0.5 liter of sweat will provoke a greater sense of thirst than loss of the same volume of blood plasma.

SECTION C. REGULATION OF POTASSIUM, CALCIUM, AND HYDROGEN ION

❐ Potassium Regulation (text pages 496-498)

55. True or false: Regulation of extracellular potassium is not particularly important because only about two percent of the total body potassium is in the extracellular fluid.

56. True or false: Cardiac arrhythmias may be a clinical sign of either potassium depletion or potassium excess.

57. True or false: Unlike sodium, potassium may be secreted by the nephron tubules.

58. True or false: Ingestion of large amounts of potassium triggers reflexes to limit the amount of potassium reabsorbed by the distal tubules.

59. True or false: Stimuli that cause increased sodium reabsorption decrease potassium secretion.

60. True or false: Stimuli that cause increased sodium excretion decrease potassium excretion.

61. True or false: A stimulus for increased aldosterone secretion is decreased plasma levels of potassium.

❒ Calcium Regulation (text pages 498-501)

62. True or false: A clinical sign of too little calcium in the extracellular fluid is muscle spasms.

63. True or false: Regulation of calcium balance, like that of sodium and potassium balance, is primarily effected by the kidneys.

64. Match the terms on the right with their correct descriptions on the left:

____	active form of vitamin D	(a) 7-dehydrocholesterol
____	necessary for conversion to active form of vitamin D	(b) parathyroid hormone
____	result of metabolism of vitamin D_3 in liver	(c) 25-OH D_3
____	converted to vitamin D_3 in the skin	(d) rickets
____	increases calcium absorption by small intestine	(e) vitamin D
____	result of vitamin D deficiency	(f) 1,25-$(OH)_2D_3$
____	found in food	

65. True or false: Parathyroid hormone stimulates resorption of calcium and phosphate from bone and reabsorption of these minerals in the kidneys.

66. True or false: Bone is a storage depot for calcium ion.

67. Hyperparathyroidism is a disease characterized by elevated parathyroid hormone levels in the blood, usually secondary to a tumor of the parathyroid glands. What symptoms would you expect in a patient with this disease?

____

____

____

❒ Hydrogen-Ion Regulation (text pages 501-504)

68. Hydrogen-ion
    (a) concentration in body fluids is decreased by diarrhea.
    (b) concentration in body fluids is increased by a fat-rich diet.
    (c) in blood is mainly in the form of weak acids.
    (d) concentration in body fluids at physiological pH is zero.

69. In the equation $H^+ + HCO_3^- \leftrightarrow H_2CO_3$, by the law of mass action, increasing the amount of HCO_3^- will ____________________ (increase/decrease) the amount of H^+ by driving the reaction to the ____________________ (left/right).

70. With regard to renal regulation of extracellular hydrogen-ion concentration,
(a) regulation is dependent upon regulating the amount of hydrogen ion filtered by the kidneys.
(b) the reabsorption of bicarbonate ion is dependent upon the reabsorption of hydrogen ion.
(c) the excretion of hydrogen ion is related to the amount of ammonia formed in tubular cells.

71. True or false: Most of the hydrogen ion excreted in the urine is bound to bicarbonate ion.

72. True or false: The kidneys work to rectify metabolic acidosis and alkalosis but have no effect on respiratory acidosis and alkalosis.

73. True or false: A consequence of an episode of severe vomiting is an increased breathing rate.

SECTION D. DIURETICS AND KIDNEY DISEASE

❐ Diuretics (text pages 505-506)

74. True or false: A diuretic is any drug that increases the volume of urine excreted as a result of inhibiting sodium reabsorption.

75. Consequences of heart failure include
(a) tissue edema.
(b) decreased plasma volume.
(c) decreased venous pressure.
(d) increased renin secretion.
(e) increased secretion of atrial natriuretic factor.

76. True or false: People taking diuretics that do not spare potassium should increase their ingestion of sodium.

❐ Kidney Disease (text pages 506-507)

77. Kidney damage by a number of different agents leads to a disease called uremia. What is uremia and why is it called this?

____

____

____

____

78. In some cases of kidney disease, the glomerular membranes become much more permeable to large molecules than normal. This condition results in tissue edema. Explain.

____

____

____

____

79. Reduced kidney function may be diagnosed by detection of reduced
    (a) GFR.
    (b) plasma creatinine.
    (c) creatinine clearance.

80. People with acute or chronic kidney failure can be kept alive by undergoing one of two kinds of ____________________. The more efficient of these methods is called ____________________ and is also known as the "artificial kidney." The other method, ______________________________, utilizes the patient's own ________________________ as the selectively permeable membrane. This latter method has the advantage of allowing ______________________________ during the treatment. Neither method offers a cure for chronic kidney failure. The only known cure is ____________________________.

ANSWERS

Boldface type indicates the answers you should have given. Words in medium-face (ordinary) type explain or expand upon the answer.

1. (a) **Correct.**

   (b) **Incorrect. Waste elimination is only one of many functions.**

   (c) **Incorrect. They are not the major regulators of such ions as zinc and iron** (or even calcium, although they are important for calcium regulation, as you will see later in this chapter).

2. **False.** The adjective that means "of the kidneys" is ~~adrenal~~ **renal**. (*Ad*renal means "near the kidney.")
   Note: The adjective "nephric" is also used to describe, for example, kidney disease. "Renal" comes from the Latin word for kidney, while "nephros" is the Greek word.

3. (a) **Correct.**

   (b) **Correct.**

   (c) **Incorrect. Sodium ion is not a waste product** (although it is excreted in the urine to maintain total-body balance of sodium).

   (d) **Incorrect. Drugs are not metabolic waste products. They are foreign chemicals.**

   The purpose of this question is to reinforce the point made in Answer 1-b: The elimination of metabolic waste is only one of many important functions of the kidneys.

4. (a) **erythropoietin**
   (b) **renin**
   (c) **1,25-dihydroxyvitamin D_3**

5. The functional unit of the kidney is the **nephron**. Each unit is composed of a **glomerulus** that acts as a filter and a **tubule** that processes the filtrate. The filter component is in turn composed of a tuft of capillaries called the **glomerular capillaries** and a capsule, called **Bowman's capsule**, that is continuous with the **tubule**. The capillary tuft is unusual in that it has **arterioles** on either side of it.

6. The nephron tubule can be divided into four distinct anatomic and functional divisions: the segment nearest the capsule, the **proximal tubule**; the long, hairpin-shaped segment, the **loop of Henle**, which is further subdivided into **ascending** and **descending** limbs; the **distal tubule**; and the **collecting ducts**, which are joined by other nephrons.

7. **True.** (This arrangement is very important for nephron function, as you will see later in this chapter.)

   Note that you have now learned about three organs that have a cortex and a medulla, and so you must be careful to specify what you are talking about. It is especially easy to confuse "renal cortex or medulla" with "*ad*renal cortex or medulla," for example.)

8. **False.** The ~~distal tubules~~ **collecting ducts** drain into the kidney pelvis, which in turn drains into the ~~urethra~~ **ureter**, a tube that carries urine to the bladder.

9. **Renal artery → afferent arteriole → glomerular capillaries**

   ↓

   Renal vein ← peritubular capillaries ← efferent arteriole

   Note: "Peritubular" means "around the tubules." These capillaries form a network in close proximity to the tubular parts of the nephron.

10. (a) **Correct.** (It is *juxtaposed* to the glomerulus — hence, the name.)
    (b) **Incorrect.** **It is composed of parts of the ascending loop of Henle and the *afferent* arteriole.**
    (c) **Correct.**
    (d) **Correct.**

11. (a) **Incorrect.** **It is a protein-free filtrate of plasma.**
    (b) **Incorrect.** **It is very different from urine.**
    (c) **Correct.**

12. **False.** The three basic processes of kidney function are filtration from the glomerulus to the capsule, secretion from the ~~tubule to the peritubular capillaries~~ **peritubular capillaries into the tubule**, and reabsorption from the ~~capillaries into the tubular lumen~~ **tubular lumen into the capillaries**. Renal nomenclature can be very confusing. When we speak of "tubular secretion," we mean "secretion *into* the tubule" instead of the more familiar route of secretion *from* something. As the text mentions, the problem stems from the fact that when secretion is accomplished by mediated transport rather than by passive processes, it is the epithelial cells of the tubular wall that do the secreting. Passage of substances out from the peritubular capillaries is always by passive diffusion. (In case it is unclear to you what is meant by "tubular epithelial cells" — that is, are they cells in the tubule or in the peritubular capillaries? — recall that capillaries consist of a specialized epithelium called *endo*thelium. Further, any reference to "tubular" is to the tubule, *not* to the peritubular capillaries.) Similarly, tubular reabsorption means "reabsorption *from* the tubule back into the capillaries" (the "re" meaning "to absorb again"), but we speak of the tubule doing the

reabsorbing because, again, any mediated transport occurs in the tubular epithelium — that is, from the lumen of the tubule through the plasma membrane of the tubular epithelial cell. If you keep in mind that glomerular filtration and tubular secretion *add* substances to the tubule, and that tubular reabsorption *subtracts* substances from the tubule, then you should be able to avoid confusion.

13. **True.**

14. **False.** Large amounts of protein in a person's urine indicates that the person ~~is eating a high protein diet~~ **has kidney disease**.

15. (a) **Incorrect.** **It occurs by bulk flow.**
    (b) **Incorrect.** **See answer (c).**
    (c) **Correct.** (Note, in text Figure 15-7, that the value of π — the protein osmotic force — is 30 mm Hg. This number is higher than the one we saw in Chapter 13 for "ordinary" systemic capillary beds. One reason for this higher osmotic force is that the fluid in the capsule, unlike interstitial fluid, is virtually protein-free and thus does not have a countering osmotic force [favoring filtration] of its own.)

16. (a) **Correct.** (Incidentally, there are about 1.25 million nephrons in each kidney.)
    (b) **Incorrect.** **Cardiac output in a 70-kg man at rest = 5 L/min = 300 L/h = 7200 L/day. GFR is 180 L/day, or about 2.5% of CO** (or about 4% if you consider only plasma in the cardiac output equation).
    (c) **Correct.** (Obviously there is a big difference between "output" and "volume.")
    (d) **Incorrect.** **It varies according to the filtration pressure.** (This fact was not stated explicitly in this section of the text, but you should have been able to reason out the correct answer because you know that blood flow to the kidneys and thus kidney blood pressure are regulated. Decreased pressure in the afferent arteriole will decrease glomerular filtration, and increased pressure will increase it. We shall come back to this subject later in this chapter.)

17. **True.**

18. **False.** Water, sodium, glucose, and **urea** all undergo tubular reabsorption, but ~~urea does~~ not **to the same extent**.

19. **False.** The kidneys regulate the plasma concentrations of water and sodium, **but not that of** glucose.

20. **False.** When a substance is actively reabsorbed from the tubule into the peritubular capillary, the active transport systems are located in the ~~capillary wall~~ **luminal or basolateral membrane of the tubular epithelium**.

21. (a) **Correct.**
    (b) **Correct.** (It has a T_m.)
    (c) **Correct.**
    (d) **Correct.**

22. (a) **Correct.**
    (b) **Correct.**
    (c) **Incorrect.** **Protein is not secreted into the lumen.**

23. **False.** The spinal reflex for micturition involves stretch receptors in the wall of the bladder that send messages about distension to ~~sympathetic~~ **parasympathetic** nerves in the spinal cord.

24. **True.**

25. (a) **Correct.** (Note in text Table 15-3 that insensible loss of water is greater than metabolic production.)
    (b) **Incorrect.** **In this case, she or he would be in positive balance for water.**
    (c) **Correct.**

26. **True.**

27. **False.** Water will move by diffusion from a compartment with ~~high~~ **low** osmolarity to a compartment with ~~low~~ **high** osmolarity.

28. **True.**

29. **False.** The ~~concentration~~ **volume** of water in the extracellular fluid is dependent upon the concentration of sodium in the extracellular fluid. (Water concentration does vary with increasing or decreasing sodium concentration, but only temporarily.)

30. (a) **Correct.**
    (b) **Incorrect.** **Sodium is actively transported, and water follows after sodium along the osmotic gradient created by the sodium transport.**
    (c) **Incorrect.** **Neither is secreted.**

31. (a) **Incorrect. The pumps are in the basolateral membrane.**
    (b) **Correct.**
    (c) **Incorrect. Hydrogen ion is countertransported in the proximal tubule, while glucose and amino acids are cotransported.** Recall from Chapter 6 that cotransport refers to the transport of a solute against its concentration gradient but in the same direction as sodium is moving along *its* gradient — that is, into the cell. In countertransport, the actively transported molecule goes in the opposite direction from sodium; in this case, out of the cell and into the lumen.

32. **False.** Water movement passively follows the active transport of sodium into the interstitial fluid outside the tubule, where both kinds of molecules then ~~diffuse~~ **move** into the peritubular capillaries **by bulk flow.** (Thus, the final step of reclaiming water, ions, and other valuable molecules from the tubular fluid is the same as the filtration step that put them there — bulk flow.)
    Note that the reabsorption of ions and water in the proximal tubule is such that their concentrations in the tubular interstitial fluid are not different from that of plasma. Therefore, it is fortunate that the pressure and osmotic differences favor bulk flow because diffusion would not be efficient under these conditions.

33. (a) **Incorrect. The permeability to water varies along different segments of the tubule.**
    (b) **Correct.**
    (c) **Correct.**

34. (a) **Incorrect. It is secreted by the posterior pituitary.**
    (b) **Correct.**
    (c) **Correct.**
    (d) **Incorrect. It acts on cells of the late distal tubule and the collecting ducts.**
    (e) **Incorrect. It opens water channels in the luminal membrane.**

35. **False.** A consequence of lack of ADH is excretion of ~~sugar in the~~ **huge amounts of** urine, diabetes ~~mellitus~~ **insipidus**. (Diabetes means "inordinate and persistent urination." Insipidus means "insipid" or "tasteless." This term distinguishes the disease from the more common "sugar diabetes," diabetes mellitus ["mellitus" means "honeyed" in Latin], which is a consequence of too much glucose in the blood, as we shall see in Chapter 17.)

36. **False.** A person lacking ADH would have to drink up to ~~180~~ **25** L of water per day to make up for the water lost in the urine. (Most of the water that is filtered is

reabsorbed in the proximal tubule, which is fully permeable to water even in the absence of ADH.)

37. **True.**

38. (a) **Correct.** (The ducts must share the same interstitial osmolarity as the loops. Recall that several nephrons contribute to each collecting duct. All of the loops are in the vicinity of their collecting duct.)

    (b) **Correct.** (The capillaries must carry the reabsorbed water away quickly, or else it would dilute the hyperosmotic interstitium.)

    (c) **Incorrect.** **It requires active transport out of the ascending limb.** (The mechanism would not work in reverse; in fact, such a mechanism would be a diluting system, not a concentrating system.)

    (d) **Incorrect.** **It requires that the descending limb be permeable to water.**

    (e) **Correct.**

39. **False.** Movement of water out of the collecting ducts takes place by ~~bulk flow~~ **diffusion**.

40. **False.** In the absence of ADH, urine is ~~isosmotic with plasma~~ **hyposmostic**. (The fluid entering the distal tubule is hyposmostic. In the presence of ADH, water can leave the tubule along its concentration gradient, but if ADH is absent, the fluid entering the collecting duct is hyposmostic. It becomes diluted further there by the active transport of sodium and chloride out of the duct, which takes place regardless of water diffusion.)

41. **An animal living in an arid environment must conserve body water very efficiently. Long loops of Hcnle allow for the establishment of a large concentration gradient by the countercurrent multiplier system and thus a very concentrated urine.** (The maximal osmolarity of the interstitial fluid in a kidney's medulla is directly related to the length of the loops: the longer they are, the greater the concentration of solute that is possible. In the case of the kangaroo rat, the osmolarity is almost four times greater than that in a human, and the urine can be four times as concentrated. As a result, the animal can survive on water ingested with food and metabolic water.)

42. **True.**

43. **True.**

44. (a) **Incorrect.** **It increases sodium reabsorption by the late distal tubules and the collecting ducts.**

    (b) **Incorrect.** **It stimulates synthesis of the proteins forming sodium channels.**

(c) **Incorrect.** **It is required for reabsorbing only about 2% of the sodium that is filtered.** (Obviously, in light of the fact that aldosterone is necessary for life, this is a most important 2%.)

45. The primary signal for the release of aldosterone from the **adrenal cortex** is the hormone **angiotensin II**, which is cleaved from a molecule called **angiotensin I** in the presence of **converting** enzyme. This transformation takes place primarily in the capillaries of the **lung**. The initial prohormone for the reaction, **angiotensinogen**, is produced by the liver and cleaved by the enzyme **renin**, which is secreted by the **granular** cells of the afferent arterioles of the kidney. The rate-limiting factor in these reactions is the concentration of **renin**.

46. (a) **Incorrect.** **This stimulus would decrease plasma renin.**
    (b) **Incorrect.** **This stimulus would also decrease plasma renin.**
    (c) **Correct.** (The granular cells are themselves baroreceptors.)

47. **False.** Atrial distension is a stimulus for the secretion of atrial natriuretic factor, which ~~stimulates~~ **inhibits** the reabsorption of sodium by the kidneys.
    Note that *diuresis* means "increased excretion of urine." "Natrium" is the Latin word for sodium (hence its chemical abbreviation Na), and so *natriuresis* means "increased excretion of sodium in the urine." Thus, a natriuretic factor is one that stimulates sodium excretion.

48. (a) **Correct.**
    (b) **Correct.**
    (c) **Correct.**
    (d) **Incorrect.** **A drug that inhibits the activity of this enzyme would decrease blood pressure.** (Enhancing the activity of converting enzyme would not have much effect on blood pressure, since renin is the rate-limiting enzyme in the angiotensin-II pathway.)
    (e) **Correct.**
    Note that in every case of drugs that decrease blood pressure, the ultimate effect is on kidney reabsorption of sodium. Choice (c) also decreases blood pressure by increasing GFR, and (e) decreases concentration of angiotensin-II, a potent vasoconstrictor. Most drugs prescribed for hypertension act primarily on the kidneys, although in only a few cases is kidney malfunction the cause of the hypertension. We shall discuss this topic again later in this chapter.

49. **False.** Following hemorrhage, ADH secretion increases because of ~~increased~~ **decreased** firing of ~~hypothalamic osmoreceptors~~ **atrial baroreceptors**.

50. **True.**

51. **False.** Because fluid lost as sweat is ~~isosmotic~~ **hyposmostic**, the baroreceptor **and osmoreceptor** reflexes are ~~the most~~ **both** important in restoring fluid volume. (Note that loss of *hyposmostic* fluid means that extracellular fluid will be *hyper*osmotic, unlike the case with hemorrhage discussed in Question 49.)

52. (a) **Incorrect. Sweating triggers reflexes that decrease GFR.**
 (b) **Correct.**
 (c) **Correct.**
 (d) **Incorrect. Water permeability in the ducts is increased** (by increased ADH).
 (e) **Correct.**

53. **True.**

54. **True.** (In both sweating and blood loss, decreased baroreceptor activity and increased angiotensin II stimulate thirst. It is only in sweating, however, that the osmoreceptors that also stimulate thirst are activated.)

55. **False.** Regulation of extracellular potassium is ~~not~~ particularly important ~~because~~ **even though** only about two percent of the total body potassium is in the extracellular fluid.

56. **True.**

57. **True.**

58. **False.** Ingestion of large amounts of potassium triggers reflexes to ~~limit~~ **increase** the amount of potassium ~~reabsorbed~~ **secreted** by the distal tubules.

59. **False.** Stimuli that cause increased sodium reabsorption ~~decrease~~ **increase** potassium secretion.

60. **True.**

61. **False.** A stimulus for increased aldosterone secretion is ~~decreased~~ **increased** plasma levels of potassium.

62. **True.**

63. **False.** Regulation of calcium balance, **un**like that of sodium and potassium balance, is ~~primarily~~ effected by the kidneys, **gastrointestinal tract, and bone**.

64. f active form of vitamin D
b necessary for conversion to active form of vitamin D
c result of metabolism of vitamin D_3 in liver
a converted to vitamin D_3 in the skin
f increases calcium absorption by small intestine
d result of vitamin D deficiency
e found in food

Note: Rickets is a result of both too little vitamin D in the diet and too little exposure to sunlight. Symptoms of rickets were first observed in children in England, where there is relatively little sunlight. The children had diets poor in fat, the source of vitamin D. (Cod liver oil is the single best source of dietary vitamin D. Today's children are fortunate that milk, bread, and other foods are now artificially fortified with the vitamin, so that children no longer have to force down an unpalatable fish oil to ensure strong, healthy bones.)

Second note: Vitamin D deficiency in adults is called osteomalacia. It results in weak bones that fracture easily and is a critical problem for the elderly. Many doctors recommend that elderly people try to spend about 15 minutes a day exposed to the sun, without sunscreen, to activate 7-dehydrocholesterol. Of course, they should also be sure to eat foods that are good sources of calcium.

65. **False.** Parathyroid hormone stimulates resorption of calcium and phosphate from bone and reabsorption of ~~these minerals~~ **calcium** in the kidneys. **It also decreases renal phosphate reabsorption.**

Note the difference between *resorption* of calcium and phosphate from bone, where they are stored, and *reabsorption* of the minerals from the tubular fluid. (See text Figure 15-27.)

66. **True.** (Of course, is also serves to support the body.)

67. **High levels of parathyroid hormone stimulate excessive movement (resorption) of calcium and phosphate from bone, causing it to weaken. Elevated plasma calcium levels cause cardiac arrhythmias and muscle weakness.**
(The alert reader may be wondering about the hormone *calcitonin* that was listed in text Table 10-1 as having an effect on plasma calcium. This hormone, which is secreted by specialized cells in the thyroid gland, inhibits resorption of minerals from bone and thus inhibits the action of parathyroid hormone on bone. Calcitonin is not a particularly important regulator of plasma calcium in adults, but in growing children and in women during pregnancy and lactation, calcitonin is thought to be important for its bone-sparing effects.)

68. (a) **Incorrect. It is increased** (because of loss of bicarbonate ion).
    (b) **Incorrect. It is increased by a protein-rich diet.**
    (c) **Correct.**
    (d) **Incorrect. It is very low** (0.00004 mM), **but not zero.** ($[H^+]$ is zero at pH 7.0, which is *not* a physiological pH.)

69. In the equation $H^+ + HCO_3^- \leftrightarrow H_2CO_3$, by the law of mass action, increasing the amount of HCO_3^- will **<u>decrease</u>** the amount of H^+ by driving the reaction to the **<u>right</u>**.

70. (a) **Incorrect. It is dependent upon regulating the amount of hydrogen ion secreted by the tubules.**
    (b) **Incorrect. It is dependent upon the secretion of hydrogen ion.**
    (c) **Correct.**

71. **False.** Most of the hydrogen ion excreted in the urine is bound to ~~bicarbonate ion~~ **either ammonia or HPO_4^{2-}**. (When hydrogen ion binds to filtered bicarbonate ion, the resulting carbonic acid breaks down into $CO_2 + H_2O$, which diffuse into the tubular cells. As text Figure 15-30 shows, this process provides for further formation of H^+ in the tubular cells. The hydrogen ion is actively transported out into the tubular lumen, where it can "rescue" another molecule of bicarbonate.)

72. **False.** The kidneys (and lungs) work to rectify ~~metabolic~~ acidosis and alkalosis ~~but have no effect on respiratory acidosis and alkalosis~~ **of whatever cause**. (Thus, breathing rate increases during metabolic acidosis, and excretion of hydrogen ion in the urine increases during respiratory acidosis.)

73. **False.** A consequence of an episode of severe vomiting is ~~an increased~~ **a decreased** breathing rate. (Vomiting causes alkalosis, which is detected by brain chemoreceptors that in turn inhibit ventilation.)

74. **False.** A diuretic is any drug that increases the volume of urine excreted. ~~as a result of inhibiting sodium reabsorption.~~ (Recall that lack of ADH results in water diuresis. A drug that inhibits ADH secretion — alcohol, for instance — is a diuretic.) Alternatively, the answer below is also correct:

    False. A ~~diuretic~~ **natriuretic** is any drug that increases the volume of urine excreted as a result of inhibiting sodium reabsorption.

75. (a) **Correct.**
    (b) **Incorrect. Plasma volume is increased.**
    (c) **Incorrect. Venous pressure is increased.**
    (d) **Correct.**

(e) **Correct.** (This secretion tends to balance other factors that lead to increased reabsorption of salt and water and worsen the edema and other symptoms of heart failure. By itself, ANF secretion cannot wholly compensate, however; thus, diuretic drugs are necessary to counteract these symptoms.)

76. **False.** People taking diuretics that do not spare potassium should increase their ingestion of ~~sodium~~ **potassium**. (It makes no sense to take a drug that enhances urinary excretion of a substance and at the same time increase ingestion of that substance.)

77. **Severely damaged kidneys cannot filter the plasma adequately. As a consequence, substances that would normally be excreted in urine are not and they collect in the plasma, causing uremia — "urine in blood."**

78. **If the glomeruli are permeable to plasma proteins, the proteins will be excreted in urine because there is no mechanism to reabsorb them. This protein loss would decrease the osmotic pressure associated with the plasma and thus increase the net filtration pressure, and filtration, at capillary beds, resulting in edema.**

79. (a) **Correct.** (This test would not be diagnostic unless the patient is normally hydrated, however. Remember that GFR will vary according to hydration and also with emotional stimuli.)

(b) **Incorrect.** **Increased plasma creatinine is diagnostic of decreased GFR.** (But again, someone who had just engaged in strenuous exercise would have elevated creatinine.)

(c) **Correct.**

Some people find the determination of clearance confusing. If we rewrite the equation on text page 507 in the form

$$\underset{(mg/min)}{\text{Creatinine excreted}} = \underset{(mg/mL)}{\text{plasma creatinine conc.}} \times \underset{(mL/min)}{\text{GFR}}$$

then it is easy to see why dividing the amount of creatinine excreted per unit time by the plasma concentration gives an estimate of the filtration rate of the creatinine, and thus of the plasma.

80. People with acute or chronic kidney failure can be kept alive by undergoing one of two kinds of **dialysis**. The more efficient of these methods is called **hemodialysis** and is also known as the "artificial kidney." The other method, **peritoneal dialysis**, utilizes the patient's own **peritoneum** as the selectively permeable membrane. This latter method has the advantage of allowing **some normal activities** during the treatment. Neither method offers a cure for chronic kidney failure. The only known cure is **a kidney transplant**.

PRACTICE

True or false (correct the false statements):

P1. The nephron has both a vascular component and a tubular component.

P2. The glomerulus is involved in filtration, secretion, and reabsorption; the peritubular capillaries serve only to exchange O_2 and CO_2 with the cells.

P3. Renal reabsorption of glucose is a T_m-limited active process.

P4. Only substances that are filtered by the kidneys can be excreted by them.

P5 Urine at the end of the collecting ducts is similar in composition to the urine in the urinary bladder.

P6. Unlike water, sodium is filtered, reabsorbed, and secreted.

P7. Under normal conditions, the concentration of glucose in the distal tubular fluid is about one-third that in Bowman's capsule.

P8. The countercurrent mechanism of the kidney allows for the formation of hypertonic urine.

P9. The walls of the ascending limb of the loop of Henle are freely permeable to water.

P10. The fluid entering the distal tubule is hypotonic with respect to plasma.

P11. At the bend in the loop of Henle, the osmolarity of the tubular fluid is several times that of the glomerular filtrate.

P12. The main force responsible for water reabsorption from the collecting ducts is the low hydrostatic pressure in the surrounding interstitial space.

P13. Excessive secretion of renin by the kidneys causes excess water retention and an increase in blood pressure.

P14. High levels of sodium chloride in the tubular fluid of the macula densa leads to increased secretion of aldosterone.

P15. A fall in the osmolarity of the blood supplying the hypothalamus is a powerful stimulus for thirst.

P16. A substance that interferes with the active transport of sodium in the distal tubules will also interfere with potassium reabsorption.

P17. The primary stimulus for parathyroid hormone secretion is increased plasma potassium levels.

P18. One response to increased hydrogen ion production in the body is decreased reabsorption of bicarbonate ion by the kidneys.

P19. Bicarbonate ions are secreted into the urine by the tubular cells.

P20. Because most of the filtered NaCl and H_2O is reabsorbed in the proximal tubule, a substance that interferes with active Na^+ reabsorption there is an effective diuretic.

Multiple choice (correct each incorrect choice:)

P21-23. With regard to sodium and water reabsorption,

P21. water reabsorption is passive and depends entirely on sodium reabsorption.

P22 the excretion of large amounts of sodium always results in the excretion of large amounts of water.

P23. the amount of sodium excreted is equal to the amount filtered plus the amount secreted.

P24-26. Drinking two liters of isotonic saline would result in

P24. increased sympathetic activity to the renal arterioles.

P25. increased GFR.

P26. stimulation of volume receptors in the atria of the heart.

P27-29. Urine volume will increase following

P27. a fall in plasma volume.

P28. inhibition of tubular sodium reabsorption.

P29. ingestion of alcohol.

P30-32. During severe exercise in a hot climate, a person may lose several liters of body fluid as hypotonic sweat, resulting in

P30. decreased plasma volume.

P31. decreased plasma osmolarity.

P32. increased circulating ADH.

P33-35. Calcium homeostasis

P33. is dependent on the renin-angiotensin system.

P34. is regulated by the actions of hormones on the gastrointestinal tract, bone, and liver.

P35. requires parathyroid hormone and vitamin C.

16: THE DIGESTION AND ABSORPTION OF FOOD

Serendipity — the accidental or unexpected discovery of something not looked for — has often played an important role in science. One example of serendipity that has increased our understanding of the functions of the stomach was the accidental shooting, in 1822, of a young French Canadian named Alexis St Martin. Happily for him, a United States Army surgeon named William Beaumont was nearby. Beaumont was able to remove the musket shot from the large hole in St Martin's lower chest and push back the protruding organs including the stomach, which had been pierced by the shot. Even more happily for St Martin, he recovered from his wounds and within a week or so began to eat — but the food came tumbling out of the hole in his stomach!

Beaumont devised a bandage that could substitute for the missing stomach wall, and St Martin could eat and drink. But the hole — termed "Beaumont's window" by science journalist Anthony Smith — allowed Beaumont a close-up view of gastric physiology that answered many questions. Beaumont took samples of gastric juice at various times before and after different kinds of meals, and established for the first time that hydrochloric acid is secreted by cells in the lining of the stomach, both during digestion and in anticipation of eating. He also noticed that acid secretion occurred when his young patient became angry. Thus, reasoned Beaumont, our digestive functions can be altered by emotional states.

In this chapter you will learn much about how the stomach and the other organs forming the gastrointestinal system — the body's "disassembly line" — function in their very important roles of bringing nutrients from the external environment into the internal one.

Instructions for answering questions in this Study Guide

1. True or false: Correct each false statement. Whenever you must correct a false statement, there will almost always be more than one way to do so. In the Guide, the answer requiring the least correction will generally be given, with the understanding that other correct answers are possible. It is usually not sufficient, in terms of demonstrating understanding, to simply insert a "no" or "not," however.

2. Multiple choice: Any, all, or none of the responses may be correct. Explain why each incorrect response is incorrect.

3. Fill in the blanks: Choose the best word or words to complete each statement.

4. Directions will be given for other types of questions as they appear.

❒ Introduction (text pages 513-515)

1. True or false: The lumen of the gastrointestinal tract is continuous with the external environment.

2. True or false: The liver and pancreas are components of the gastrointestinal tract.

3. List the four fundamental processes of the gastrointestinal system:
   ______________________, ______________________,
   ______________________, and ______________________.

4. From inspection of text Figure 16-2,
   (a) most digestion occurs in the esophagus and stomach.
   (b) most absorption occurs across the large intestine.
   (c) the absorbed products of carbohydrate and protein digestion go directly to the liver.

❒ Overview: Functions of the Gastrointestinal Organs (text pages 515-518)

5. Saliva
   (a) is produced by endocrine glands.
   (b) is important for the sensation of taste.
   (c) aids the swallowing of food.
   (d) begins the digestion of protein.

6. The functions of the stomach include
   (a) the complete digestion of protein.
   (b) dissolving all the food that enters it.
   (c) killing bacteria.
   (d) storing partially digested food.

7. True or false: The pH of the contents of the stomach lumen is considerably higher than that of interstitial fluid.

8. The small intestine has three anatomically distinct segments: the ________________ leading from the stomach, the ________________ in the middle, and the ________________, which connects to the large intestine. Most digestion and absorption occur in the ________________ and the ________________. The products of protein, carbohydrate, and fat digestion — ________________, ________________, and ________________, respectively — cross the epithelial cells and enter the ________________ or the ________________.

9. True or false: Much of the material secreted into the small intestine comes from the liver and pancreas.

10. In the small intestine,
   (a) acidic chyme is neutralized.
   (b) mineral ions are actively absorbed.
   (c) water molecules are actively absorbed.
   (d) fatty acids are actively absorbed.

11. Match the substance or structure on the right with the correct description on the left:

____ digest(s) protein	(a) bile salts
____ digest(s) carbohydrate	(b) gallbladder
____ dissolve(s) fat	(c) liver
____ dissolve(s) proteins and polysaccharides	(d) pancreas
____ secrete(s) bile	(e) amylase
____ store(s) bile	(f) pepsin
____ secrete(s) bicarbonate	(g) hydrochloric acid
	(h) large intestine

12. True or false: A greater volume of material enters the lumen of the gastrointestinal tract from the body than is ingested as food and water.

13. True or false: The digestion and absorption of food are highly inefficient processes because so much secreted material is lost in the feces.

❐ Structure of the Gastrointestinal Tract Wall (text pages 518-520)

14. The four layers of the gastrointestinal tract wall, beginning on the luminal side, are the ________, the ________, the ________, and the ________. Nerve networks called ________ are found in two layers: the ________ and the ________. Smooth muscle layers are found in the ________ and ________. The mucosa is lined by a layer of ________ that includes two kinds of secretory cells: ________ cells that secrete into the lumen and ________ cells that secrete into the blood.

15. True or false: Contraction of the circular smooth muscle in the gastrointestinal tract wall decreases the diameter of the lumen, while contraction of the longitudinal smooth muscle shortens the tract.

16. Villi
   (a) are found throughout the gastrointestinal tract.
   (b) increase the surface area of the intestinal mucosa.
   (c) are covered by connective tissue.
   (d) are in close proximity to capillaries and lacteals.

❐ Digestion and Absorption (text pages 520-526)

17. Regarding the digestion and absorption of carbohydrates,
   (a) polysaccharides are broken down to monosaccharides by amylase.
   (b) amylase is activated by acid.
   (c) sucrose and maltose are actively transported across the small intestinal epithelium.
   (d) glucose absorption is accomplished by facilitated diffusion.

18. True or false: Following a "meal" that consists solely of carbohydrate (for example, sugar cubes), no amino acids will be absorbed by the small intestine.

19. True or false: Short chains of amino acids and some intact proteins are absorbed from the small intestine.

20. True or false: An aminopeptidase is an enzyme that catalyzes the cleavage of amino acids from the carboxyl terminus of a peptide chain.

21. True or false: Three enzymes secreted by the pancreas and important for protein digestion are trypsin, chymotrypsin, and pepsin.

22. Lipase
    (a) catalyzes the breakdown of triacylglycerol to glycerol and three fatty acids.
    (b) emulsifies lipids.
    (c) is secreted by the endocrine pancreas.

23. The emulsification of fats
    (a) results in the formation of small fat droplets.
    (b) results in the formation of micelles.
    (c) depends upon the amphipathic structure of bile salts.
    (d) allows fats to become water-soluble.

24. True or false: The products of lipid digestion diffuse into the intestinal epithelium as micelles.

25. True or false: The function of micelles is to store the products of lipid digestion in water-soluble form.

26. The products of lipid digestion
    (a) are resynthesized into triacylglycerol in intestinal epithelial cells.
    (b) are packaged for secretion by the granular endoplasmic reticulum of intestinal epithelial cells.
    (c) diffuse into capillaries in the vicinity of the intestinal epithelium.
    (d) travel in the circulatory system in the form of micelles.

27. Fat-soluble vitamins
    (a) must be digested to fatty acids and monoglyceride.
    (b) include vitamins A, B_{12}, and D.
    (c) circulate in chylomicrons.

28. Intrinsic factor
    (a) is necessary for digestion of vitamin B_{12}.
    (b) is secreted by cells in the gastric mucosa.
    (c) lack is associated with anemia.

29. True or false: The most abundant solute in chyme is iron.

30. True or false: The intestinal epithelium is permeable to water only in the presence of ADH.

31. The absorption of iron
    (a) is active.
    (b) is generally 100% of the amount ingested.
    (c) results in storage as ferritin in intestinal epithelial cells.
    (d) is regulated by the amount of ferritin in mucosal cells.

❐ **Regulation of Gastrointestinal Processes** (text pages 526-545)

32. True or false: Regulation of digestive processes depends upon reflex arcs.

33. True or false: Neural regulation of digestive processes is accomplished exclusively by the enteric nervous system of the gastrointestinal tract.

34. Receptors for digestive reflexes
    (a) are located in the gastrointestinal tract wall.
    (b) include chemoreceptors, osmoreceptors, and mechanoreceptors.
    (c) may relay information to integrative centers in the CNS or the enteric plexuses.
    (d) may be endocrine cells.

35. True or false: The long neural digestive reflex arcs bypass the enteric nervous system.

36. Regarding endocrine regulation of gastrointestinal processes,
    (a) the hormone secreted by the stomach is secretin.
    (b) chemoreceptors for glucose and fat in the small intestine secrete GIP.
    (c) CCK is the primary efferent pathway for a reflex arc controlling the levels of acid in the small intestine.
    (d) gastrin secretion is controlled in part by parasympathetic nerves.

37. The three phases of gastrointestinal control
    (a) refer to the places in the gastrointestinal tract where the digestion or absorption is occurring.
    (b) include phases in the mouth, stomach, and intestines.
    (c) include short and long reflexes.

38. True or false: The presence of acid and peptides in the stomach stimulates the gastric phase of gastrointestinal control.

39. True or false: The presence of food in the mouth stimulates salivary gland secretion.

40. Swallowing
    (a) is a reflex.
    (b) is coordinated by centers in the brainstem.
    (c) requires cessation of respiration.
    (d) includes closure of the glottis.
    (e) includes contraction of the upper esophageal sphincter.

41. True or false: Food travels down the esophagus primarily under the influence of gravity.

42. True or false: The presence of food in the esophagus is a stimulus for secondary peristalsis.

43. Regurgitation of food from the stomach into the esophagus
    (a) is ordinarily inhibited by the upper esophageal sphincter.
    (b) causes irritation of the esophageal mucosa.
    (c) causes heartburn.
    (d) is a common problem during pregnancy.

44. The exocrine glands of the gastric mucosa
    (a) are most elaborate in the fundus and body of the stomach.
    (b) contain cells that secrete protective mucus onto the mucosa.
    (c) contain chief cells that secrete HCl.
    (d) contain parietal cells that secrete pepsinogen.

45. Regarding the secretion of HCl,
    (a) hydrogen ion is actively transported out of parietal cells by ATPase pumps in the mucosal membrane.
    (b) it would cease if bicarbonate ions were not simultaneously secreted into the interstitial fluid.
    (c) the source of the secreted hydrogen ions is carbonic acid.
    (d) the pH of the blood leaving the area of the parietal cells is lower than normal.

46. HCl secretion is stimulated
    (a) during the gastric and intestinal phases of gastrointestinal control and inhibited during the cephalic phase.
    (b) by acetylcholine and gastrin.
    (c) by the smell of food.

47. True or false: The amount of HCl secreted during a meal is independent of the type of food ingested.

48. Pepsin
    (a) is secreted in the form of an inactive precursor.
    (b) molecules activate other pepsin molecules.
    (c) is inactivated in the presence of acid.
    (d) indirectly stimulates secretion of acid.

49. True or false: The receptive relaxation of the stomach smooth muscle that allows for increased volume with little pressure occurs during the gastric phase of digestive control.

50. The pyloric sphincter
   (a) is a muscle separating the esophagus from the fundus.
   (b) is generally closed when no food is in the stomach.
   (c) closes at the approach of a peristaltic wave.

51. With regard to gastric motility,
   (a) the basic electrical rhythm of the gastric smooth muscle is three depolarizations per minute regardless of whether or not food is present.
   (b) gastric smooth muscle contracts three times a minute regardless of whether or not food is present.
   (c) the force of contraction is increased by gastrin and inhibited by enterogastrones.
   (d) the rate of contraction is increased by gastrin and inhibited by enterogastrones.

52. True or false: In general, sympathetic stimulation increases gastric secretion and motility, while parasympathetic stimulation decreases these two processes.

53. Vomiting
   (a) is a reflex coordinated by the enteric nervous system.
   (b) may be initiated by chemicals in the gastrointestinal tract.
   (c) has adaptive value.

54. The secretions of the exocrine pancreas include ______________________________ by cells in the glands and ______________________________ by cells lining the ducts. Secretion of the former is stimulated primarily by ______________________________, while that of the latter is stimulated primarily by ______________________________. The secretions are carried to the lumen of the gastrointestinal tract by the pancreatic duct, which joins the ______________________________ duct just before it enters the ______________________________.

55. True or false: Normally the amount of bicarbonate ions secreted into the intestine about equals the amount of acid secreted into the stomach.

56. True or false: Severe diarrhea can lead to metabolic alkalosis.

57. Enterokinase
   (a) is an enzyme produced by the pancreas.
   (b) is activated by trypsin.
   (c) mediates the first step in activating digestive enzymes.

58. A person who lacked the exocrine portion of the pancreas would not be able to
   (a) digest fats.
   (b) digest proteins.
   (c) digest starches.
   (d) neutralize gastric acid.

59. In the liver,
   (a) portal-vein blood and hepatic artery blood mix in the hepatic capillary beds.
   (b) bile in the canaliculi mixes with blood in the hepatic capillary beds.
   (c) bile canaliculi merge to form the common bile duct.

60. A person without a gall bladder
   (a) cannot secrete bile.
   (b) cannot store bile.
   (c) will have difficulty digesting a large, fat-rich meal.

61. True or false: About 95% of the bile secreted by the liver is recycled back to the liver by the enterohepatic circulation.

62. True or false: The same hormone that stimulates secretion of bicarbonate ion by the pancreas also stimulates bicarbonate ion secretion into the bile.

63. True or false: The same hormone that stimulates pancreatic enzyme secretion stimulates hepatic secretion of bile salts.

64. Bile pigments
   (a) are important for fat digestion.
   (b) are formed from catabolism of the globin part of hemoglobin.
   (c) are yellowish.
   (d) impart color to the bile, feces, and urine.

65. True or false: The presence of high-osmolarity chyme in the small intestine stimulates both water movement into the intestinal lumen from the blood and the discharge of more chyme from the stomach into the duodenum.

66. Regarding the motility of the small intestine,
   (a) it varies depending upon whether a meal is being digested and absorbed.
   (b) during a meal, intestinal motility functions primarily to mix chyme and bring it into contact with intestinal mucosa.
   (c) it is the same down the entire length of the small intestine.
   (d) segmentation is the primary means of propelling unabsorbed matter toward the large intestine.

67. True or false: The intestino-intestinal reflex coordinates the emptying of chyme from the ileum into the colon with the emptying of chyme from the stomach into the duodenum.

68. True or false: The primary function of the large intestine is to store and dilute unabsorbed fecal material.

69. True or false: Because the large intestine has a greater diameter than the small intestine, the large intestine has a greater surface area.

70. True or false: Flatus is a result of air being swallowed with food.

71. True or false: Voluntary control of defecation requires learning to keep the external anal sphincter smooth muscle contracted.

❑ **Pathophysiology of the Gastrointestinal Tract** (text pages 545-547)

72. Describe two properties of gastric mucus that enable it to help protect the gastric mucosa from damage by acid and pepsin.
   (a) ______________________________
   (b) ______________________________

73. Ulcers are
 (a) most common in the gastric mucosa.
 (b) always caused by hypersecretion of gastric acid.
 (c) treated by methods that inhibit acid secretion.

74. Gallstones
 (a) are crystallized bile salts.
 (b) may cause pain.
 (c) may cause impaired fat digestion.
 (d) may cause general nutritional deficiencies.
 (e) may cause jaundice.

75. Lactose intolerance
 (a) is an inability to digest milk sugar.
 (b) is most common in very young children.
 (c) can cause painful bloating and diarrhea.

76. Constipation
 (a) is associated with symptoms caused by accumulation of toxins in the stool.
 (b) is caused by failure to defecate at least once a day.
 (c) is most often a problem of the elderly.
 (d) may be relieved by ingestion of foods with a high proportion of undigestible carbohydrates.

77. True or false: Any agent that interferes with water absorption or that causes water secretion into the gastrointestinal tract can cause diarrhea.

ANSWERS

Boldface type indicates the answers you should have given. Words in medium-face (ordinary) type explain or expand upon the answer.

1. **True.**

2. **False.** The liver and pancreas are components of the gastrointestinal ~~tract~~ **system**.

3. **digestion**, **secretion**, **absorption**, and **motility**

4. (a) **Incorrect.** **Most digestion occurs in the stomach and** (early part of the) **small intestine**.
   (b) **Incorrect.** **Most absorption occurs across the small intestine.**
   (c) **Correct.** (The hepatic [adjective for liver] portal vein is the second portal system — that is, one set of capillary beds draining into a vein that then breaks up into another set of capillary beds — we have come across. The other is the hypothalamo-pituitary portal system.)

5. (a) **Incorrect.** **It is produced by exocrine glands** (glands with ducts).
   (b) **Correct.**
   (c) **Correct.**
   (d) **Incorrect.** **It begins the digestion of starch.**

6. (a) **Incorrect.** **Protein is only partially digested by the pepsin in the stomach.**
   (b) **Incorrect.** **It dissolves everything but fat.**
   (c) **Correct.**
   (d) **Correct.** (Release of the partially digested food — chyme — into the small intestine is regulated, as we shall see.)

7. **False.** The pH of the contents of the stomach lumen is considerably ~~higher~~ **lower** than that of interstitial fluid. (The pH of gastric fluid is about 2.0 compared to 7.4 for blood.)

8. The small intestine has three anatomically distinct segments: the **duodenum** leading from the stomach, the **jejunum** in the middle, and the **ilium**, which connects to the large intestine. Most digestion and absorption occur in the **duodenum** and the **jejunum**. The products of protein, carbohydrate, and fat digestion — the **amino acids**, **monosaccharides**, and **fatty acids**, respectively, — cross the epithelial cells and enter the **blood** or **lymph**.

9. **True.** (See text Figure 16-4.)

10. (a) **Correct.**
    (b) **Correct.**
    (c) **Incorrect.** **Water is absorbed passively, along an osmotic gradient.**
    (d) **Incorrect.** **Fatty acids diffuse across the epithelial cell membranes.**

11. **f** digest(s) protein
e digest(s) carbohydrate
a dissolve(s) fat
g dissolve(s) proteins and polysaccharides
c secrete(s) bile
b store(s) bile
d (and c) secrete(s) bicarbonate

12. **True.**

13. **False.** The digestion and absorption of food are highly ~~in~~**efficient** processes because so ~~much~~ **little** secreted material is lost in the feces.

14. The four layers of the gastrointestinal tract wall, beginning on the luminal side, are the **mucosa**, the **submucosa**, the **muscularis externa,** and the **serosa**. Nerve networks called **plexuses** are found in two layers: the **submucosa** and the **muscularis externa**. Smooth muscle layers are found in the **mucosa** and **muscularis externa**. The mucosa is lined by a layer of **epithelium** that includes two kinds of secretory cells: **exocrine** cells that secrete into the lumen and **endocrine** cells that secrete into the blood.

15. **True.** (Coordination of the contractions of these smooth muscle layers produces gastrointestinal motility, which we shall learn more about presently.)

16. (a) **Incorrect. They are found only in the small intestine.**
(b) **Correct.**
(c) **Incorrect. They are covered by epithelial cells having microvilli.**
(d) **Correct.**

17. (a) **Incorrect. They are broken down to smaller polysaccharides and to disaccharides by amylase.** (Each disaccharide has its own specific enzyme. That is, maltose is broken down to glucose by malt*ase*, lactose to glucose and galactose by lact*ase*, and so on.)
(b) **Incorrect. It is inactivated by acid.**
(c) **Incorrect. Sucrose and maltose are disaccharides and are not absorbed until they are digested to monosaccharides.**
(d) **Incorrect. Glucose absorption is accomplished by secondary active transport.**

18. **False.** Following a "meal" that consists solely of carbohydrate (for example, sugar cubes), ~~no~~ amino acids **from the digestion of secreted digestive enzymes and mucus** will be absorbed by the small intestine.

19. **True.**

20. **False.** An aminopeptidase is an enzyme that catalyzes the cleavage of amino acids from the ~~carboxyl~~ **amino** terminus of a peptide chain.

21. **False.** Three enzymes secreted by the pancreas and important for protein digestion are trypsin, chymotrypsin and ~~pepsin~~ **carboxypeptidase**.

22. (a) **Incorrect. The breakdown is to a monoglyceride and two fatty acids.**
    (b) **Incorrect. Bile salts emulsify lipids.**
    (c) **Incorrect. It is secreted by the exocrine pancreas.**

23. (a) **Correct.**
    (b) **Incorrect. It results in an emulsion.** (Micelle formation is a separate process, but it would not proceed efficiently without emulsification.)
    (c) **Correct.**
    (d) **Incorrect. Only fats in micelles are water-soluble.**

24. **False.** The products of lipid digestion diffuse into the intestinal epithelium as ~~micelles~~ **free fatty acids and monoglycerides.**

25. **True.**

26. (a) **Correct.**
    (b) **Incorrect. They are packaged by the Golgi apparatus of intestinal epithelial cells.**
    (c) **Incorrect. They diffuse into lacteals in the vicinity of the intestinal epithelium.**
    (d) **Incorrect. They travel in the form of chylomicrons.**

27. (a) **Incorrect. Fat-soluble vitamins must be dissolved in micelles in order to be absorbed, but they are not digested.**
    (b) **Incorrect. Vitamin B_{12} is water-soluble.**
    (c) **Correct.**

28. (a) **Incorrect. It is necessary for *absorption* of vitamin B_{12}.**
    (b) **Correct.**
    (c) **Correct.**

29. **False.** The most abundant solute in chyme is ~~iron~~ **sodium**.

30. **False.** The intestinal epithelium is **always** permeable to water ~~only in the presence of ADH~~. (It is fortunate that this is so, because people lacking ADH can absorb enough water to replace what is lost in urine. As a general rule, molecules whose plasma

concentrations are regulated by the kidneys do not have their absorption regulated. Absorption of trace elements, however, is generally regulated.)

31. (a) **Correct.**

    (b) **Incorrect.** **Only about 10% of ingested iron is absorbed** (except in cases of iron deficiency, and even then, absorption does not reach 100%).

    (c) **Correct.**

    (d) **Correct.** (As the text mentions, the feedback mechanisms are unclear, but iron uptake and release into blood are clearly correlated with cellular iron stores.)

32. **True.**

33. **False.** Neural regulation of digestive processes is accomplished ~~exclusively~~ **in part** by the enteric nervous system of the gastrointestinal tract. (Do not be puzzled by this use of the term "nervous system" to describe the digestive-tract neural plexuses. Some investigators include the enteric nervous system as a third branch of the autonomic nervous system, but others prefer to treat it as a special local regulatory system — in other words, to ignore it until gastrointestinal physiology is discussed.)

34. (a) **Correct.**
    (b) **Correct.**
    (c) **Correct.**
    (d) **Correct.**

35. **False.** The long neural digestive reflex arcs ~~bypass~~ **interact with** the enteric nervous system.

36. (a) **Incorrect.** **It is gastrin.** (This is easy to remember because "gastro-" refers to the stomach.)

    (b) **Correct.** (The GIP story illustrates how problems with nomenclature can arise in physiology. When this peptide, which resembles secretin in structure, was first identified, its function was thought to be inhibition of gastric secretion and motility. Accordingly, it was named "gastric inhibitory peptide." Other scientists questioned the physiological relevance of GIP's gastric-inhibitory activity, however, because supraphysiological amounts of it were required to produce an effect. Thus, GIP may be just a weak agonist for secretin in terms of its inhibitory effects [see text Table 16-2, page 528], but the acronym is here to stay. GIP does stimulate the secretion of insulin by endocrine cells of the pancreas, however, which has prompted the awkward-sounding "glucose insulinotropic peptide." The best bet is simply to call it GIP and let it go at that.)

(c) **Incorrect.** **Secretin is the primary efferent pathway for acid in the small intestine.** (Note that the stimulus for secretin secretion is acid in the small intestine. Secretin decreases that acid by stimulating bicarbonate-rich secretion by the pancreas and liver, and also by inhibiting both acid secretion by the stomach and the rate of gastric emptying.)

(d) **Correct.**

37. (a) **Incorrect.** **They refer to the location of the stimuli that reflexly initiate the control.**

(b) **Correct.** (Note that the "cephalic phase" includes all stimuli in the head, including those in the mouth.)

(c) **Correct.**

38. **True.**

39. **True.**

40. (a) **Correct.**

(b) **Correct.**

(c) **Correct.**

(d) **Correct.**

(e) **Incorrect.** **It includes relaxation of the upper sphincter.**

41. **False.** Food travels down the esophagus primarily under the influence of ~~gravity~~ **peristaltic waves**.

42. **True.**

43. (a) **Incorrect.** **It is ordinarily inhibited by the *lower* esophageal sphincter.**

(b) **Correct.**

(c) **Correct.**

(d) **Correct.**

44. (a) **Correct.**

(b) **Correct.**

(c) **Incorrect.** **The chief cells secrete pepsinogen.**

(d) **Incorrect.** **Parietal cells secrete HCl.** (This is a case where remembering "P stands for parietal and pepsin" will help you to remember it *wrong*. It may be helpful to remember that this is a "backwards" case.)

45. (a) **Correct.**

(b) **Correct.** (Examine text Figure 16-17. Note that the transport of bicarbonate ions out of the cell allows the reaction catalyzed by carbonic anhydrase to proceed to the right, that is, to the formation of carbonic acid. If transport of bicarbonate ion stopped, mass action would drive the reaction to the left, and there would no longer be hydrogen ion available to neutralize the hydroxyl ion left when water is broken down to hydrogen ion [which is secreted] and hydroxyl ion [which must be neutralized]. A second reason bicarbonate ion secretion is important is that this secretion is linked with the uptake of chloride ions, which are secreted into the lumen along with the hydrogen ions.)

(c) **Incorrect. The source is water.**

(d) **Incorrect. The pH is higher than normal** (less acidic) **because of the high levels of bicarbonate buffer.**

46. (a) **Incorrect. It is stimulated during the gastric and cephalic phases of gastrointestinal control and inhibited during the intestinal phase.**

(b) **Correct.**

(c) **Correct.**

47. **False.** The amount of HCl secreted ~~during a meal~~ is ~~independent of the type of food ingested~~ **greater during a protein-rich meal than during a protein-poor one.**

48. (a) **Correct.** (These precursor enzymes are often called "proenzymes.")

(b) **Correct.** (They are autocatalytic.)

(c) **Incorrect. Pepsin is inactive in the *absence* of acid.** (Pepsin is a very unusual enzyme in that it requires a pH of about 2 for optimal catalytic activity. This requirement is in addition to the need for acid to alter the shape of pepsinogen so that it can cleave other pepsinogen molecules to the active pepsin.)

(d) **Correct.** (By digesting proteins to peptides, which then stimulate the acid secretion.)

49. **False.** The receptive relaxation of the stomach smooth muscle that allows for increased volume with little pressure occurs during the ~~gastric~~ **cephalic** phase of digestive control.

50. (a) **Incorrect. It separates the antrum from the duodenum.**

(b) **Incorrect. It is generally open.**

(c) **Correct.**

51. (a) **Correct.**
    (b) **Incorrect.** **Depolarizations are strong enough to trigger action potentials and contractions only when stimulated by hormones or neural signals** (as happens when food is either present or sensed).
    (c) **Correct.**
    (d) **Incorrect.** **The rate of contraction stays the same.**

52. **False.** In general, sympathetic stimulation ~~increases~~ **decreases** gastric secretion and motility, while parasympathetic stimulation ~~decreases~~ **increases** these two processes.

53. (a) **Incorrect.** **It is coordinated by the medullary vomiting center.**
    (b) **Correct.**
    (c) **Correct.** (As the text points out, removing toxic chemicals before they can be absorbed is beneficial, and the intensely unpleasant experience of nausea is helpful for learning not to eat the offending food again. Some people unwisely use the vomiting reflex intentionally as a form of dieting. They induce vomiting by tickling the back of their throat to expel the contents of a meal before it can be absorbed and therefore experience the joy of eating without the side effect of gaining weight. As the text points out, however, frequent vomiting causes dangerous salt, water, and circulatory imbalances. It can even be fatal.)

54. The secretions of the exocrine pancreas include **digestive enzymes** by cells in the glands and **bicarbonate ion** by cells lining the ducts. Secretion of the former is stimulated primarily by **CCK,** while that of the latter is stimulated primarily by **secretin**. The secretions are carried to the lumen of the gastrointestinal tract by the pancreatic duct, which joins the **common bile** duct just before it enters the **duodenum**.

55. **True.**

56. **False.** Severe diarrhea can lead to metabolic ~~alkalosis~~ **acidosis due to the net loss of bicarbonate ion**.

57. (a) **Incorrect.** **It is produced by intestinal epithelial cells**.
    (b) **Incorrect.** **It activates trypsin, not the reverse.**
    (c) **Correct.**

58. (a) **Correct.** (The pancreas is the only source of lipase.)
    (b) **Incorrect.** **He or she would have impaired protein digestion, but it would not be lacking entirely.** (Recall that the stomach secretes pepsin, which breaks down proteins to smaller peptide chains, and that the intestinal epithelium makes aminopeptidase.)

(c) **Incorrect.** **As with (b), starch digestion would be impaired but not absent.** (Recall the salivary amylase and the maltase, lactase, and sucrase secreted by the intestinal epithelial cells.)

(d) **Incorrect.** **The liver also secretes some bicarbonate ion into the gastrointestinal tract lumen.**

59. (a) **Correct.** (Examine text Figure 16-27. The circular and oval structures — the portal vein, hepatic artery, bile duct, and central vein — should be viewed as cross-sections of three-dimensional structures perpendicular to the plane of the page. The "blood" indicated is in a capillary. According to this arrangement, all hepatocytes are in close proximity to blood from the heart and to blood from the gastrointestinal system.)

(b) **Incorrect.** **Bile and blood do not mix.**

(c) **Incorrect.** **They merge to form the common hepatic duct, which becomes the common bile duct beyond the gallbladder.**

60. (a) **Incorrect.** **Bile is continuously secreted by the liver, with or without a gallbladder present.**

(b) **Correct.**

(c) **Correct.** (The adaptive value of a gallbladder is to allow relatively large amounts of bile to be secreted at once. Animals that lack a gallbladder or humans who have it surgically removed cannot digest large quantities of fat efficiently.)

61. **False.** About 95% of the bile **salts** secreted by the liver ~~is~~ **are** recycled back to the liver by the enterohepatic circulation. (Most of the bile pigments and trace metals are excreted in the feces, as is some portion of the cholesterol and lecithin. The largest component of bile, like all digestive secretions, is water, which is absorbed all along the intestine.)

62. **True.** (Secretin)

63. **False.** The same hormone that stimulates pancreatic enzyme secretion stimulates ~~hepatic secretion of bile salts~~ **contraction of the gallbladder and relaxation of the sphincter of Oddi**.

64. (a) **Incorrect.** **They have nothing to do with fat digestion.** (They are solely excretory products.)

(b) **Incorrect.** **They are formed from catabolism of heme.** (The globin, being a protein, is broken down to amino acids. The iron part of the heme is recycled for new hemoglobin formation.)

(c) **Correct.** (Except for the brown pigments that have been further metabolized by bacteria in the intestines.)

(d) **Correct.**

65. **False.** The presence of high-osmolarity chyme in the small intestine stimulates ~~both~~ water movement into the intestinal lumen from the blood and **inhibits** the discharge of more chyme from the stomach into the duodenum.

66. (a) **Correct.**

(b) **Correct.**

(c) **Incorrect.** **The rate of segmentation movements gets lower moving from duodenum to jejunum to ileum.**

(d) **Incorrect.** **The peristaltic waves of the migrating motility complex are the primary means of moving material toward the large intestine.**

67. **False.** The ~~intestino-intestinal~~ **gastroileal** reflex coordinates the emptying of chyme from the ileum into the colon with the emptying of chyme from the stomach into the duodenum.

68. **False.** The primary function of the large intestine is to store and ~~dilute~~ **concentrate** unabsorbed fecal material.

69. **False.** ~~Because~~ **Even though** the large intestine has a greater diameter than the small intestine, the large intestine has a ~~greater~~ **lesser** surface area.

70. **False.** Flatus is a result of ~~air being swallowed with food~~ **gas produced by bacteria in the large intestine** (primarily from undigested carbohydrates).

71. **False.** Voluntary control of defecation requires learning to keep the external anal sphincter ~~smooth~~ **skeletal** muscle contracted. (Recall that only skeletal muscles can be contracted voluntarily.)

72. (a) **It is slightly alkaline, which helps neutralize acid and inactivate pepsin.**

(b) **It contains protein mucins that also bind and neutralize acid.**

73. (a) **Incorrect.** **They are most common in the duodenum.**

(b) **Incorrect.** **Many ulcer patients have normal acid secretion.**

(c) **Correct.**

74. (a) **Incorrect.** **They are either crystallized cholesterol or crystallized bile pigments.**

(b) **Correct.**

(c) **Correct.**

(d) **Correct.**

(e) **Correct.**

(For all these reasons, people who experience gallstones frequently have their gallbladders removed. This surgery combined with a low-fat diet generally solves the problem.)

75. (a) **Correct.**

(b) **Incorrect. It is most common in adults.** (It is fortunate that it is very *un*common in very young children, who rely on milk for their sole source of nourishment.)

(c) **Correct.**

Note: Although the text mentions the percentage of white Americans who have diminished lactose tolerance, it fails to state that this percentage is far higher in nonwhite Americans and in most people who do not live in western Europe. In only a few countries in Africa, for example, cattle are valued highly because people can drink cow's milk. In other countries, no one drinks milk.

76. (a) **Incorrect. It is associated with symptoms caused by distension of the rectum.**

(b) **Incorrect. It is caused by failure to defecate before the rectum becomes distended, usually after several days.**

(c) **Correct.**

(d) **Correct.**

77. **True.**

PRACTICE

True or false (correct the false statements):

P1. Most food absorption occurs in the first quarter of the small intestine.

P2. The volume of fluids secreted by the gastrointestinal tract in a day is far greater than the volume of food and drink ingested.

P3. Glucose absorption occurs by active transport.

P4. Polysaccharides must be broken down to disaccharides in order to be absorbed.

P5. In patients lacking exocrine pancreas secretion, fat digestion is normal provided bile is still produced.

P6. The breakdown products of dietary fats are resynthesized into fat by intestinal cells and pass from these cells into lacteals.

P7. Removal of the stomach would interfere with absorption of vitamin B_{12}.

P8. "Heartburn" following a large meal is usually due to pressure of the stomach against the heart.

P9. Gastric chief cells secrete pepsin.

P10. The secretion of gastrin by cells in the stomach lining is stimulated by increasing gastric acidity.

P11. Prolonged vomiting results in dehydration and reduced acidity of the blood.

P12. Secretion of secretin is stimulated by the presence of fat in the duodenum, whereas CCK release is triggered by the presence of acid.

P13. Bile secreted by the liver contains the major enzymes for digesting fats.

P14. Bile pigments are essential for the adequate digestion of fat.

P15. The breakdown products of hemoglobin are eliminated from the body as bile salts in the feces.

P16. Reabsorption of bile salts leads to further secretion of bile.

P17. During a meal, peristalsis is the predominant form of movement in the small intestine.

P18. Lactose intolerance is caused by an amylase deficiency.

Multiple choice (correct each incorrect choice):

P19-21. Hydrochloric acid in the stomach

- P19. kills bacteria.
- P20. breaks down proteins into amino acids.
- P21. converts pepsinogen into pepsin.

P22-24. The pancreas

- P22. contains both endocrine and exocrine glands.
- P23. secretes large quantities of enzymes in response to the hormone secretin.
- P24. neutralizes the hydrochloric acid secreted by the stomach.

P25-27. Hydrochloric acid secretion by the stomach is

- P25. stimulated when a hungry person smells food.
- P26. stimulated by increased osmolarity of the intestinal contents.
- P27. accompanied by bicarbonate secretion into the blood.

P28-30. CCK

- P28. stimulates HCl secretion by the stomach.
- P29. stimulates secretion of enzymes by the pancreas.
- P30. inhibits contraction of the sphincter of Oddi.

17: REGULATION OF ORGANIC METABOLISM, GROWTH, AND ENERGY BALANCE

In 1918, a young girl named Elizabeth Hughes was diagnosed with diabetes mellitus — a fearful, fatal disease for which there is still no cure and, at that time, no treatment beyond slow starvation. Diabetes is an ancient disease, and its symptoms — excessive urination, thirst, hunger, weakness, gradual wasting, and finally coma and death — had been known for centuries. But no one guessed its cause until the late nineteenth century, when scientists studying the role of the pancreas in fat digestion found that removal of the pancreas produced diabetes in dogs. Various early trials to find the pancreatic compound missing in diabetics failed, however, and the best doctors could do for patients like Elizabeth was to severely restrict their intake of food, which prolonged the course of the disease but did not reverse it. By August of 1922, when she was 15 years old, Elizabeth weighed only 45 pounds and was near death.

But Elizabeth Hughes was lucky. Just one year earlier a physician in Toronto, Frederick Banting, had come upon the idea of making pancreatic extract from pancreas that had been tied off so that the exocrine part had essentially digested itself, leaving "islet" (endocrine) cells intact. Using this method, Banting and his assistant, graduate student Charles Best, produced extracts of pancreas that could reverse the disease when injected in dogs. With the help of physiologist John J. MacLeod and biochemist James Collip, these Canadian pioneers purified their extract and discovered the hormone *insulin* — a discovery that today allows insulin-dependent diabetics to live full and reasonably healthy lives. Banting and MacLeod received the Nobel Prize for Physiology or Medicine in 1923 for their efforts, and Elizabeth Hughes, who was one of the first human recipients of the hormone, went on to live to the age of 74, nearly 60 years longer than would have been possible without what she herself called her "miracle" drug.

In this chapter you will learn why insulin is so vitally important and why its absence causes so much misery, as well as many other aspects of the regulation of metabolism and growth.

Instructions for answering questions in this Study Guide

1. True or false: Correct each false statement. Whenever you must correct a false statement, there will almost always be more than one way to do so. In the Guide, the answer requiring the least correction will generally be given, with the understanding that other correct answers are possible. It is usually not sufficient, in terms of demonstrating understanding, to simply insert a "no" or "not," however.

2. Multiple choice: Any, all, or none of the responses may be correct. Explain why each incorrect response is incorrect.

3. Fill in the blanks: Choose the best word or words to complete each statement.

4. Directions will be given for other types of questions as they appear.

SECTION A. CONTROL AND INTEGRATION OF CARBOHYDRATE, PROTEIN, AND FAT METABOLISM

❒ Events of the Absorptive and Postabsorptive States (text pages 554-559)

1. True or false: During the absorptive state, the body makes use of stored nutrients for energy.

2. True or false: A person who eats three meals a day spends about half of each day in the absorptive state.

3. True or false: The liver gets first crack at all the absorbed nutrients.

4. Absorbed glucose is
   (a) stored as fat in skeletal muscle.
   (b) stored as glycogen in adipose tissue.
   (c) converted to fat in the liver.
   (d) utilized by most cells for energy.

5. True or false: The source of the triacylglycerol used for synthesis of very-low-density lipoproteins in the liver is absorbed fatty acids and glycerol.

6. The triacylglycerol in the VLDL complexes is broken down to __________________ and __________________ by the enzyme __________________, found in adipose-tissue capillary walls. The __________________ formed by the reaction diffuse into adipose cells to be stored as __________________, and the __________________ molecules are recycled back to the liver. Thus, triacylglycerol formed in the liver from absorbed __________________ is ultimately stored as __________________ in adipose tissue.

7. True or false: Most of the glucose stored after a meal is stored in the form of glycogen.

8. True or false: Most of the triacylglycerol absorbed as chylomicrons is ultimately stored as fat in the adipose tissue.

9. During the absorptive phase, amino acids
   (a) are primarily stored as protein in most body cells.
   (b) are used for protein synthesis in the liver.
   (c) are used as energy sources by the liver.
   (d) may be converted to fat or glucose in the liver.

10. True or false: During the absorptive phase, there is net synthesis of fat, glycogen, and protein, but this process is reversed during the postabsorptive phase.

11. True or false: During the postabsorptive phase, metabolic reactions are carried out by the liver and other tissues to maintain relative constant levels of fatty acids in the blood.

12. True or false: The brain and other nervous tissue absolutely require glucose for energy.

13. The formation of "new" glucose in the body is called ____________. Most of this glucose is normally formed in the ____________(what organ?), from which it is secreted into the blood. The most important sources of new glucose are the breakdown of ____________ in the liver (called ____________), formation of ____________ and ____________ by muscle cells, breakdown of ____________ in adipose tissue (called ____________), and catabolism of ____________ in muscle and other tissues.

14. True or false: Most of the energy used by the body during fasting is provided by gluconeogenesis.

15. True or false: Fatty acids released by lipolysis of triacylglycerol in adipose tissue travel in the plasma as VLDL.

16. True or false: The major energy sources for non-nervous tissue during the postabsorptive period are fatty acids and keto acids.

❒ Endocrine and Neural Control of the Absorptive and Postabsorptive States (text pages 559-568)

17. The four hormones of the endocrine pancreas, an organ also called the ____________________, are ____________________ from the A cells, ____________________ from the B cells, and ____________________ and ____________________, whose functions are less clearly understood at this time. The most important of the four is ____________________, which controls virtually all the metabolic reactions of the ____________________ state.

18. Upon binding to its receptor on target cells, insulin
    (a) causes its receptor to autophosphorylate.
    (b) activates cyclic AMP.
    (c) stimulates active transport of glucose into cells.
    (d) stimulates active transport of amino acids into cells.

19. True or false: Insulin has no effect on the carrier-mediated transport of glucose in the liver; therefore, insulin does not stimulate the uptake of glucose by that organ.

20. True or false: The effects of insulin on membrane transport and on enzyme activity are mediated by the same receptor.

21. The effects of insulin on glycogen synthesis in muscle cells are "triple-barreled:" It ____________________ (stimulates/inhibits) glucose uptake, ____________________ (stimulates/inhibits) the activity of the rate-limiting enzyme in glycogen breakdown, and ____________________ (stimulates/inhibits) the enzyme catalyzing the rate-limiting step in glycogen synthesis. Simultaneously, enzymes catalyzing glycolysis are ____________________ (stimulated/inhibited).

22. With regard to fat metabolism, insulin
    (a) increases facilitated fatty acid diffusion across target cell membranes.
    (b) increases activity of lipoprotein lipase.
    (c) increases activity of intracellular lipase.
    (d) stimulates formation of precursors for triacylglycerol synthesis.
    (e) increases triacylglycerol synthesis by increasing glucose transport in adipose tissue cells.

23. True or false: Insulin stimulates protein anabolism by increasing amino acid transport, but it has no other effects on protein metabolism.

24. True or false: Insulin inhibits the enzymes in the liver that catalyze reactions leading to gluconeogenesis.

25. True or false: The metabolic events characteristic of the absorptive state can be attributed to the presence of large amounts of insulin in the blood, whereas the events of the postabsorptive phase can be attributed largely to decreased insulin levels.

26. The B cells of the pancreas
   (a) are chemoreceptors for glucose.
   (b) increase secretion of insulin at the start of a meal before blood glucose levels have had a chance to rise.
   (c) increase insulin secretion in response to elevated amino acid levels in blood.

27. True or false: The most important of the glucose-counterregulatory controls is epinephrine.

28. Pancreatic glucagon
   (a) utilizes cyclic AMP as a second messenger.
   (b) secretion is stimulated by elevated blood glucose and amino acid levels.
   (c) inhibits glucose uptake by cells.
   (d) stimulates the activity of most of the enzymes inhibited by insulin.
   (e) stimulates glycogenolysis in liver and muscle cells.

29. True or false: During absorption of a high-protein, low-carbohydrate meal, net synthesis of protein and storage of glucose will occur and blood glucose levels will fall sharply.

30. Epinephrine and sympathetic nervous stimulation
   (a) inhibit insulin secretion.
   (b) inhibit glucagon secretion.
   (c) inhibit glycogenolysis in liver and skeletal muscle.
   (d) promote gluconeogenesis in liver.

(e) promote lipolysis in adipose tissue.

(f) inhibit glucose uptake by liver cells.

31. True or false: The chemoreceptors for the reflexes that stimulate sympathetic activity and epinephrine secretion in response to hypoglycemia are located in the carotid and aortic bodies.

32. Cortisol

(a) secretion is elevated during fasting.

(b) is required for cells to respond normally to hormones promoting gluconeogenesis and lipolysis.

(c) is permissive for insulin actions.

(d) in high amounts increases gluconeogenesis and inhibits glucose uptake.

33. True or false: Pituitary growth hormone has effects on protein metabolism similar to those of insulin, but its effects on carbohydrate and lipid metabolism are similar to those of elevated cortisol.

❒ **Fuel Homeostasis in Exercise and Stress** (text pages 568-569)

34. During exercise,

(a) blood glucose levels fall dramatically.

(b) decreased blood glucose and increased epinephrine stimulate insulin secretion.

(c) epinephrine stimulates glucagon secretion and inhibits skeletal muscle uptake of glucose.

(d) skeletal muscles increase their rate of utilization of fatty acids.

(e) epinephrine and glucagon stimulate glycogenolysis in muscle and liver cells.

35. True or false: During stress, elevated levels of plasma cortisol stimulate the catabolism of muscle protein and the conversion of keto acids to glucose in the liver.

❒ Diabetes Mellitus (text pages 569-572)

36. Of the two types of diabetes mellitus, the more serious but less common is __________ (type 1/type 2), or insulin-__________________ (dependent/independent) diabetes. This form is caused by destruction of the ______________________ cells of the pancreas. The other form of the disease is characterized by abnormalities in the insulin ________________________.

37. True or false: In an insulin-dependent diabetic, liver metabolism of glucose is normal because the liver does not depend on insulin for its uptake of glucose.

38. True or false: The primary fuel source for most cells in untreated type 1 diabetes is keto acids.

39. True or false: Metabolic acidosis caused by excessive blood levels of ketones is one of the most serious consequences of untreated type 1 diabetes.

40. True or false: One serious consequence of insulin lack is the decreased ability of the brain to utilize glucose efficiently, a condition that can lead to brain dysfunction and death.

41. Clinical signs of severe diabetes include
   (a) glucose in the urine.
   (b) ketones in the urine.
   (c) hypoglycemia.
   (d) expiration of acetone.
   (e) decreased respiratory rate.

42. True or false: The excretion of glucose in the urine of a diabetic patient is a result of the inability of the kidney tubules to reabsorb glucose in the absence of insulin.

43. Without checking the text, replace the missing arrows signifying "increase" or "decrease" from this copy of text Figure 17-11:

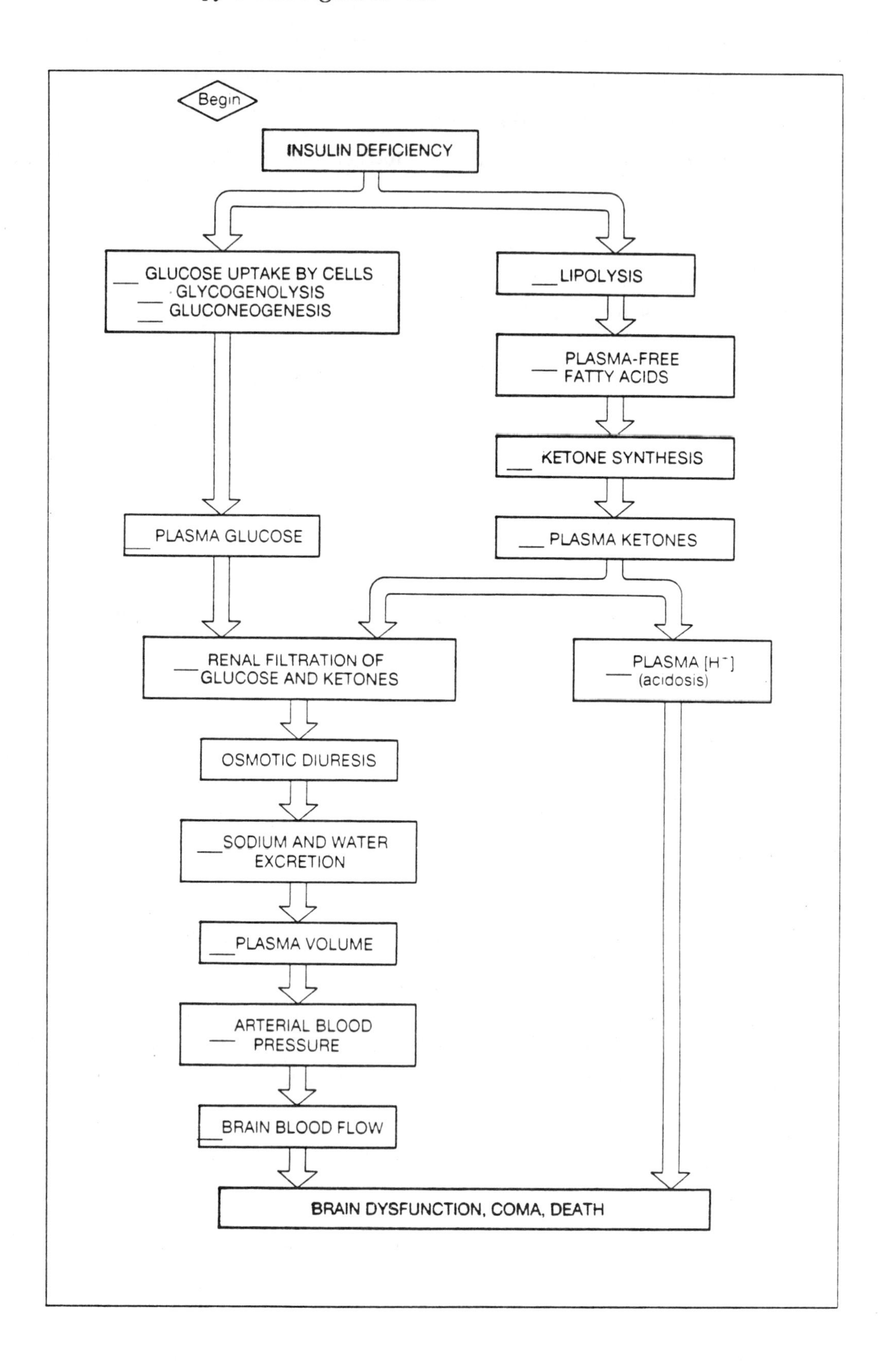

44. Insulin-dependent diabetics need to monitor their blood and urine carefully during the day (and even sometimes at night) to make sure their glucose levels remain stable. Explain how the following would affect a diabetic's blood glucose levels and the consequent dose of insulin he or she would need:

(a) Exercise

___

___

___

(b) Chronic stress

___

___

___

(c) A glucagon-secreting tumor

___

___

___

45. Prior to the discovery of insulin in 1921, the only known treatment for diabetes mellitus was a near-starvation diet with almost no carbohydrates. Explain why this diet was helpful in delaying fatal coma.

___

___

___

___

___

46. Type 2 diabetes

(a) is most common among overweight adults.

(b) may not require insulin therapy.

(c) may be associated with abnormal insulin receptors.

(d) may be associated with too few insulin receptors.

(e) is helped by dieting and exercise.

47. True or false: High levels of glucose in the blood are detrimental to many body structures and functions other than causing blood volume and chemistry abnormalities.

❐ Hypoglycemia as a Cause of Symptoms (text pages 572-573)

48. Hypoglycemia
    (a) may result from a glucagon-secreting tumor.
    (b) may result from administration of too much insulin to a diabetic.
    (c) may be a result of liver disease.
    (d) is not a particularly common problem in people other than insulin-dependent diabetics.
    (e) may be fatal.

❐ Regulation of Plasma Cholesterol (text pages 573-575)

49. Cholesterol is
    (a) a precursor of steroid hormones.
    (b) a component of plasma membranes.
    (c) a precursor of bile pigments.
    (d) an important fuel source for nonnervous tissue.
    (e) an essential nutrient in the diet.

50. Most of the cholesterol in the blood circulates associated with ________________________ (low-density/high-density) lipoproteins. These complexes carry cholesterol ____________(to/from) cells. The other complexes, ____________________(LDL/HDL), carry cholesterol____________(to/from) cells and promote its secretion into the ________________________________. The cholesterol carried by the former complexes is known as ________________("good"/"bad") cholesterol because it is associated with ______________________(increased/decreased) deposition of cholesterol on arterial walls and thus _________________(increased/decreased) likelihood of atherosclerosis.

51. In the blood, the ratio of LDL cholesterol to HDL cholesterol is
    (a) positively correlated with risk of atherosclerotic heart disease.
    (b) decreased in people who smoke cigarettes.
    (c) decreased in people who exercise regularly.
    (d) generally lower in young women than in young men.

52. True or false: Cholesterol is metabolized to bile acids in the liver.

53. True or false: Plasma levels of cholesterol are not homeostatically regulated, but instead depend solely on the intake of cholesterol in the diet.

SECTION B. CONTROL OF GROWTH

❐ Bone Growth (text page 575)

54. True or false: In bone, osteoblasts secrete a protein matrix upon which calcium salts are deposited.

55. Bone growth
    (a) occurs as a result of osteoblast division.
    (b) follows a straight line from infancy to adulthood.
    (c) of the long bones requires growth of the epiphyseal cartilage.
    (d) of the long bones ceases after closure of the epiphyseal growth plate.

❐ Environmental Factors Influencing Growth (text pages 575-576)

56. True or false: A high-protein, low-calorie diet is the best for achieving maximal growth in infants and children.

57. All other things being equal, the most damage to a child's growth and development would be caused by malnutrition during (choose one)
    (a) the teenage growth spurt.
    (b) the first six months after birth.
    (c) prenatal development.

❐ Hormonal Influences on Growth (text pages 576-579)

58. True or false: Growth regulation can only be positive. That is, growth regulating hormones can only stimulate growth, not suppress it.

59. True or false: Oncogenes can be aberrant receptors for growth factors.

60. Pituitary growth hormone
    (a) directly promotes protein anabolism in many cells.
    (b) stimulates insulin secretion by many target cells.
    (c) causes differentiation of precursor cells that then respond to IGF-I by proliferating.
    (d) is absent or deficient in African Pygmies.
    (e) hypersecretion in adults leads to giantism.
    (f) is absent or deficient in pituitary dwarfs.

61. True or false: A hypothalamic tumor that secretes somatostatin may cause dwarfism in a young child.

62. True or false: Growth hormone is secreted in largest amounts during exercise.

63. True or false: The actions of thyroid hormones in growth are permissive to the actions of growth hormone.

64. True or false: Insulin is important for growth both before and after birth.

65. Estrogen and testosterone in females and males, respectively,
    (a) are responsible for the pubertal growth spurt.
    (b) are responsible for epiphyseal plate closure.
    (c) cause increased strength and mass of skeletal muscle.

66. True or false: Cortisol is permissive for the growth-promoting actions of growth hormone.

❐ **Compensatory Growth** (text page 579)

67. True or false: Removal of one adrenal gland leads to compensatory growth of the remaining adrenal gland.

SECTION C. SUMMARY OF LIVER FUNCTIONS (text pages 579-580)

68. A person who suffers significant damage to his or her liver will have a variety of impairments, including those listed below. For each, cite the diminished liver function(s) that is/are primarily responsible:

(a) impaired intestinal absorption of calcium

____

____

(b) impaired digestion of fat

____

____

(c) impaired cholesterol metabolism

____

____

(d) decreased levels of total circulating thyroid and steroid hormones

____

____

(e) tissue edema

____

____

(f) jaundice

____

____

(g) accumulation of ammonia (NH_3) in blood

____

____

(h) hyperglycemia during the absorptive phase of metabolism

____

____

(i) hypoglycemia during the fasting phase of metabolism

____

____

(j) impaired reabsorption of sodium in response to decreased blood volume

____

____

SECTION D. REGULATION OF TOTAL-BODY ENERGY BALANCE

❒ Basic Concepts of Energy Expenditure and Caloric Balance (text pages 580-588)

69. According to the energy-balance equation for chemical reactions in the body:
$\Delta E = H + W$
    (a) ΔE is ______________________________
    (b) H is ______________________________
    (c) W is ______________________________

70. In the body,
    (a) most of the energy released during metabolism is used for doing work.
    (b) most of the energy used for doing work is transferred to ATP.
    (c) heat energy can be used for doing work.
    (d) external work refers to work done by skeletal muscles in moving objects outside the body.

71. True or false: Some of the energy liberated by metabolism may be stored in the form of ATP.

72. The metabolic rate of a person
    (a) is the total energy expenditure of that person in a unit of time.
    (b) who is fasting and at rest is equal to the heat generated by that person.
    (c) cannot be measured directly.
    (d) can be measured indirectly by measuring the expiration of CO_2 per unit time.
    (e) varies according to activity and physiological state.

73. The basal metabolic rate
    (a) is measured in a fasting person at rest.
    (b) does not include work done by the skeletal muscles.
    (c) is greater in a child than in an adult.
    (d) is generally greater in women than in men.
    (e) is greater in a pregnant woman than in one who is not pregnant.
    (f) is greater in someone who is swimming than in someone at rest.

74. True or false: All other things being equal, a person who is hypothyroid will have a higher BMR than a person whose thyroid function is normal.

75. True or false: The calorigenic effect of thyroid hormones refers to the rate at which the body synthesizes organic molecules.

76. A person who is hyperthyroid
    (a) is frequently cold.
    (b) is usually hungry.
    (c) is mentally sluggish.
    (d) tends to be irritable and restless.

77. True or false: Stimuli that activate the sympathetic nervous system increase the metabolic rate.

78. Food-induced thermogenesis is
    (a) an increased metabolic rate associated with the absorptive phase of metabolism.
    (b) primarily a result of increased activity by the stomach.
    (c) greater with protein-rich meals than with protein-poor meals providing the same total calories.

79. Text Table 17-9 lists the energy expended by a 70-kg person involved in different activities. "Lying still, awake" accounts for 77 kcal/h, which add up to 1848 kcal/24 h. Below the table the text says, "[T]otal energy expenditure may vary for a normal young adult from a value of 1500 kcal/24 h to..." Which of the following might account for the discrepancy between the calculated value of 1848 kcal/24 h and the 1500 kcal/24 h value?
    (a) The value in the table is for a person smaller than the "normal young adult."
    (b) The value in the table is for a male, and the "normal young adult" is a female.
    (c) The metabolic rate during sleep is less than "basal."

80. Regarding total-body energy balance,
    (a) in most normal adults the energy taken in as food is about equal to the energy expended in heat production and external work.
    (b) in normal adults metabolic rate is regulated more closely than is food intake.
    (c) in growing children energy intake must exceed internal heat production plus external work.

81. True or false: Insulin is a satiety signal that suppresses appetite, whereas glucagon stimulates appetite.

82. Obesity
    (a) is defined as being 20% or more above "desirable" body weight.
    (b) may cause problems with self-image but will not cause problems related to physical health.
    (c) is entirely hereditary.
    (d) is usually caused by too little secretion of thyroid hormones.
    (e) is best treated with a combination of sensible dieting and exercise.

83. Calculate your own body mass index: One inch = 2.54 centimeters and one kilogram = 2.2 pounds. (The answer is given using data from a woman five feet tall who weighs 110 pounds.)

84. True or false: The thinner one is, the healthier one is. One can never be "too thin."

85. Anorexia nervosa is
    (a) an eating disorder.
    (b) a psychological disorder.
    (c) a biological disorder.

❒ Regulation of Body Temperature (text pages 588-597)

86. The two basic physiological mechanisms for maintaining normal body temperature in the face of decreasing environmental temperature are increasing ______________________________ and decreasing ______________________________.

87. True or false: In response to cold stimuli, human beings increase their rate of heat production primarily by increasing basal metabolic rate.

88. A person standing outside in the sun on a cold, clear day will gain heat by ____________________ from the sun and lose body heat by ____________________ through the air around him. Any wind would increase this loss by the process of ____________________. Additional heat would be lost by ____________________ of water lost from the skin or during expiration, called ____________________ water loss.

89. Regarding the regulation of body temperature,
    (a) the temperature of the skin is regulated to stay about 37°C.
    (b) behavioral adjustments are important components of body temperature regulation.
    (c) any variation in body core temperature is abnormal.

90. True or false: The body's most effective mechanism for reducing heat loss is vasoconstriction of blood vessels in the skin.

91. The thermoneutral zone is
    (a) that area of the body that is midway between body core temperature and environmental temperature.
    (b) the temperature of the body core.
    (c) the environmental temperature at which the body can regulate core temperature without either increasing or decreasing heat production.
    (d) about 72°F.

92. True or false: The reason high humidity plays such a significant role in increasing the discomfort felt on very hot days is that it decreases the cooling properties of conduction.

93. Which of the following are body responses to decreases in environmental temperature below the thermoneutral zone?
    (a) increased sympathetic activity to vasoconstrictor smooth muscles in the skin.
    (b) increased sympathetic activity to vasodilator smooth muscles in the skin.
    (c) decreased motor activity.
    (d) increased secretion of epinephrine.

94. True or false: Receptors for regulation of body temperature are heat and cold receptors in the skin, the CNS, and internal organs.

95. Acclimatization to heat includes
   (a) increased production of sweat for any given increase in environmental temperature.
   (b) secretion of more concentrated sweat than before acclimatization.
   (c) increased production of aldosterone by the sweat glands.

96. True or false: Fever differs from other forms of hypothermia in that it results from resetting the hypothalamic "thermostat" to a lower level.

97. Endogenous pyrogens
   (a) are released by infection-fighting leukocytes.
   (b) act on the temperature-integrating centers of the hypothalamus.
   (c) have actions that are enhanced by aspirin.
   (d) include interleukin 1 and 6, and tumor necrosis factor.

98. Exercise in a hot environment
   (a) is always accompanied by a rise in body core temperature.
   (b) is always accompanied by an increased rate of heat loss.
   (c) may lead to heat exhaustion.
   (d) may be fatal.

ANSWERS

Boldface type indicates the answers you should have given. Words in medium-face (ordinary) type explain or expand upon the answer.

1. **False.** During the **post**absorptive state, the body makes use of stored nutrients for energy.

   Alternatively, the answer given below is also correct:

   False. During the absorptive state, the body makes use of ~~stored~~ **ingested** nutrients for energy.

2. **True.** (Four hours for each meal = twelve hours)

3. **False.** The liver gets first crack at all the absorbed nutrients **except fat**.

4. (a) **Incorrect. It is stored as glycogen in skeletal muscle.**
   (b) **Incorrect. It is stored as fat in adipose tissue.**
   (c) **Correct.**
   (d) **Correct.**

5. **False.** The source of the triacylglycerol used for synthesis of very-low-density lipoproteins in the liver is absorbed ~~fatty acids and glycerol~~ **glucose**. (Just as in the adipose tissue.)

6. The triacylglycerol in the VLDL complexes is broken down to **fatty acids** and **glycerol** by the enzyme **lipoprotein lipase**, found in adipose-tissue capillary walls. The **fatty acids** formed by the reaction diffuse into adipose cells to be stored as **triacylglycerol**, and the **glycerol** molecules are recycled back to the liver. Thus, triacylglycerol formed in the liver from absorbed **glucose** is ultimately stored as **triacylglycerol** in adipose tissue.

7. **False.** Most of the glucose stored after a meal is stored in the form of ~~glycogen~~ **triacylglycerol**.

8. **True.** (Even though the glycerol released during breakdown of triacylglycerol by lipoprotein lipase is not stored in the adipose tissue, it does function to bring more fatty acids to the adipose tissue in the form of VLDL. Therefore, the generalization that "absorbed fat is stored as fat in the adipose tissue" is valid.)

9. (a) **Incorrect. They are used for protein synthesis but not stored as protein.**
   (b) **Correct.**
   (c) **Correct.** (The liver is an exception to the "rule" that glucose is used for energy by most tissues during the absorptive phase.)
   (d) **Incorrect. They are converted to fat** (via keto and fatty acids) **but not to glucose during the absorptive phase.**

10. **True.**

11. **False.** During the postabsorptive phase, metabolic reactions are carried out by the liver and other tissues to maintain relative constant levels of ~~fatty acids~~ **glucose** in the blood.

12. **True.** (Even though prolonged starvation does bring about changes that allow the brain to utilize ketones, this adjustment does not change the basic fact that the brain needs glucose, and all of the intricate metabolic shifts you will learn in this chapter take place because of this fact.)

13. The formation of "new" glucose in the body is called **gluconeogenesis**. Most of this glucose is normally formed in the **liver**, from which it is secreted into the blood. The most important sources of new glucose are the breakdown of **glycogen** in the liver (called **glycogenolysis**), formation of **lactate** and **pyruvate** by muscle cells, breakdown of **triacylglycerol** in adipose tissue (called **lipolysis**), and catabolism of **protein** in muscle and other tissues. (Many people find the terms "gluconeogenesis" and "glycogenolysis" confusing, or at least difficult to differentiate. When you look carefully at the words, their meanings are clear — the new formation of glucose, and the lysis, or breakdown, of glycogen, respectively. Perhaps the problem stems from the fact that one avenue to gluconeogenesis is by way of glycogenolysis. But you must remember that the terms are not synonyms. You may well wonder why we don't simply say "glucose synthesis" instead of gluconeogenesis. We do not use that phrase because of the implication it leaves that we can synthesize glucose "from scratch," that is, from CO_2 and H_2O. Of course you remember that only green plants can do that.

14. **False.** Most of the energy used by the body during fasting is provided by ~~gluconeogenesis~~ **fat utilization**. (As the text points out, gluconeogenesis can account for only about 720 kcal/day, less than half the minimal caloric requirement of a normal individual.)

15. **False.** Fatty acids released by lipolysis of triacylglycerol in adipose tissue travel in the plasma ~~as VLDL~~ **bound to albumin**.

16. **False.** The major energy sources for non-nervous tissue during the postabsorptive period are fatty acids and keto**nes** ~~acids~~. (As the text mentions, keto acids and ketones are not the same. The liver — and to some extent the kidneys — are the only sources of keto acids because they are the only organs with enzymes that catalyze protein deamination. One fact that increases the problem of correctly identifying ketones and keto acids is that two of the ketones are acids, and these two acidic ketones contribute to metabolic acidosis in untreated diabetics, as you will see later in this chapter.)

17. The four hormones of the endocrine pancreas, an organ also called the **islets of Langerhans**, are **glucagon** from the A cells, **insulin** from the B cells, and **somatostatin** and **pancreatic polypeptide**, whose functions are less clearly understood at this time. The most important of the four is **insulin**, which controls virtually all the metabolic reactions of the **absorptive** state.

18. (a) **Correct.**
    (b) **Incorrect. The activated insulin receptor inhibits adenylate cyclase.**
    (c) **Incorrect. It stimulates the facilitated diffusion of glucose** (but not in liver or brain).
    (d) **Correct.**

19. **False.** Insulin has no effect on the carrier-mediated transport of glucose in the liver; ~~therefore~~ insulin **nevertheless** ~~does not~~ stimulate**s** the uptake of glucose by that organ (by increasing glucose phosphorylation and thus decreasing the concentration of free glucose in the cytoplasm, creating a concentration gradient favorable to the inward diffusion of glucose).

20. **True.**

21. The effects of insulin on glycogen synthesis in muscle cells are "triple-barreled:" It **stimulates** glucose uptake, **inhibits** the activity of the rate-limiting enzyme in glycogen breakdown, and **stimulates** the enzyme catalyzing the rate-limiting step in glycogen synthesis. Simultaneously, enzymes catalyzing glycolysis are **stimulated**. (So long as glucose levels in the cells are high, both the anabolic pathway to glycogen storage and the catabolic pathway to Krebs cycle intermediates and energy production will be active.)

22. (a) **Incorrect. Fatty acids enter cells by simple diffusion**. (Insulin increases fatty acid uptake indirectly by increasing the activity of lipoprotein lipase, which catalyzes breakdown of blood triacylglycerol to fatty acids and glycerol.)
    (b) **Correct.**
    (c) **Incorrect. It inhibits the activity of this enzyme.** (This enzyme catalyzes fat breakdown in cells.)
    (d) **Correct.** (It stimulates activity of the enzymes that convert glucose to fatty acids and glycerophosphate.)
    (e) **Correct,** (It also increases the activity of fatty acid-synthetic enzymes and inhibits those that promote triacylglycerol breakdown.)

23. **False.** Insulin stimulates protein anabolism by increasing amino acid transport~~,~~ ~~but it has no other effects on protein metabolism~~ **and increasing the activity of protein anabolic enzymes, and by decreasing the activity of protein catabolic enzymes.**

24. **True.**

25. **True.** (Even though other hormones do play a role in the postabsorptive state, low insulin is probably the most important component of control of metabolism during fasting.)

26. (a) **Correct.** (The cells that secrete insulin detect glucose levels in the blood entering the pancreas. Incidentally, like liver and brain cells but unlike adipose tissue and muscle cells, the B islet cells do not require insulin for glucose uptake. It would make no sense for detectors of blood glucose to be dependent upon insulin in order to measure it.)

    (b) **Correct.** (This is the feedforward mechanism mentioned by the text, and is mediated by GIP and parasympathetic stimulation.)

    (c) **Correct.**

27. **False.** The most important of the glucose-counterregulatory controls is ~~epinephrine~~ **glucagon**.

28. (a) **Correct.**

    (b) **Incorrect.** **Its secretion is stimulated by lowered blood glucose levels and elevated blood amino acid levels.**

    (c) **Incorrect.** **It has no effect on glucose uptake.**

    (d) **Correct.** (However, the converse of this statement is not true. That is, glucagon does not inhibit the activity of the enzymes stimulated by insulin. Thus, glucagon cannot act as a complete insulin antagonist.)

    (e) **Incorrect.** **It stimulates glycogenolysis in liver but not in muscle.** (With the exception of stimulating lipolysis in adipose tissue, the major effects of glucagon are all on the liver. This fact makes good teleological sense — remember teleology from Chapter 1? — because the hormonal secretions of the pancreas are carried directly to the liver by the hepatic portal vein. Thus, the liver gets first crack at those hormones as well as at the glucose and amino acids absorbed after digestion.)

29. **False.** During absorption of a high-protein, low-carbohydrate meal, net synthesis of protein ~~and storage of glucose~~ will occur **and storage of glucose will be offset by catabolic processes and gluconeogenesis, allowing** blood glucose levels ~~will fall sharply~~ **to remain quite stable**.

30. (a) **Correct.**

    (b) **Incorrect.** **They increase glucagon secretion.**

    (c) **Incorrect.** **They stimulate glycogenolysis in these tissues.**

    (d) **Correct.**

(e) **Correct.**

(f) **Incorrect. They inhibit glucose uptake by skeletal-muscle cells.**

31. **False.** The chemoreceptors for the reflexes that stimulate sympathetic activity and epinephrine secretion in response to hypoglycemia are located in the ~~carotid and aortic bodies~~ **brain** (and liver. The liver receptors have not been completely characterized.) Note that we have another "-emia" word. "Hypoglycemia" means "too little glucose in blood." Conversely, "hy*per*glycemia" refers to too much blood glucose. This latter condition is the hallmark of diabetes mellitus.

32. (a) **Incorrect. It is usually not elevated during fasting.**

    (b) **Correct.**

    (c) **Incorrect. It is permissive for the actions of the counterregulatory hormones, and for the effects of decreased insulin on enzyme activity.** (In other words, the enzymes that mediate the gluconeogenic and glucose-sparing reactions require basal levels of cortisol for their activity.)

    (d) **Correct.**

33. **True.**

34. (a) **Incorrect. During prolonged exercise, blood glucose levels decline but not dramatically.**

    (b) **Incorrect. These stimuli both decrease insulin secretion.**

    (c) **Incorrect. Epinephrine stimulates glucagon secretion but cannot inhibit glucose uptake by exercising muscle.** (As your text mentions, the mechanism for this phenomenon of increased muscle uptake of glucose without insulin stimulation is not clear, but it has important implications for people who are marginally diabetic — that is, who secrete some but not enough insulin — as we shall see in the next section.)

    (d) **Correct.**

    (e) **Incorrect. Both hormones stimulate glycogenolysis in the liver, but only epinephrine increases glycogen breakdown in muscle.**

35. **True.**

36. Of the two types of diabetes mellitus, the more serious but less common is **type 1**, or insulin-**dependent** diabetes. This form is caused by destruction of the **B** cells of the pancreas. The other form of the disease is characterized by abnormalities in the insulin **receptor**.

37. **False.** In an insulin-dependent diabetic, liver metabolism of glucose is **ab**normal ~~because~~ **even though** the liver does not depend on insulin for its uptake of glucose. (The enzymes governing glycogen synthesis and glycolysis are inhibited, and those that mediate glycogen breakdown are stimulated; thus, glucose levels are relatively high in the liver, decreasing the amount of glucose taken up. Whatever glucose is transported into the liver just diffuses right back out.)

38. **False.** The primary fuel source for most cells in untreated type-I diabetes is **ketones** ~~acids~~.

39. **True.**

40. **False.** One serious consequence of insulin lack is the decreased ~~ability of the brain to utilize glucose efficiently~~ **flow of blood to the brain**, a condition that can lead to brain dysfunction and death. (The purpose of this question is to point out that the body's redundant controls, which ensure that blood glucose levels remain high enough for the brain to have an adequate supply of its obligatory fuel source, function just fine in this regard. The only problem with such redundancy in keeping glucose levels high enough for the brain is that different mechanisms break down so that other, equally vital homeostatic parameters, such as blood pressure, cannot be controlled.)

41. (a) **Correct.**
    (b) **Correct.**
    (c) **Incorrect.** **The finding would be hyperglycemia.**
    (d) **Correct.** (Acetone is breathed out in expired air.)
    (e) **Incorrect.** **Respiratory rate is increased, as a result of metabolic acidosis.**

42. **False.** The excretion of glucose in the urine of a diabetic patient is a result of the ~~inability~~ **high levels of plasma glucose, exceeding the transport maximum for glucose** of the kidney tubules ~~to reabsorb glucose in the absence of insulin~~. (Glucose is actively cotransported into the tubular epithelium secondary to the active transport of sodium ion. This process, like the active transport of glucose across the intestinal epithelium, is independent of insulin.)

43.

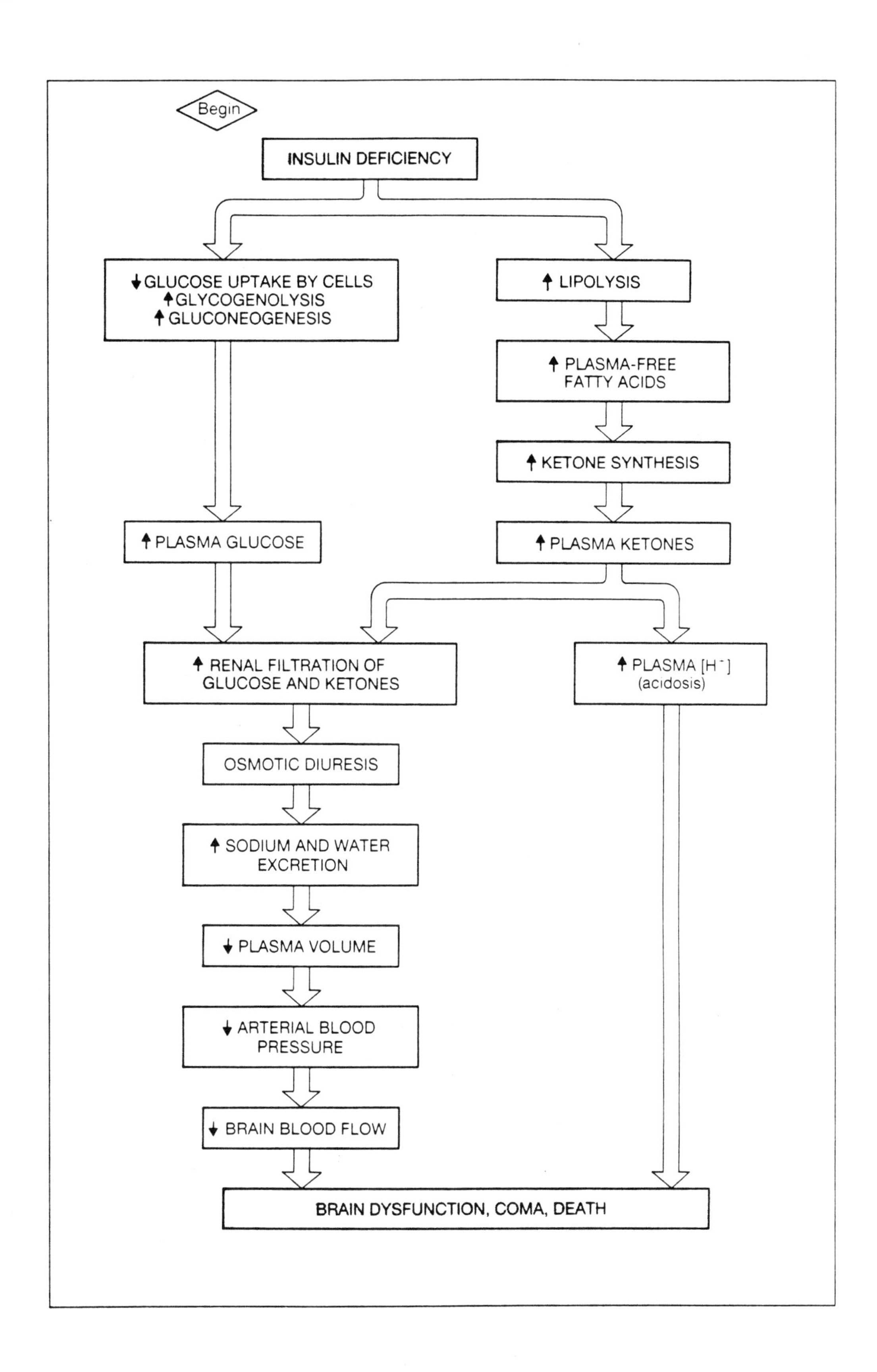
Begin
INSULIN DEFICIENCY
↓ GLUCOSE UPTAKE BY CELLS
↑ GLYCOGENOLYSIS
↑ GLUCONEOGENESIS
↑ LIPOLYSIS
↑ PLASMA-FREE FATTY ACIDS
↑ KETONE SYNTHESIS
↑ PLASMA GLUCOSE
↑ PLASMA KETONES
↑ RENAL FILTRATION OF GLUCOSE AND KETONES
↑ PLASMA [H⁺] (acidosis)
OSMOTIC DIURESIS
↑ SODIUM AND WATER EXCRETION
↓ PLASMA VOLUME
↓ ARTERIAL BLOOD PRESSURE
↓ BRAIN BLOOD FLOW
BRAIN DYSFUNCTION, COMA, DEATH

44. (a) **Exercise stimulates glucose uptake by muscle cells even in the absence of insulin, and it also increases the affinity of insulin receptors for the hormone. Therefore, an exercising diabetic would require a somewhat lower dose of insulin than normal** (or one who is sedentary).

(b) **Stress worsens diabetes because it triggers increased secretion of cortisol (resulting from increased CRF and ACTH secretion) and increased sympathetic activity. Cortisol, epinephrine, and sympathetic nerve stimulation increase glycogenolysis, gluconeogenesis, protein catabolism, lipolysis, and ketone formation and inhibit glucose uptake by nonbrain tissues. Thus, a larger dose of insulin would be required.**

(c) **A glucagon-secreting tumor would mimic stress in worsening diabetes** (in fact, some cases of diabetes are caused by inappropriately elevated glucagon secretion without a defect in insulin secretion or its receptors). **Glucagon does not interfere with insulin-mediated transport of glucose in nonbrain tissue, however.** (A glucagon-secreting tumor should be removed if possible.)

45. **In the absence of insulin, any ingested carbohydrate would add to the circulating glucose levels (and thus worsen the hyperglycemia) and would increase water loss. Therefore, a low-calorie, nearly carbohydrate-free diet could provide some fuel without aggravating the hyperglycemia as much as a normal diet. Furthermore, a starving patient would have very little stored fat to mobilize to contribute to the hyperglycemia and excessive levels of ketones in blood (ketonemia).**

46. (a) **Correct.**

(b) **Correct.** (In some cases insulin is required.)

(c) **Correct.**

(d) **Correct.** (But in many cases the receptors are abnormal as well as reduced in number. Although down-regulation of insulin receptors is part of the problem in this type of diabetes, it is usually not the whole problem.)

(e) **Correct.** (This is always true, no matter what the source of the defect.)

47. **True.**

48. (a) **Incorrect.** **It may result from an insulin-secreting tumor.** (A glucagon-secreting tumor would cause hyperglycemia.)

(b) **Correct.**

(c) **Correct.** (Gluconeogenesis and ketone formation would be impaired.)

(d) **Correct.** (And in those people the hypoglycemia results from incorrect monitoring of blood glucose and an inappropriate insulin dose for the activity or diet.)

(e) **Correct.** (Insulin overdose can lead to insulin *shock*, where the patient loses consciousness because of too little glucose to fuel the brain cells. Thus, an insulin-dependent diabetic might be found in a coma for two reasons: 1) too little insulin, leading to dehydration, lowered blood pressure, and acidosis; and 2) too much insulin, leading to dangerous hypoglycemia and brain dysfunction. One way for the amateur without blood-testing equipment to tell the difference is to observe the breathing. If it is rapid, then acidosis and too little insulin are the problem. One can also smell the breath for the presence of acetone — a plasma ketone —which also indicates too little insulin.)

49. (a) **Correct.**
    (b) **Correct.**
    (c) **Incorrect.** **It is a precursor for bile salts.**
    (d) **Incorrect.** **It is not catabolized for energy.**
    (e) **Incorrect.** **It is not necessary to consume cholesterol. The liver synthesizes all the body needs.**

50. Most of the cholesterol in the blood circulates associated with **low-density** lipoproteins (LDL). These complexes carry cholesterol **to** cells. The other complexes, **HDL**, carry cholesterol **from** cells and promote its secretion into the **bile**. The cholesterol carried by the former complexes is known as **"bad"** cholesterol because it is associated with **increased** deposition of cholesterol on arterial walls and thus **increased** likelihood of atherosclerosis.

51. (a) **Correct.**
    (b) **Incorrect.** **The ratio is increased in people who smoke.**
    (c) **Correct.**
    (d) **Correct.** (Although this was not stated explicitly in the text, premenopausal women have both lower total blood cholesterol and lower LDL/HDL ratios than do men the same age. You should have been able to arrive at the correct answer by tying together the text's statements that the best indicator of atherosclerotic heart disease is the ratio of plasma LDL-cholesterol to HDL-cholesterol, and that premenopausal women have a lower incidence of heart disease than do men.)

52. **True.** (In case you are wondering, "bile *acids*" is just another name for bile *salts* — the amphipathic emulsifiers of fat during digestion.)

53. **False.** Plasma levels of cholesterol are ~~not~~ homeostatically regulated, ~~but instead~~ **and** depend ~~solely~~ **little** on the intake of cholesterol in the diet. (However, as the text mentions, the relative saturation of the fatty acids in the diet does affect plasma cholesterol, perhaps by altering its synthesis in the liver or its recovery from bile.

Foods rich in cholesterol, like red meat, also have relatively high levels of saturated fats and so tend to increase blood cholesterol levels if consumed in large amounts. In addition, some people may not regulate plasma cholesterol as closely as is normal; for them, a diet high in cholesterol would increase blood levels of the steroid.)

54. **True.**

55. (a) **Incorrect.** **It occurs as a result of chondrocyte division.**
    (b) **Incorrect.** **It follows an S-shaped curve with two growth spurts.**
    (c) **Correct.** ("Growth plate," "epiphyseal cartilage," and "cartilage plate" are all synonymous.)
    (d) **Correct.** (Some bones, such as those in the hands and skull, do not have epiphyses that close and so can continue growing indefinitely. They do so only under unusual circumstances, however, as you will see later in this chapter.)

56. **False.** A high-protein, ~~low~~ **high**-calorie diet is the best for achieving maximal growth in infants and children. (Obviously there can be too much of a good thing, and children who overeat to extremes will become obese. However, for growth to occur there must be adequate protein (and other vital nutrients) as well as sufficient calories in the diet to ensure net anabolism. We shall come back to "caloric balance sheets" later in this chapter.)

57. (a) **Incorrect.**
    (b) **Incorrect.**
    (c) **Correct.** (These choices are ordered in reverse order of danger to normal growth and development.)

58. **False.** Growth regulation ~~can only be~~ **is either** positive **or negative**. That is, growth regulating hormones can ~~only~~ **either** stimulate growth~~, not~~ **or** suppress it.

59. **False.** Oncogenes can ~~be~~ **code for** aberrant receptors for growth factors. (Remember that genes are DNA and code for protein synthesis. Receptors for all chemical messengers are proteins.)

60. (a) **Correct.**
    (b) **Incorrect.** **It stimulates secretion of insulin-like growth factor I by target cells.** (IGF-I is, as the text mentions, also called somatomedin C. The term "somatomedin" refers to the molecule's role as a mediator of the somatotropic actions [stimulating growth of the body, or soma] of growth hormone. The "IGF" designation refers to the molecule's structure, which closely resembles that of insulin. In fact, in high

enough doses insulin molecules can bind to the IGF-I receptor and thus can act as a weak agonist for IGF-I. Unfortunately for type 1 diabetics, the reverse is not true.)

(c) **Correct.**

(d) **Incorrect.** **Pygmies have elevated growth hormone with very low IGF-I.**

(e) **Incorrect.** **It leads to acromegaly.** Giants have growth hormone hypersecretion before closure of their epiphyseal plates. The tallest known human being is known as the Alton giant after the name of his home town in Illinois. He was normal in height at birth (in 1928) but grew to be nearly four feet tall at one year of age. He continued growing until his death at age 22 (of an infection), when he towered at 8 feet 11 inches. The cause of his remarkable growth was a growth-hormone-secreting tumor.

Note that not all unusually tall people have pituitary tumors. Most if not all very tall athletes, such as basketball players, simply are genetically programmed to be tall. They can reach this potential, of course, only with adequate nutrition, as mentioned earlier.

(f) **Correct.** Pituitary dwarfs (so-called because of deficiency in secretion of pituitary growth hormone) are the other side of the coin from giants, or acromegalics. Dwarfs may be treated with injections of growth hormone and can achieve nearly normal stature.

61. **True.**

62. **False.** Growth hormone is secreted in largest amounts during ~~exercise~~ **sleep**.

63. **False.** The actions of thyroid hormones in growth are permissive to the actions of growth hormone **except for the action of thyroid hormones on brain development**. (The dwarfism associated with hypothyroidism at birth is much worse than that associated with growth hormone deficiency because of the mental retardation that occurs with the former.)

64. **True.** (As the text mentions, insulin is important for fetal growth. So, apparently, is a related molecule called IGF-*II*. The importance of insulin postnatally is shown in the fact [not mentioned in the text] that diabetic people have low levels of IGF-I in their plasma and that diabetic children do not grow as well as their nondiabetic siblings.)

65. (a) **Correct.**

(b) **Correct.** (If it seems confusing that the same hormones can stimulate both growth and cessation of growth, remember that the processes are separate: growth involves stimulation of chondrocyte mitosis;

cessation of growth involves the laying down of bone [ossification] and increased activity of osteoblasts. The sex steroids speed up both processes, but cessation of growth is stimulated just slightly more than is growth, so that ossification catches up with cartilage proliferation and the plate closes.)

(c) **Incorrect. Estrogen has no effect on skeletal muscle growth.**

66. **False.** Cortisol is ~~permissive for the growth-promoting actions of growth hormone~~ **a growth-inhibiting hormone**. (If you answered: **False.** ~~Cortisol is~~ **Thyroid hormones are**..., that would also be correct.)

67. **True.** (In the adrenals, this compensation is caused in large part by the increased ACTH secretion that results from the diminution of the negative feedback signal for its release. Sympathetic nerves supplying the gland also seem to play a role, as they may in the case of compensatory growth of a remaining kidney.

68. (a) **Decreased levels of the enzyme that converts vitamin D_3 to 25-OH D_3** (see text Figure 15-28).
(b) **Decreased secretion of bile salts.**
(c) **Decreased synthesis of cholesterol and lipoproteins, and decreased catabolism of cholesterol to bile salts and secretion of cholesterol into bile.**
(d) **Decreased synthesis of thyroid-binding and steroid-binding proteins.**
(e) **Decreased production of albumin and other plasma proteins.**
(f) **Decreased extraction of bilirubin from blood and decreased secretion of bilirubin into bile.**
(g) **Decreased conversion of ammonia to urea.**
(h) **Decreased glycogen and triacylglycerol synthesis.**
(i) **Decreased glycogenolysis and** (more important) **decreased gluconeogenesis and ketone production.**
(j) **Decreased angiotensinogen production** (which decreases the amount of angiotensin produced and thus decreases aldosterone secretion).

69. (a) **energy released by catabolism of organic molecules**
(b) **heat energy**
(c) **energy used for doing work**

70. (a) **Incorrect. Most of the energy released during metabolism is heat energy.**
(b) **Incorrect. All of the energy used for doing work is transferred to ATP.**
(c) **Incorrect. Heat energy cannot be used for doing work** (but it keeps us warm).
(d) **Correct.**

71. **False.** Some of the energy liberated by metabolism may be stored in the form of ~~ATP~~ **organic molecules**. (ATP is *not* a storage molecule for energy. The energy transferred to ATP during catabolism is used essentially immediately to do work.)

72. (a) **Correct.**
    (b) **Correct.** (This is true because, as the text points out, energy used to perform internal work other than the net synthesis of organic molecules shows up eventually as heat.)
    (c) **Incorrect.** **It can be measured directly in a calorimeter.**
    (d) **Incorrect.** **It can be measured indirectly by measuring the consumption of O_2 per unit time.**
    (e) **Correct.**

73. (a) **Correct.**
    (b) **Incorrect.** **The work done by the skeletal muscles involved in breathing accounts for some of the BMR.**
    (c) **Correct.**
    (d) **Incorrect.** **It is generally greater in men than in women.**
    (e) **Correct.**
    (f) **Incorrect.** **One cannot speak of having a basal metabolic rate while engaging in a physical activity.** (The metabolic rate of the swimmer would almost certainly be greater than the metabolic rate of the resting person, however.)

74. **False.** All other things being equal, a person who is hypothyroid will have a ~~higher~~ **lower** BMR than a person whose thyroid function is normal.

75. **False.** The calorigenic effect of thyroid hormones refers to the rate at which the body ~~synthesizes~~ **catabolizes** organic molecules.

76. (a) **Incorrect.** **A hyperthyroid person is intolerant of** (made unusually uncomfortable by) **heat.**
    (b) **Correct.**
    (c) **Incorrect.** **A hyperthyroid person is mentally alert.**
    (d) **Correct.**

77. **True.**

78. (a) **Correct.**
    (b) **Incorrect.** **It is primarily a result of the increased metabolic processing of amino acids and glucose in the liver.**
    (c) **Correct.**

79. (a) **Incorrect.** **The reverse may be correct.**
(b) **Correct.** (This might account for the discrepancy.)
(c) **Correct.**

80. (a) **Correct.** (In other words, most people who do not consciously try to reduce or gain weight maintain a relatively constant body weight.)
(b) **Incorrect.** **The opposite is correct.** (This question is easy to answer for those who have read the text but would probably not be answered correctly by someone relying on intuition, especially someone who had read the text only up to page 584. We tend to think of metabolic rate as being closely regulated by hormones while food intake is determined in part by hunger but also by habit and emotional state. One problem with this assumption is that it is *basal* metabolic rate that is affected by hormones, not total metabolic rate. The other problem is that food intake also has controls, which operate through hormones and other signals, that we do not automatically think of.)
(c) **Correct.** (If this is not the case, children will not grow.)

81. **False.** Insulin **and glucagon** ~~is a~~ **are** satiety signal**s** that suppress~~es~~ appetite~~, whereas glucagon stimulates it~~. (This answer also runs contrary to what you might expect, based on the fact that glucagon levels rise as blood glucose levels fall, and only in the case of a high-protein, low-carbohydrate meal would glucagon be present in appreciable amounts during the absorptive phase.)

82. (a) **Correct.** (As the text mentions, there is some controversy about which weight table is the better to use in determining "desirable" weight.)
(b) **Incorrect.** **Obese people suffer higher risk of several diseases, including hypertension, atherosclerosis, and diabetes.**
(c) **Incorrect.** **It is largely, but not entirely, determined by heredity.**
(d) **Incorrect.** **Patients who are hypothyroid are not necessarily obese, and most obese people have normal thyroid function.** (This correlation, or lack of one, between hypothyroidism and obesity was not mentioned in the text, but many people think there is one. Some obese people use it as an excuse. The only problem is that hypothyroidism is easily treated, while obesity is not.)
(e) **Correct.** ("Sensible" of course means no crash or fad diets. Such diets lead people on a roller coaster of weight loss followed by rapid reaccumulation of the lost pounds. And, as the text mentions, contrary to popular belief, the moderate increase of metabolic rate through exercise does not increase appetite in obese individuals as it would in people who are not obese.)

83. 5 feet = 60 inches = 2.54 cm/inch x 60 inches = 152.4 cm
= 1.52 m
110 pounds/(2.2 pounds/kg) = 50 kg
BMI = kg/m^2 = $50/(1.52)^2$ = 50/2.31 = **21.6 kg/m^2**
(The woman in the example is not obese. How about you?)

84. **False.** The thinner one is, the healthier one is **likely to be, up to a point**. ~~One can never be "too thin."~~ **Being too thin can be even more dangerous than being obese.**

85. (a) **Correct.**
(b) **Correct.**
(c) **Correct.**

Anorexia is also very dangerous. When accompanied by binge eating followed by induced vomiting (which is called bulemia) or by induced diarrhea (from laxatives), anorexia can cause death from "simple" starvation or from salt, water, and hydrogen ion imbalances. The hypothalamus seems to be involved with this disorder, but we do not yet know its basis. A cure almost always must involve psychological counseling as well as physiological therapy. We do know that overemphasis on "Be thin!" can do more harm than good, especially in adolescent girls.

86. The two basic physiological mechanisms for maintaining normal body temperature in the face of decreasing environmental temperature are increasing **heat production** and decreasing **heat loss**.

87. **False.** In response to cold stimuli, human beings increase their rate of heat production primarily by increasing ~~basal metabolic rate~~ **shivering thermogenesis**. (Although changes in the levels of thyroid hormones are thought to play minimal roles in the regulation of body temperature in normal adult humans, recall that hypothyroid individuals are intolerant of cold and hyperthyroid patients are intolerant of heat. Thus, normal levels of thyroid hormones *are* important for the setting of normal body temperature and for the body's adjustments to changes in environmental temperature even though *changes* in thyroid hormone levels are not thought to play a major role in those adjustments.)

88. A person standing outside in the sun on a cold, clear day will gain heat by **radiation** from the sun and lose body heat by **conduction** through the air around him. Any wind would increase this loss by the process of **convection**. Additional heat would be lost by **evaporation** of water lost from the skin or during expiration, called **insensible** water loss.

89. (a) **Incorrect. Body core temperature is regulated to stay about 37°C.**
(b) **Correct.**

(c) **Incorrect.** **There is a normal circadian variation in body temperature, and body temperature also varies with the menstrual cycle.**

90. **True.**

91. (a) **Incorrect.** **See answer (c).**
(b) **Incorrect.** **See answer (c).**
(c) **Correct.**
(d) **Incorrect.** **It is between 75 and 86°F.**

92. **False.** The reason high humidity plays such a significant role in increasing the discomfort felt on very hot days is that it decreases the cooling properties of ~~conduction~~ **sweat evaporation**.

93. (a) **Correct.**
(b) **Incorrect.** **It would be decreased.**
(c) **Incorrect.** **It would be increased** (to increase muscle tone and shivering thermogenesis, and perhaps voluntary motor activity).
(d) **Correct.**

94. **True.**

95. (a) **Correct.**
(b) **Incorrect.** **The sweat is more dilute.**
(c) **Incorrect.** **Aldosterone production by the adrenal cortex is increased.**

96. **False.** Fever differs from other forms of ~~hypo~~**hyper**thermia in that it results from resetting the hypothalamic "thermostat" to a ~~lower~~ **higher** level.

97. (a) **Correct.**
(b) **Incorrect.** **They act on the thermoreceptors in the hypothalamus** (which are near the integrating centers).
(c) **Incorrect.** **Their actions are mediated by prostaglandins, whose synthesis is inhibited by aspirin.**
(d) **Correct.**

98. (a) **Correct.**
(b) **Correct.**
(c) **Correct.**
(d) **Correct.** (If it leads to untreated heat stroke. Cooling a victim of heat stroke is very important, particularly splashing him or her with cool water and fanning to help increase heat loss by convection.)

PRACTICE

True or false (correct the false statements):

P1. Plasma glucose concentrations are higher in the postabsorptive period than in the absorptive period because of gluconeogenesis in the liver.

P2. Insulin secretion is stimulated by elevated levels of blood glucose or amino acids, whereas glucagon secretion is inhibited by these stimuli.

P3. Epinephrine is produced by the A cells of the pancreas.

P4. Insulin is the most important hormone in the regulation of metabolism.

P5. Cortisol is permissive for the actions of insulin.

P6. During exercise, glucose uptake by muscle cells is increased because of increased insulin secretion.

P7. Type 1 diabetes is often associated with obesity.

P8. A baby with untreated growth hormone deficiency will have impaired growth and mental development.

P9. The resting energy requirement of a person is independent of body size.

P10. Basal metabolic rate is increased by epinephrine and decreased by thyroid hormones.

P11. When total-body energy balance is positive in an adult, energy is being stored as fat.

P12. Anorexia nervosa is excessive thinness usually caused by hyperthyroidism.

P13. Heat loss from evaporation occurs only when one is sweating.

P14. The integrating centers for temperature-regulating reflexes are located in the brainstem.

P15. The cause of hyperthermia is endogenous pyrogens.

Multiple choice (correct each incorrect choice):

P16-20. Following a meal,

P16. insulin facilitates the diffusion of glucose into all cells except those of the liver.

P17. the conversion of triacylglycerol to glycerol and fatty acids by lipoprotein lipase is increased.

P18. glucose is the primary energy source for most body cells except those of the liver.

P19. glucose is stored as glycogen in adipose tissue.

P20. glucose is stored as fat in skeletal muscle.

P21-25. During fasting,

P21. liver glycogen stores are quickly depleted.

P22. the liver is the source of most of the glucose entering the blood.

P23. epinephrine stimulates lipolysis in adipose tissue.

P24. glucagon stimulates gluconeogenesis in muscle cells.

P25. low levels of insulin promote glycogenolysis in the liver.

P26-30. Compared to a normal person, a person with untreated diabetes mellitus would have

P26. decreased glycogen stores in muscle.

P27. increased blood ketone levels.

P28. increased blood pH.

P29. increased urine volume.

P30. increased triacylglycerol stores in adipose tissue.

P31-36. Growth hormone

P31. stimulates epiphyseal growth of long bones.

P32. stimulates protein synthesis.

P33. increases fat mobilization.

P34. increases glucose utilization.

P35. stimulates secretion of IGF-I.

P36. is the most important hormone for fetal growth.

P37-40. Cholesterol

P37. is synthesized by the liver.

P38. is carried to cells by HDL.

P39. synthesis is directly correlated with levels of cholesterol in the diet.

P40. is recycled from bile.

18: REPRODUCTION

It's a girl! Or is it???

One aspect of life most of us take for granted is that there are two sexes — male and female. A baby is born one or the other and (usually) remains the same sex throughout life. As is true of all other aspects of physiology, however, sexual development can be abnormal and can result in a baby that appears at birth to be one sex but in fact is actually the other.

One of the most interesting and unusual examples of mistaken gender identification occurred in the Dominican Republic. There, over a period of four generations, more than thirty seemingly normal girls have been spontaneously "transformed" into boys at puberty. The local name for this spontaneous sex-change is "guevodoces," which means "penis at twelve." The affected children's genitals, which appeared female at birth, grow and change to become male. The explanation of how and why this change can happen — what goes wrong and how — is enlightening about how sex development normally occurs.

Sexual differentiation, like all other life processes, depends fundamentally upon genetic information. Male sex-determining genes, if present, "program" the developing gonad to become a testis. If the male sex-determining genes are not present, the gonad will form an ovary. Testosterone secreted by the fetal testis causes formation of male genitals, including the penis and scrotum. If the gonad is an ovary or is a defective testis, then the genitals will appear female, with a vagina, a clitoris, and so forth.

The guevodoces children have normal testes, which secrete normal amounts of testosterone during fetal life. The defect in these children is the lack of an enzyme that converts testosterone to a more active form called *dihydro*testosterone (DHT) in many target tissues. It is DHT that causes development of the male reproductive tract in normal male fetuses. Testosterone is an weak *agonist* of DHT, however, and is secreted in very high levels during puberty. In pubertal guevodoces children, the high levels of testosterone interact with the DHT receptor well enough to accomplish the changes at puberty that normally go on *in utero* (in the womb).

One of the most interesting facets of this story is that the boys who start out in life as girls apparently make a smooth transition to the idea of being a boy. Gender identity (that is, the sex one thinks of one's self as being or having) thus must be, at least in their case, quite plastic.

In this chapter you will learn more about the role of hormones in sex differentiation and in other aspects of reproductive physiology.

Instructions for answering questions in this Study Guide

1. True or false: Correct each false statement. Whenever you must correct a false statement, there will almost always be more than one way to do so. In the Guide, the answer requiring the least correction will generally be given, with the understanding that other correct answers are possible. It is usually not sufficient, in terms of demonstrating understanding, to simply insert a "no" or "not," however.

2. Multiple choice: Any, all, or none of the responses may be correct. Explain why each incorrect response is incorrect.

3. Fill in the blanks: Choose the best word or words to complete each statement.

4. Directions will be given for other types of questions as they appear.

❐ **Introduction** (text page 602)

1. The primary reproductive organs are not the genitals, but rather the ______________ — the ______________ in males and the ______________ in females. Two fundamental processes occur in these organs: (1) ______________, which refers to the production of germ cells, or ______________ (______________ in the male and ______________ in the female); and (2) ______________. Both processes are controlled by the anterior pituitary hormones ______________ and ______________, whose secretion is controlled by ______________ from the hypothalamus. ______________ (Short /Long)-loop negative feedback by ______________ and ______________ also affects secretion of the pituitary gonadotropic hormones.

2. True or false: Secretion of the gonadal sex hormones does not begin until the onset of puberty.

❐ **General Principles of Gametogenesis** (text pages 602-604)

3. True or false: One way in which gametogenesis in males differs from gametogenesis in females is that mitosis of primordial germ cells occurs in males during fetal life only.

4. Meiosis
   (a) differs from mitosis in that in meiosis cell division occurs without chromosome replication.
   (b) differs from mitosis in that in meiosis two cell divisions follow one chromosomal replication.
   (c) allows for mixing of paternal and maternal genes on homologous chromosomes.
   (d) generally allows for mixing of paternal and maternal genes for the same trait on the same chromosome.
   (e) results in germ cells with 23 maternal chromosomes in one daughter cell and 23 paternal chromosomes in the other.
   (f) results in four daughter cells, each with one-half the number of chromosomes of the primordial germ cell.

SECTION A. MALE REPRODUCTIVE PHYSIOLOGY

❒ Anatomy (text pages 604-606)

5. The primary male sex organs, the ________________________, are located outside the body in a sac called the ________________________. In addition to these organs, the male reproductive tract includes ____________________________ that store and convey sperm to the exterior of the body. This system begins at the site of sperm formation, the __________________________________, followed by the ____________________, which is attached to the outer part of each testis. From there, sperm are conveyed in the ______________________________, which, together with nerves and blood vessels forming the ______________________________, enters the abdominal cavity via the ______________________________. At the back of the bladder two glands, called the ________________________________, empty into each duct, which at this point are called the ____________________________________. The two ducts meet within the ____________________ gland and join the ____________________ from the urinary bladder. As the ____________________ enters the penis, two paired glands, called the ____________________________, drain into it. The glands, ducts, and penis constitute the male ____.

6. True or false: The testes are located outside the abdomen because testosterone secretion requires a temperature lower than core body temperature.

7. True or false: In the testis, spermatogenesis takes place in the seminiferous tubules and testosterone is produced by the Sertoli cells.

8. Semen consists of the fluid secretions of the ______________________________, ____________________________, and ________________________________ and the ____________________ cells suspended in that fluid. One function of the seminal fluid is to provide __________________________ against acidic vaginal secretions, and another is to provide ______________________ for the cells.

9. True or false: Prostaglandins were given their name because they are secreted in high concentration into semen by the prostate gland.

❒ Spermatogenesis (text pages 606-608)

10. Spermatogenesis
    (a) begins with the mitotic division of a single spermatogonium.
    (b) depletes the number of spermatogonia available for later spermatogenesis.
    (c) results in four primary spermatocytes for every spermatogonium.
    (d) results in four spermatozoa for every primary spermatocyte.
    (e) normally continues from puberty throughout the lifetime of a man.

11. True or false: The products of the first meiotic division of spermatogenesis are spermatids.

12. True or false: Conversion of spermatids to spermatozoa involves cell division.

13. The process of spermatogenesis from primary spermatocyte to sperm
    (a) takes about three weeks in the human.
    (b) takes place in an environment different from that of the mitotic cell divisions that precede it.
    (c) requires participation of Sertoli cells.

14. Compared with a spermatid, a spermatozoan
    (a) is longer.
    (b) has more cytoplasm.
    (c) has fewer chromosomes.
    (d) has a head and tail.

15. True or false: The part of the spermatozoan that enables it to swim through fluid is the acrosome.

16. The blood-testis barrier
    (a) is formed primarily by Leydig cells.
    (b) protects the spermatogonia from elements in blood that might damage them.
    (c) is formed by tight junctions between cells.
    (d) separates the spermatogonia from the rest of the germ cells.

17. True or false: The only endocrine cells of the testis are the Leydig cells.

❐ Transport of Sperm (text pages 608-611)

18. Sperm
    (a) move from the lumen of the seminiferous tubule into the epididymis by self-propulsion.
    (b) are stored in the epididymis and vas deferens.
    (c) mature in the vas deferens.
    (d) become concentrated in the epididymis.

19. Erection of the penis
    (a) is a spinal reflex.
    (b) is a result of sympathetic stimulation of vascular smooth muscle in the erectile tissue of the penis.
    (c) can be triggered by stimulation of tactile receptors in the penis.
    (d) is a result of arteriolar vasodilation and venous constriction of blood vessels in the penis.
    (e) can be inhibited by input to the autonomic nerves from higher brain centers.
    (f) is required for penetration of the vagina.

20. Ejaculation
    (a) is a spinal reflex.
    (b) is triggered when afferent stimulation reaches a threshold of synaptic input to parasympathetic nerves innervating smooth muscle in the ducts and glands.
    (c) is divided into two phases — erection and orgasm.
    (d) is accomplished by contraction of smooth muscles only.
    (e) can occur in a man who has suffered spinal cord damage and has paralysis and no feeling from the waist down.
    (f) normally results in expulsion of about three million sperm from the urethra.

21. True or false: Vasectomy causes impotence because it interferes with normal blood flow to the penis.

❐ Hormonal Control of Male Reproductive Functions (text pages 611-614)

22. The hormone from the anterior pituitary gland that directly stimulates spermatogenesis is ______________________. This hormone exerts its effects through

________________ cells, which respond to the hormone by secreting ________________ (endocrine/paracrine) factors. The other pituitary hormone, ________________, stimulates spermatogenesis indirectly by increasing the secretion of ________________ by ________________ cells. Secretion of both pituitary hormones is stimulated by pulsatile secretion of ________________ from the ________________. Negative feedback control of pituitary hormone secretion is exerted by ________________, which acts both at the pituitary and at the hypothalamus, and by ________________, which selectively inhibits secretion of ________________ by the ________________.

23. A man with undescended testes (that is, testes in the abdominal cavity instead of the scrotum) is sterile (that is, he cannot make sperm). How would his hormone levels be affected, compared with a those of a normal man? Indicate whether they would be increased (I), decreased (D), or unchanged (U).

(a) ______ testosterone (b) ______ FSH (c) ______ LH

(d) ______ inhibin (e) ______ GnRH

24. True or false: In normal men, the blood levels of reproductive hormones and the production of sperm vary markedly from one day to the next.

25. Testosterone
   (a) is the most potent androgen the body secretes.
   (b) promotes spermatogenesis by actions on Leydig cells.
   (c) stimulates secretion of prostatic fluid.
   (d) stimulates growth of body and facial hair.
   (e) stimulates growth of scalp hair.
   (f) has effects on behavior.
   (g) is a prohormone.

26. Dihydrotestosterone is
   (a) secreted by the testis.
   (b) an estrogen.
   (c) required for development of the male reproductive tract.

27. True or false: Prolactin is a hormone that has no function in the male.

SECTION B. FEMALE REPRODUCTIVE PHYSIOLOGY

❐ Anatomy (text pages 614-616)

28. The uterine tubes
    (a) are part of the female external genitalia.
    (b) connect the uterus to the ovaries.
    (c) are the same as the fallopian tubes.
    (d) are the same as the oviducts.

29. The uterus
    (a) is largely composed of smooth muscle.
    (b) is hollow.
    (c) joins the vagina at the cervix.
    (d) is also called the womb.
    (e) is the source of menstrual bleeding.

30. The vagina is
    (a) part of the female external genitalia.
    (b) a muscular tube.
    (c) the birth canal.
    (d) the female organ of copulation.

31. True or false: The female external genitalia are collectively called the vestibule.

32. Menstrual cycles
    (a) refer only to events occurring in the uterus.
    (b) are caused by changes in hormone levels.
    (c) in normal women repeat every 28 days.

❐ Ovarian Function (text pages 616-619)

33. Oogenesis
    (a) refers to the production of gametes in the female.
    (b) begins at puberty.
    (c) results in the formation of four ova, each with 23 chromosomes.
    (d) takes place entirely in the ovary.

34. True or false: At birth, the ovaries of a female baby contain about one million oogonia.

35. Oocyte development in the ovary occurs in structures called follicles. The simplest of these, the ________________ follicle, consists of an immature ________________ surrounded by a single flattened layer of ________________ cells. As the follicle begins to develop, the cells surrounding it change shape and the structure is now called a ________________ follicle. At the next stage of development, the ________________ increases in size and the ________________ increase in number. The two cell types both secrete components of a layer of clear material called the ________________, but they maintain contact with each other by means of cytoplasmic processes from the ________________ that make gap junctions with the ________________. As the follicle continues to grow, a new layer of differentiated connective-tissue cells, called the ________________, surrounds the follicle. After the follicle reaches a certain diameter, fluid secreted by the ________________ cells starts to form a space called the ________________. At this point the ________________ has reached about maximal size, but the follicle continues to grow, primarily because of expansion of the ________________.

36. The dominant follicle
    (a) undergoes atresia.
    (b) continues to develop after other antral follicles have begun to degenerate.
    (c) undergoes ovulation.
    (d) is visible to the naked eye.

37. True or false: The dominant follicle normally takes about two weeks to grow from a small-antral follicle to a mature follicle.

38. True or false: Most of the primordial follicles a woman is born with eventually are ovulated.

39. At ovulation,
    (a) the follicle ruptures.
    (b) the second meiotic division of the oocyte occurs.
    (c) the oocyte, along with its surrounding thecal layer, is expelled onto the ovarian surface.

40. The corpus luteum
    (a) is composed of granulosa and theca cells from the ovulated follicle.
    (b) is an exocrine gland.
    (c) has a lifespan of ten to twelve days unless it is "rescued" by fertilization of the ovulated oocyte.

41. True or false: In the average 28-day menstrual cycle, days 1 through 13+ constitute the luteal phase and days 14 through 28 constitute the follicular phase.

42. True or false: Both the follicle and the corpus luteum secrete only steroid hormones.

❐ **Control of Ovarian Function** (text pages 619-624)

43. The accompanying figure shows the plasma concentrations of the pituitary gonadotropins and the ovarian steroids during an idealized, 28-day menstrual cycle. Name the eight hormones/events/phases indicated on the drawing.

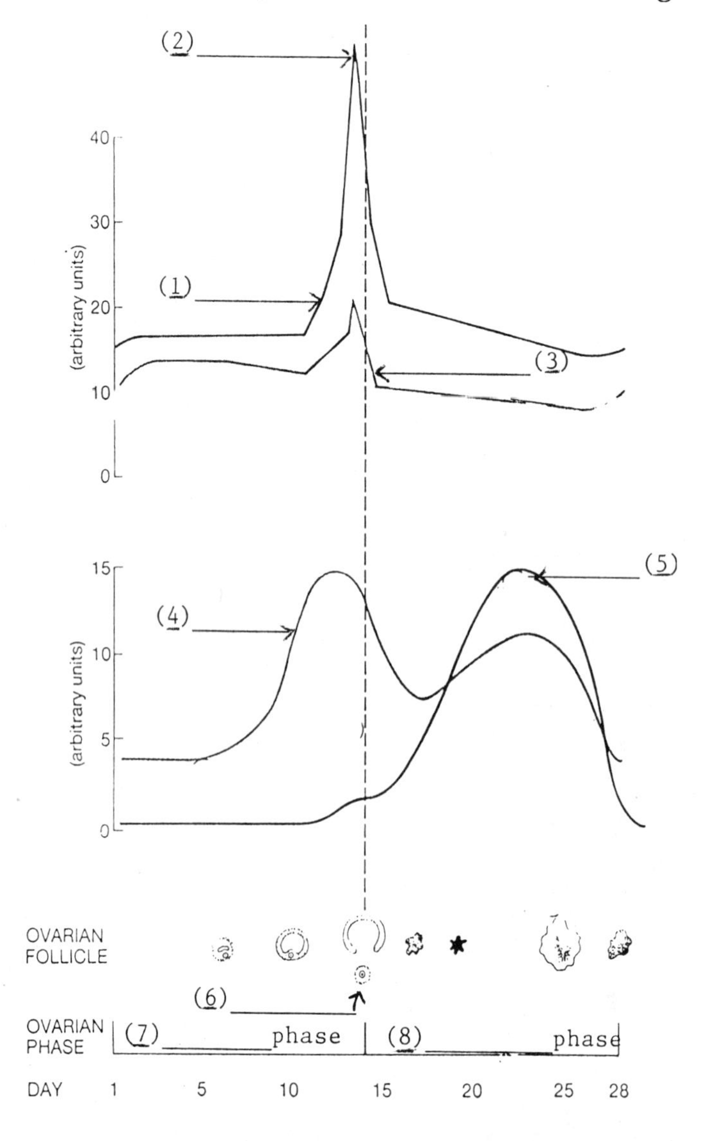

44. FSH
  (a) stimulates development from primordial follicles to primary follicles.
  (b) stimulates growth of small-antral follicles to larger follicles.
  (c) secretion increases at the beginning of the follicular phase and then declines.
  (d) stimulates growth of theca cells.
  (e) stimulates secretion of estrogen.

45. LH
  (a) secretion during the follicular phase is steady until it surges at the time of ovulation.
  (b) receptors are present in high numbers on granulosa cells during the early part of the follicular phase.
  (c) stimulates androgen production by theca cells.
  (d) is required for estrogen secretion by granulosa cells.
  (e) secretion during the latter part of the follicular phase is decreased by inhibin secreted by granulosa cells.

46. True or false: With respect to hormonal control and secretion, theca cells in the ovary are analogous to Leydig cells in the testis .

47. True or false: Atresia of all but the dominant follicle is a result of the diminished secretion of FSH during the second week of the follicular phase.

48. Maturation of one dominant follicle requires
  (a) FSH.
  (b) LH.
  (c) inhibin.
  (d) estrogen.
  (e) progesterone.

49. Estrogen
  (a) in low concentrations inhibits pituitary secretion of LH and FSH.
  (b) in high concentrations stimulates pituitary sensitivity to GnRH.
  (c) in the presence of high concentrations of progesterone stimulates increased secretion of GnRH.
  (d) helps FSH induce LH receptors on theca cells.
  (e) secretion is required for the LH surge.

50. Effects of the LH surge on the mature follicle include
 (a) stimulation of estrogen secretion.
 (b) stimulation of oocyte maturation.
 (c) stimulation of prostaglandin synthesis.
 (d) decreased antral fluid secretion.
 (e) luteinization of the unovulated granulosa cells.

51. During the luteal phase of the menstrual cycle,
 (a) pre- and early-antral follicles begin to grow.
 (b) FSH stimulates secretion of estrogen and progesterone.
 (c) the corpus luteum degenerates if the ovulated oocyte is not fertilized.
 (d) LH and FSH levels reach their lowest levels.

52. In terms of events in the ovary and plasma levels of reproductive hormones, what would happen if the corpus luteum did not self-destruct?

____

____

____

____

____

____

❐ **Uterine Changes in the Menstrual Cycle** (text pages 624-626)

53. The proliferative phase of the menstrual cycle
 (a) refers to the regrowth, in the uterus, of the myometrial layer that was sloughed during menstruation.
 (b) corresponds to the early follicular phase of the ovarian cycle.
 (c) is a result of the effects of estrogen secretion.

54. The secretory phase of the menstrual cycle
 (a) corresponds to the luteal phase of the ovarian cycle.
 (b) refers to the endocrine function of the endometrium.
 (c) prepares the uterus for the arrival of the fertilized ovum.
 (d) requires both estrogen and progesterone.

55. True or false: Cervical mucus secreted under the influence of estrogen alone is thick and sticky, while that secreted under the influence of both estrogen and progesterone is clear and nonviscous.

56. The sloughing of the endometrial lining, called ________________, is a direct result of the action of ________________, which are released as a result of withdrawal of ____________ and ____________. The agents that cause the sloughing are also responsible for the pain that is sometimes associated with it, called ________________.

❒ **Other Effects of Estrogen and Progesterone** (text pages 626-628)

57. True or false: Most of the hormonal effects on female accessory sex organs and secondary sex characteristics are attributable to progesterone.

58. Actions of estrogen include
   (a) stimulation of fat deposition under the skin in the breasts and hips.
   (b) stimulation of skeletal muscle growth.
   (c) vaginal cornification.
   (d) stimulation of myometrial contractions.
   (e) stimulation of bone growth and strength.
   (f) stimulation of acne-causing fluid from sebaceous glands.
   (g) stimulation of growth of pubic hair.
   (h) stimulation of prolactin secretion.
   (i) potentiation of milk synthesis by prolactin.

59. The actions of progesterone
   (a) on the female reproductive tract are often antagonistic to those of estrogen.
   (b) are limited to the female reproductive tract and breasts.
   (c) are mediated through plasma membrane receptors.

60. True or false: Dysmenorrhea refers to a combination of physical and psychological symptoms of uncertain cause occurring in the days preceding the onset of menstruation in some women.

❐ Androgens in Women (text page 628)

61. Androgens in women
    (a) have identifiable effects only in disease states in which they are secreted in abnormally large amounts.
    (b) are secreted in increased amounts during and after puberty, relative to prepubertal levels.
    (c) are important for maintaining sex drive.

❐ Female Sexual Response (text page 628)

62. The female sexual response includes
    (a) erection of the clitoris.
    (b) engorgement of blood vessels in the genitals and breasts.
    (c) orgasm.
    (d) ejaculation.

63. True or false: The mechanisms underlying the physiological response of females to sexual stimulation are similar to those resulting in erection in males.

❐ Pregnancy (text pages 628-643)

64. True or false: The "fertile period" (that period of time in a woman's menstrual cycle during which intercourse is likely to lead to pregnancy) normally lasts about one week.

65. Regarding transport of the ovum,
    (a) during ovulation the secondary oocyte is ejected from the ovary directly into the uterine tube.
    (b) movement of the oocyte is facilitated by cilia beating toward the ovary.
    (c) movement of the oocyte toward the uterus is facilitated by contractions of the smooth muscle in the uterine tube wall.

66. Regarding transport of sperm in the female reproductive tract,
    (a) sperm swim unaided through most of the tract.
    (b) most of the ejaculated sperm arrive in the vicinity of the ovum (if there is one).

(c) sperm transport through the cervix is made possible by actions of progesterone on the cervical mucus.
(d) during this transport sperm acquire the ability to fertilize an ovum.

67. Activation of sperm
(a) requires prior capacitation in the epididymis.
(b) is required for sperm to be motile.
(c) occurs in the vicinity of the ovum.
(d) includes changes in the acrosome and plasma membrane.

68. In general terms, ________________________ is the fusion of a sperm with an ovum. Early events in this process include binding of proteins on the ____________________________ to receptors on the __________________________. After moving through the ______________________________ with the aid of enzymes from the ____, the sperm fuses to the __________________________________, triggering several events in this order: 1) activation of ____________________________________ in the periphery of the ovum that secrete _______________________ into the ______________________________, causing changes in its structure and blocking ______________________; 2) drawing in of the ____________________________________ into the ovum; 3) completion of the second ____ and extrusion of the second ____________________________; and 4) formation of __________________________ from the sperm and ovum that migrate to the center of the fertilized cell, now called a ________________________, and replicate.

69. Can fertilization occur in the uterus? Explain.

____

____

____

____

70. True or false: Monozygotic twins are two individuals who originated from a single zygote.

71. True or false: While passing down the uterine tube, the zygote undergoes several meiotic cell divisions.

72. Cleavage of the zygote yields, after a few days, a multicellular, fluid-filled structure called the ______________________________. This structure has two general types of differentiated cells: the ___________________________________, which will develop into the embryo and some of its associated membranes, and the ______________________________, which initiates the embedding of the structure into the endometrium, a process called _______________________________. These latter cell types also eventually provide the fetal portion of the ______________________.

73. Implantation
   (a) outside the uterus is said to be myopic.
   (b) occurs midway through the luteal phase.
   (c) requires preparation of the endometrium by estrogen and progesterone.
   (d) is a result of digestion of myometrial cells by enzymes in the trophoblast.

74. The placenta
   (a) is an organ of nourishment for the embryo/fetus.
   (b) is composed solely of tissues derived from the trophoblast.
   (c) functions as the lungs and kidneys of the fetus prior to birth.
   (d) allows for mixing of maternal and fetal blood.
   (e) screens out all potentially harmful or noxious agents so that they cannot affect the fetus.

75. Describe the blood supply of the placenta (both maternal and fetal components).

____

____

____

____

76. True or false: Blood in the umbilical arteries is well-oxygenated, whereas that in the umbilical vein is poorly oxygenated.

77. True or false: During pregnancy, the fetus is suspended by the umbilical cord in the lumen of the uterus.

78. True or false: Chorionic villus sampling is the removal of a sample of amniotic fluid for diagnosis of genetic or chromosomal disorders of the fetus.

79. Chorionic gonadotropin
   (a) is a hormone.
   (b) is similar in action to FSH.
   (c) is secreted by endometrial cells.
   (d) rescues the corpus luteum.
   (e) is the basis for pregnancy tests.

80. Functions of estrogen during pregnancy include
   (a) stimulation of myometrial growth.
   (b) maintenance of the endometrium.
   (c) stimulation of prolactin secretion.
   (d) stimulation of breast development.
   (e) stimulation of milk synthesis.

81. Functions of progesterone during pregnancy include
   (a) stimulation of myometrial growth.
   (b) inhibition of myometrial contractility.
   (c) inhibition of GnRH secretion.
   (d) stimulation of breast development.
   (e) stimulation of milk synthesis.

82. During pregnancy,
   (a) estradiol is the major estrogen produced.
   (b) the placenta secretes steroid and peptide hormones.
   (c) the placenta requires precursors from the maternal adrenal cortex for estrogen synthesis.

83. True or false: Symptoms of diabetes mellitus worsen during pregnancy.

84. True or false: Symptoms of toxemia of pregnancy include glucose in the urine, abnormal fluid retention, and hypertension.

85. Examine text Table 18-9 on page 635. Note the changes during pregnancy in kidney function, blood volume, and circulation. Starting with the decrease in peripheral resistance that occurs as blood flows through the high-volume, low-resistance placenta, describe how this decreased resistance leads to an increased blood volume. (A quick review of the reflexes learned in Chapters 13 and 15 may be helpful.)

____

____

____

____

____

____

____

____

86. Parturition
    (a) refers to delivery of the infant and placenta.
    (b) normally occurs between the 27th and 28th weeks of pregnancy.
    (c) requires hormonal and sympathetic stimulation of the myometrium.
    (d) is aided by an increase in the progesterone/estrogen ratio.
    (e) is an example of positive feedback.

87. True or false: Relaxin is a tranquilizer given to pregnant women to make the birth process less stressful.

88. Regarding the control of parturition,
    (a) oxytocin secretion is stimulated by cervical dilation.
    (b) cervical dilation precedes coordinated uterine contractions.
    (c) oxytocin increases myometrial estrogen receptors.
    (d) prostaglandins stimulate myometrial contractions.
    (e) stretching of the myometrium is one signal that begins uterine contractions.

89. True or false: Mammary gland development is not complete until mid-to-late pregnancy.

90. Mammary glands consist of ducts that end in clusters of ____________________, the site of milk synthesis. At their other end, the ducts open onto the ____________________ of the breast. The stimulus of suckling reflexly inhibits secretion of ________________ and/or stimulates secretion of ________________

by the hypothalamus, and thus increases secretion of ____________________ by the anterior pituitary. This latter hormone acts on ________________________ cells, causing them to ______________________________. The same suckling stimulus causes ________________ to be released from the posterior pituitary. This hormone acts on ______________________ cells, causing them to ____________________ and resulting in ________________________.

91. True or false: The major constituents of milk are water, protein, fat, and glucose.

92. True or false: A woman cannot become pregnant while she is lactating.

93. For each description on the left, indicate which contraceptive method(s) listed on the right fit(s) that description.

________ require(s) a doctor's prescription for purchase	(a) condom
________ most effective reversible method	(b) intrauterine device
________ has highest failure rate	(c) minipill
________ often irreversible	(d) rhythm method
________ provide(s) barrier to sperm	(e) vasectomy
________ inhibit(s) ovulation	(f) vaginal sponge
________ require(s) yearly replacement	(g) diaphragm
________ should not be used by women over 35 or by those who smoke	(h) combination pill
	(i) vaginal spermicide
________ kill(s) sperm	(j) tubal ligation
________ protect(s) against sexually transmitted diseases	
________ has/have protective effects against endometrial cancer	
________ reliability improved by keeping track of basal body temperature	

94. True or false: A compound that inhibited the binding of progesterone to its receptor would be an effective contraceptive.

95. Would inhibin be an effective contraceptive for men? Why or why not?

____

____

____

____

SECTION C: THE CHRONOLOGY OF REPRODUCTIVE FUNCTION

❒ Sex Determination (text page 643)

96. Regarding the sex chromosomes and sex determination,
    (a) individuals who have two X chromosomes are genetic females.
    (b) individuals who have a Y chromosome are genetic males.
    (c) X chromosomes contain more genes than Y chromosomes.
    (d) cells in normal females contain sex chromatin whereas cells in normal males do not.

97. True or false: For every primary spermatocyte, two sperm bearing an X chromosome and two sperm bearing a Y chromosome will be produced.

❒ Sex Differentiation (text pages 644-646)

98. True or false: The gene on the Y chromosome that determines whether gonads will be testes or ovaries codes for a protein called Mullerian inhibiting hormone.

99. In the differentiation of a normal male,
    (a) undifferentiated gonads develop into testes during the fifth week of embryonic life.
    (b) the Wolffian ducts regress.
    (c) pituitary LH causes fetal Leydig cells to secrete testosterone.
    (d) testosterone directly stimulates development of the epididymis, vas deferens, and seminal vesicles.
    (e) testosterone directly stimulates development of the penis and scrotum.

100. In the differentiation of a normal female,
    (a) Mullerian inhibiting hormone causes the Mullerian ducts to regress.
    (b) the uterus and uterine tubes are formed from the Wolffian ducts.
    (c) female genitalia form in the absence of hormonal stimulation.
    (d) the gonads remain undifferentiated throughout fetal life.

101. People with androgen insensitivity syndrome
    (a) have female genitalia.
    (b) have ovaries.
    (c) have differentiated Mullerian duct structures.

(d) lack the enzyme that converts testosterone to DHT.

(e) are capable of spermatogenesis.

(f) are genetic males who appear female.

102. Masculinization by androgens of a genetic female during fetal life

(a) can occur when the mother ingests androgens during her pregnancy.

(b) results in formation of testes and masculine external genitalia.

(c) results in regression of Mullerian duct structures.

(d) can occur as a result of defective enzymes in the fetal adrenal cortex.

(e) will have pronounced effects on her behavior in adult life.

❐ **Puberty** (text page 646)

103. The onset of puberty

(a) is triggered by an increase in sensitivity of the gonads to pituitary gonadotropins.

(b) is triggered by an increase in sensitivity of the pituitary to GnRH.

(c) is triggered by an increase in GnRH secretion.

(d) in females is marked by the menarche.

(e) in males is marked by growth of the testes and scrotum.

104. True or false: The onset of puberty may be delayed indefinitely in girls who are very thin.

❐ **Menopause** (text pages 646-647)

105. Menopause

(a) is the cessation of menstrual cycles.

(b) occurs as a result of cessation of gonadotropin secretion.

(c) has a counterpart in men called the climacteric.

(d) results in low plasma levels of estrogens.

106. True or false: Postmenopausal women are at increased risk, compared to premenopausal women, of osteoporosis, atherosclerosis, and hypertension.

ANSWERS

Boldface type indicates the answers you should have given. Words in medium-face (ordinary) type explain or expand upon the answer.

1. The primary reproductive organs are not the genitals, but rather the **gonads** — the **testes** in males and the **ovaries** in females. Two fundamental processes occur in these organs: (1) **gametogenesis**, which refers to the production of germ cells, or **gametes** (**spermatozoa** in the male and **ova** in the female); and (2) **secretion of sex hormones.** Both processes are controlled by the anterior pituitary hormones **follicle-stimulating hormone** and **luteinizing hormone**, whose secretion is controlled by **gonadotropin-releasing hormone** from the hypothalamus. **Long**-loop negative feedback by **sex steroids** and **inhibin** also affects secretion of the pituitary gonadotropic hormones.

2. **False.** Secretion of the gonadal sex hormones ~~does not~~ begin**s during fetal life but then ceases during infancy and remains low** until the onset of puberty.

3. **False.** One way in which gametogenesis in males differs from gametogenesis in females is that mitosis of primordial germ cells occurs in ~~males~~ **females** during fetal life only.

4. (a) **Incorrect. See answer (b).**

   (b) **Correct.** (Even though the second meiotic division does occur without further chromosome replication, answer (a) is not correct because it implies that chromosome replication does not occur at all. Although such a system might work, in that the resultant daughter cells would have one-half the chromosome complement of the primordial germ cell and so could combine with germ cells from the other sex to form a fertilized cell with 46 chromosomes, there would be much less chance for diversity among the offspring because crossing over could not occur. The mechanism of meiosis assures the greatest mixture of possible traits, which is beneficial from the standpoint of evolution.)

   (c) **Correct.** (This is the process of crossing over.)

   (d) **Incorrect. Crossing over is normally an even exchange of homologous genes on homologous chromosomes.** (If two genes for the same trait were to stay on the same chromosome, then both chromosomes — the one with too many genes and the one with too few — would be defective. This does sometimes happen, and if the resultant germ cells with defective chromosomes are involved in fertilization, then the resultant zygote would be abnormal and in some cases would die.)

   (e) **Incorrect. The maternal and paternal homologous chromosomes are randomly distributed.**

   (f) **Correct.**

5. The primary male sex organs, the **testes**, are located outside the body in a sac called the **scrotum**. In addition to these organs, the male reproductive tract includes **ducts** that store and convey sperm to the exterior of the body. This system begins at the site of sperm formation, the **seminiferous tubules**, followed by the **epididymis,** which is attached to the outer part of each testis. From there, sperm are conveyed in the **vas deferens**, which, together with nerves and blood vessels forming the **spermatic cord**, enters the abdominal cavity via the **inguinal canal**. At the back of the bladder two glands, called the **seminal vesicles**, empty into each duct, which at this point are called the **ejaculatory ducts**. The two ducts meet within the **prostate** gland and join the **urethra** from the urinary bladder. As the **urethra** enters the penis, two paired glands, called the **bulbourethral glands**, drain into it. The glands, ducts, and penis constitute the male **accessory reproductive organs**.

   Note: The plurals of Latin and Greek words are often unusual, which is one reason we tend to speak of structures with names derived from these languages in the singular, even though there are frequently two of them. The plural of the Latin "testis" — "testes"— is not hard to remember because the word is quite familiar. However, epididymis/epididymides is more difficult. The plural of "vas deferens" is "vasa deferentia." Regardless of the nomenclature, you should remember which parts of the male reproductive system are paired and which are single structures:

Paired	Single
testes	prostate
epididymides	urethra
vasa deferentia	penis
spermatic cords	scrotum
ejaculatory ducts	
seminal vesicles	
bulbourethral glands	

6. **False.** The testes are located outside the abdomen because ~~testosterone secretion~~ **spermatogenesis** requires a temperature lower than core body temperature.

7. **False.** In the testis, spermatogenesis takes place in the seminiferous tubules and testosterone is produced by the ~~Sertoli~~ **Leydig** (interstitial) cells.

8. Semen consists of the fluid secretions of the **seminal vesicles**, **prostate gland**, and **bulbourethral glands** and the **sperm** cells suspended in that fluid. One function of the seminal fluid is to provide **buffers** against acidic vaginal secretions, and another is to provide **nutrients** for the cells.

9. **False.** Prostaglandins were given their name because they ~~are~~ **were thought to be** secreted in high concentration into semen by the prostate gland. (Perhaps it is just as well that the discoverers of prostaglandins in semen were mistaken about their gland of origin. "Seminalvesiclin" is even harder to write and pronounce than prostaglandin!)

10. (a) **Correct.**

    (b) **Incorrect.** **Depletion does not occur because one spermatogonium in each clone de-differentiates back to a primitive spermatogonium.**

    (c) **Incorrect.** **There are many more than four primary spermatocytes for every initial primitive spermatogonium, and two primary spermatocytes for each differentiated spermatogonium that undergoes the final mitotic division.** (Note from this answer that spermatogonia differ from each other depending upon where they are in the differentiation pathway toward primary spermatocytes.)

    (d) **Correct.**

    (e) **Correct.**

11. **False.** The products of the first meiotic division of spermatogenesis are ~~spermatids~~ **secondary spermatocytes**.

    Alternatively, the answer below is also correct:

    False. The products of the ~~first~~ **second** meiotic division of spermatogenesis are spermatids.

12. **False.** Conversion of spermatids to spermatozoa involves cell ~~division~~ **differentiation**.

13. (a) **Incorrect.** **It takes 64 days** (more than two months).
    (b) **Correct.**
    (c) **Correct.**

14. (a) **Correct.**
    (b) **Incorrect.** **It has less cytoplasm.** (See text Figure 18-8 and its legend.)
    (c) **Incorrect.** **It has the same number of chromosomes** (23).
    (d) **Correct.**

15. **False.** The part of the spermatozoan that enables it to swim through fluid is the ~~acrosome~~ **flagellum**.

    Alternatively, the answer below is also correct:

    False. The ~~part~~ **enzyme-filled vesicle** of the spermatozoan that enables it to ~~swim through fluid~~ **penetrate the ovum** is the acrosome.

16. (a) **Incorrect.** **It is formed by Sertoli cells.**

    (b) **Incorrect.** **It protects primary spermatocytes, secondary spermatocytes, spermatids, and sperm.**

    (c) **Correct.**
    (d) **Correct.**

17. **False.** The ~~only~~ endocrine cells of the testis are the Leydig cells **and the Sertoli cells**.

18. (a) **Incorrect.** **They are conveyed from the seminiferous tubules to the epididymis by fluid movement and peristalsis**. (Sperm are not motile at this time.)
    (b) **Correct.**
    (c) **Incorrect.** **They mature in the epididymis**.
    (d) **Correct.**

19. (a) **Correct.**
    (b) **Incorrect.** **It is a result of parasympathetic stimulation.** (Sympathetic stimulation is inhibitory.)
    (c) **Correct.**
    (d) **Correct.**
    (e) **Correct.**
    (f) **Correct.**

20. (a) **Correct.**
    (b) **Incorrect.** **Sympathetic nerves are triggered.**
    (c) **Incorrect.** **The two phases are emission and expulsion of semen from the urethra.**
    (d) **Incorrect.** **Smooth-muscle contraction is responsible for movement of sperm up the vasa into the ejaculatory ducts, for secretion of glandular fluid, and to some extent for expulsion from the urethra. Skeletal-muscle contraction is important for the latter part of ejaculation, however.**
    (e) **Correct.** (Erection and ejaculation are both spinal reflexes, and so they can occur even in a man who has no feeling in his genitals. Thus, men with such spinal cord damage can father children, at least if the damage is not at the level of the synapses necessary for the reflexes.)
    (f) **Incorrect.** **It normally results in expulsion of between two and three *hundred* million sperm**.

21. **False.** Vasectomy **does not** causes impotence ~~because it interferes with~~ **and it has no effect on** normal blood flow to the penis.

    Alternatively, the answer below is also correct:

    False. Vasectomy causes ~~impotence~~ **sterility** because it interferes with ~~normal blood flow to the penis~~ **sperm transport out of the lower part of the vas deferens**.

    (Erection and ejaculation still occur normally in a vasectomized man. In fact, the sperm themselves contribute so little to the actual volume of semen ejaculated — about 5-10% — that men do not notice a difference.)

22. The hormone from the anterior pituitary gland that directly stimulates spermatogenesis is **FSH**. This hormone exerts its effects through **Sertoli** cells, which respond to the hormone by secreting **paracrine** factors. The other pituitary hormone, **LH**, stimulates spermatogenesis indirectly by increasing the secretion of **testosterone** by **Leydig** cells. Secretion of both pituitary hormones is stimulated by pulsatile

secretion of **GnRH** from the **hypothalamus**. Negative feedback control of pituitary hormone secretion is exerted by **testosterone**, which acts both at the pituitary and at the hypothalamus, and by **inhibin**, which selectively inhibits secretion of **FSH** by the **pituitary gland**.

Note: One clear effect of FSH on spermatogenesis is that this hormone stimulates androgen-binding protein secretion by the Sertoli cells. Spermatogenesis requires high levels of testosterone, higher than those present in plasma. The binding protein from Sertoli cells acts as a "sink" for the testosterone produced in the nearby Leydig cells, promoting diffusion of the steroid into the seminiferous tubules and allowing the tubules to concentrate it.

23. (a) **U**; (b) **I**; (c) **U**; (d) **D**; (e) **U**

Sertoli-cell function is inhibited by high temperatures. Therefore inhibin secretion will be diminished, which will lessen the negative feedback on FSH, causing FSH levels to rise. Leydig-cell function is normal at elevated testis temperature, however, and so testosterone, LH, and GnRH levels will be normal. (It was this hormonal picture of elevated blood FSH in the face of normal LH and testosterone levels in men with undescended testes or other conditions involving decreased Sertoli-cell function that prompted scientists to postulate the existence of inhibin. This hormone has subsequently been purified and found to be a large polypeptide.)

24. **False.** In normal men, the blood levels of reproductive hormones and the production of sperm ~~vary markedly~~ **remain quite constant** from one day to the next.

25. (a) **Correct.** (The qualifier "the body secretes" is necessary because, in some cells, testosterone is not very potent at all and must be converted either to dihydrotestosterone or to estradiol. Dihydrotestosterone is not secreted, and estradiol is an estrogen.)

(b) **Incorrect.** **It promotes spermatogenesis through actions on Sertoli cells.**

(c) **Correct.** (And other functions of other accessory sex organs.)

(d) **Correct.** (And other secondary sex characteristics.)

(e) **Incorrect.** **It stimulates loss of scalp hair.** (If the gene for baldness is present.)

(f) **Correct.** (As the text points out, testosterone is required for development of sex drive in males and may be involved in stimulating aggressive behavior in human males as it is in other animals. However, human behaviors in general are much less dependent upon hormones than are those of other animals, and more dependent upon learning.)

(g) **Correct.**

26. (a) **Incorrect.** **It is produced by enzymatic conversion of testosterone in its target cells.**

(b) **Incorrect.** **It is an androgen.**

(c) **Correct.**

27. **False.** Prolactin is a hormone that has ~~no~~ **significant** functions in the male (including potentiation of the actions of LH and testosterone on their target cells).

28. (a) **Incorrect.** **They are part of the internal genitalia.**
    (b) **Incorrect.** **They are attached to the uterus but not to the ovaries.** (The ovaries are held in place by ligaments.)
    (c) **Correct.** (The 16th-century Italian anatomist Fallopio first described them, but his guess as to their function was wrong. He thought they "ventilated" the uterus.)
    (d) **Correct.** (This name describes one of their major functions: to provide passage for the ovum toward the uterus.)

29. (a) **Correct.**
    (b) **Correct.**
    (c) **Correct.**
    (d) **Correct.**
    (e) **Correct.**

30. (a) **Incorrect.** **Only the entrance to the vagina is external.**
    (b) **Correct.**
    (c) **Correct.** (Although this fact was not stated in this section, it is logical that at birth the baby would exit the uterus through the vagina.)
    (d) **Correct.** (Again, not stated explicitly, but logical. Thus, the vagina serves the same function in the female as the penis in the male.)

31. **False.** The female external genitalia are collectively called the ~~vestibule~~ **vulva**.

32. (a) **Incorrect.** **They refer to cyclic changes in the entire female reproductive system.**
    (b) **Correct.**
    (c) **Incorrect.** **They average 28 days in length but vary considerably even in normal women.** (The name "menstrual" comes from the Latin *mensis*, which means "month." This term is used because the cycles last about a month. Menstruation is also sometimes called the *menses*.)

33. (a) **Correct.**
    (b) **Incorrect.** **It begins before birth.**
    (c) **Incorrect.** **It results in one mature ovum** (if the secondary oocyte is fertilized) **and three polar bodies, each with 23 chromosomes.** (If the secondary oocyte is <u>not</u> fertilized, then the end result is one secondary oocyte and one polar body, each with 23 duplicated chromosomes, or 46 chromatids. These cells die and are phagocytized after about one day.)
    (d) **Incorrect.** **The second meiotic division occurs, if it occurs, in the uterine tubes.**

34. **False.** At birth, the ovaries of a female baby contain about one million ~~oogonia~~ **primary oocytes**. (There are no undifferentiated oogonia left at the time of birth.)

35. Oocyte development in the ovary occurs in structures called follicles. The simplest of these, the **primordial** follicle, consists of an immature **primary oocyte** surrounded by a single flattened layer of **granulosa** cells. As the follicle begins to develop, the cells surrounding it change shape and the structure is now called a **primary** follicle. At the next stage of development, the **oocyte** increases in size and the **granulosa cells** increase in number. The two cell types both secrete components of a layer of clear material called the **zona pellucida**, but they maintain contact with each other by means of cytoplasmic processes from the **granulosa cells** that make gap junctions with the **oocyte**. As the follicle continues to grow, a new layer of differentiated connective-tissue cells, called the **theca**, surrounds the follicle. After the follicle reaches a certain diameter, fluid secreted by the **granulosa** cells starts to form a space called the **antrum**. At this point the **oocyte** has reached about maximal size, but the follicle continues to grow, primarily because of expansion of the **antrum**.

36. (a) **Incorrect.** **The dominant follicle keeps growing as the other growing follicles undergo atresia.**

    (b) **Correct.**

    (c) **Correct.**

    (d) **Correct.** (Text Figure 18-14 shows this quite clearly. The ovary, shown in actual size in part H, is probably smaller than you expected, but the follicles are easily visible, particularly the one about to ovulate. The oocyte in the middle cumulus oophorous, the largest human cell, is about the size of a printed period. Also note in part G of Figure 8-14 that the oocyte seems to have shrunk. Actually, all parts of that follicle are smaller (less enlarged) in this diagram in order to show the size of the antrum in a mature follicle relative to the size of the rest of the follicle. If the mature follicle were drawn on the same scale as the developing follicles, it would have taken most of the page.)

37. **True.** (Note in text Figure 18-15 that the dominant follicle is among the group of "multiple follicles" that begin to develop on the first day of the cycle. After about the first week, the dominant follicle is the only one that continues to grow — usually. Occasionally there are two or more *co*-dominant follicles, and multiple births (nonidentical twins, triplets, or more) will result if the ovulated oocytes are fertilized.)

38. **False.** Most of the primordial follicles a woman is born with eventually ~~are ovulated~~ **undergo atresia**.

39. (a) **Correct.**

 (b) **Incorrect. Ovulation occurs shortly after the completion of the first meiotic division. The second division does not occur until fertilization.**

 (c) **Incorrect. The oocyte is surrounded by its zona pellucida and cumulus oophorous.** (The theca remains in the ovary.)

 Note: The adhering cumulus oophorous (which means "egg cloud") of granulosa cells does not just go along for the ride. It is very important for the transport of the oocyte down the uterine tube, which we shall discuss later. The importance of the zona pellucida will be made clear then as well.

40. (a) **Correct.** (These cells are reorganized into a structure that looks quite different from the follicle. As the text mentions, the cells enlarge, and they also take on a sort of yellowish color because of accumulated lipid. Hence the structure is called the corpus luteum, which is Latin for "yellow body.")

 (b) **Incorrect. It has endocrine function, not exocrine.**

 (c) **Correct.** (You should be wondering just what that rescue signal might be. Go ahead — take a guess!)

41. **False.** In the average 28-day menstrual cycle, days 1 through 13+ constitute the ~~luteal~~ **follicular** phase and days 14 through 28 constitute the ~~follicular~~ **luteal** phase.

42. **False.** Both the follicle and the corpus luteum secrete ~~only~~ **both** steroid **and peptide** hormones.

43. (1) **LH**
 (2) **LH surge**
 (3) **FSH**
 (4) **Estrogen** (estradiol)
 (5) **Progesterone**
 (6) **Ovulation**
 (7) **Follicular** phase
 (8) **Luteal** phase

44. (a) **Incorrect. This development is independent of pituitary hormones.**

 (b) **Correct.** (It is this action that gives FSH its name — follicle-stimulating hormone.)

 (c) **Correct.**

 (d) **Incorrect. It stimulates the growth of granulosa cells.**

 (e) **Correct.**

45. (a) **Incorrect.** **The surge begins about 48 hours prior to ovulation.**

    (b) **Incorrect.** **They are present in high numbers on theca cells during the early part of the follicular phase** (and in high numbers in granulosa cells during the late part of the follicular phase).

    (c) **Correct.** (It also stimulates theca-cell proliferation.)

    (d) **Correct.**

    (e) **Incorrect.** **Inhibin decreases FSH secretion, not LH secretion** (in the female as in the male).

46. **True.**

47. **True.** (As the text points out, the dominant follicle may have enough FSH receptors to make do with the decreased, but still measurable, plasma FSH levels. You might suppose that if FSH levels were *not* decreased, then more than one follicle would be likely to mature, and you would be right. Some women have difficulty conceiving because of hormonal imbalances that interfere with normal ovulation. These women may take fertility drugs [gonadotropins] to promote ovulation. The drugs are given in an attempt to mimic the normal secretion of FSH and LH, but sometimes the doses given are too high, and so multiple follicles mature and ovulate, leading to twins, triplets, sometimes even quintuplets or more. When the number of babies is more than two, their development may be compromised, as we shall see in a later section.)

48. (a) **Correct.**

    (b) **Correct.**

    (c) **Correct.** (Without it, there would likely be more than one.)

    (d) **Correct.**

    (e) **Incorrect.** **Progesterone is not required.**

49. (a) **Correct.**

    (b) **Correct.**

    (c) **Incorrect.** **The combination of estrogen and progesterone decreases the pulse frequency of GnRH** (and thus secretion of LH and FSH by the pituitary. It also decreases the sensitivity of the pituitary cells to GnRH.)

    (d) **Incorrect.** **It helps FSH induce LH receptors on granulosa cells.**

    (e) **Correct.**

    Note: One of the most confusing aspects of the study of physiology is the one you have just learned — that the nature (negative or positive) of the feedback of a target gland hormone (estrogen and to some extent progesterone) depends on that hormone's concentration in blood. It is perfectly logical to wonder how the same hormone can have opposite effects depending on whether it is present in high or low amounts. Endocrinologists are as puzzled by this question as students are, and they are working hard to find answers. In the meantime, it may be helpful to describe some experimental evidence that led to our understanding of the dual role of estrogen in helping to regulate the menstrual cycle.

The figure below depicts the plasma levels of LH (solid line) and estrogen (dashed line) in a woman who is postmenopausal (and thus has little ovarian function) or who has had her ovaries removed.

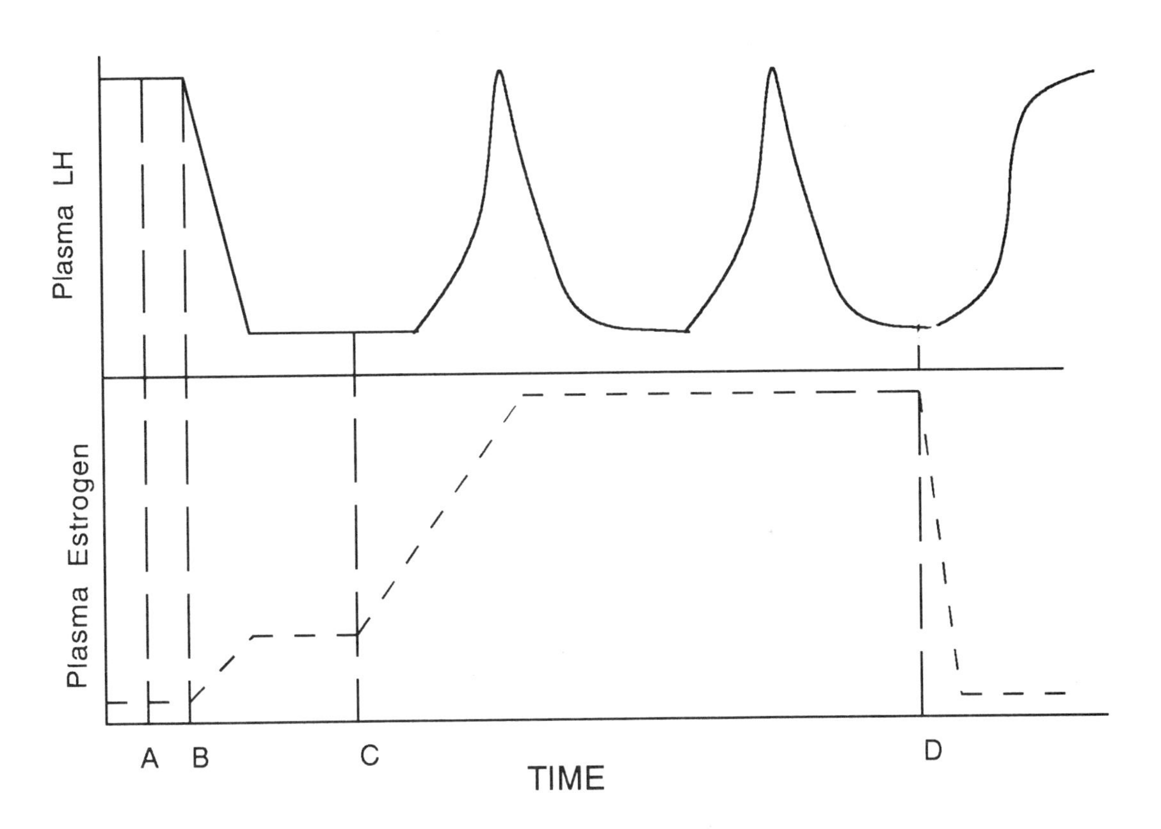

At time A, before any treatment, LH levels are high and estrogen levels are very low. (What little estrogen is present is converted from adrenal androgens.) At time B, the woman is given an infusion of estrogen which increases her plasma estrogen levels to their value in a normal early follicular phase (compare with text Figure 18-16). Simultaneously, plasma LH levels begin to fall to a level that also mimics LH levels in the early follicular phase. This decline in plasma LH is a result of negative feedback (primarily on anterior pituitary cells) by the moderately low (as opposed to the earlier *very* low) levels of estrogen in the blood. At time C, the steady state achieved in B is interrupted by increasing the estrogen infusion to mimic levels in the late follicular phase. An LH surge follows a few hours later. Note that the surge is still a surge, not a continued rise as long as estrogen is high, or a high plateau or steady state as in time period A. Therefore, positive feedback is terminated somehow (just as in the case of the action potential). In this artificial situation, we can also induce more than one surge by keeping estrogen levels elevated in the absence of progesterone.

One other point should be kept in mind: Whatever the various actions of estrogen on pituitary (or any other target) cells, the actions are all accomplished through interaction with estrogen receptors in the nucleus of the estrogen target cells.

50. (a) **Incorrect.** **The LH surge decreases the secretion of estrogen by the follicle and begins the secretion of progesterone.**

(b) **Correct.**

(c) **Correct.** (This is one of the factors that lead to follicle rupture and expulsion of the oocyte.)

(d) **Incorrect.** **Antral fluid secretion is increased.**

(e) **Correct.** ("Luteinization" means "yellowing," but it refers to the transformation of the granulosa cells remaining in the follicle into a corpus luteum. This luteinizing effect of LH prompted its name — luteinizing hormone.)

51. (a) **Incorrect.** **They do not grow during the luteal phase because plasma FSH levels are too low.**

(b) **Incorrect.** **LH continues to stimulate estrogen and progesterone secretion.**

(c) **Correct.** (No one should have missed this the second time around.)

(d) **Correct.**

52. **If the corpus luteum did not self-destruct, it would continue secreting estrogen and progesterone. High levels of these hormones would inhibit GnRH secretion and thus secretion of LH and FSH. Low levels of these latter two hormones (particularly FSH) would prevent new follicles from developing. The luteal phase would go on indefinitely.** (As we shall see, these events are not very different from what actually happens during pregnancy. Obviously, the self-destruct mechanism of the corpus luteum is very important for maintaining the cyclic pattern of female reproductive physiology.)

53. (a) **Incorrect.** **It refers to regrowth of the endometrial layer** (glandular epithelium). (The myometrium does grow somewhat under the influence of estrogen during the proliferative phase, but it is never sloughed. Note that "myo-" again refers to muscle. The ending "metrium" is from the Greek word for uterus, *metra*.)

(b) **Incorrect.** **It corresponds to the mid-to-late part of the follicular phase.**

(c) **Correct.**

54. (a) **Correct.**

(b) **Incorrect.** **It refers to exocrine secretion** (of a nourishing endometrial fluid).

(c) **Correct.**

(d) **Correct.** (Estrogen is required during the secretory phase even after it induces progesterone receptors because it is necessary to retain the proliferated endometrium.)

55. **False.** Cervical mucus secreted under the influence of estrogen alone is ~~thick and sticky~~ **clear and nonviscous**, while that secreted under the influence of both estrogen and progesterone is ~~clear and nonviscous~~ **thick and sticky**. (As the text mentions, the cervical mucous "plug" formed by progesterone seems to play a protective role for the uterus if pregnancy occurs. The kind of mucus secreted in response to estrogen alone actually helps sperm travel through the cervix into the body of the uterus, by forming straight channels for them to swim in.)

56. The sloughing of the endometrial lining, called **menstruation**, is a direct result of the action of **prostaglandins**, which are released as a result of withdrawal of **estrogen** and **progesterone**. The agents that cause the sloughing are also responsible for the pain that is sometimes associated with it, called **dysmenorrhea**. (As in other cases where prostaglandin production causes pain, inhibitors of prostaglandin synthesis, such as aspirin, are helpful in alleviating dysmenorrhea.)

57. **False.** Most of the hormonal effects on female accessory sex organs and secondary sex characteristics are attributable to ~~progesterone~~ **estrogen**. (The effects of progesterone are almost entirely related to providing a suitable environment for pregnancy, or gestation.)

58. (a) **Correct.** (Other actions of estrogen on fat include increasing the HDL/LDL ratio, which may account for the hormone's apparent ability to protect against atherosclerosis. As was pointed out in Chapter 17, women before menopause [that is, women who have relatively high plasma levels of estrogen] have a greatly reduced incidence of atherosclerotic heart disease compared to men; after menopause, however, when estrogen levels fall, the incidence of this disease becomes equal in men and women.)

    (b) **Incorrect.** **Estrogen is not anabolic for skeletal muscle.**

    (c) **Correct.** (This effect of estrogen toughens the vaginal mucosa and is protective against damage during intercourse.)

    (d) **Correct.** (This effect is important for delivery of the baby at the end of pregnancy, as we shall soon see.)

    (e) **Correct.** (The increased strength is what is protective against osteoporosis.) Other effects of estrogen on bone growth are seen in the shoulders and pelvis. Females have narrow shoulders and broad hips compared to males. More important, from the standpoint of reproduction, the size and shape of the pelvic outlet (the bottom of the pelvic girdle) are different in women — the outlet is larger and rounder — so that a baby's head can pass through it.

    (f) **Incorrect.** **Estrogen stimulates secretion of sebaceous-gland fluid that is "antiacne."** (So why do pubertal girls sometimes suffer from acne? Acne is caused by increased adrenal androgen secretion during this time.)

(g) **Incorrect.** **This is also an action of adrenal androgen** (although the female pattern — an inverted pyramid - is estrogen-stimulated.)

(h) **Correct.**

(i) **Incorrect.** **It inhibits this action of prolactin.** (The importance of these actions involving prolactin will be clear soon.)

59. (a) **Correct.** (This is especially true of effects on smooth muscle motility and contractility.)

(b) **Incorrect.** **Progesterone also has important effects on the brain** (such as the negative-feedback effects of progesterone on the hypothalamus. Progesterone also apparently resets the hypothalamic thermostat so that basal body temperature and respiratory rate are increased.)

(c) **Incorrect.** **They** (and the actions of all other steroid hormones) **are mediated through receptors in the nucleus of target cells.**

60. **False.** ~~Dysmenorrhea~~ **Premenstrual syndrome** refers to a combination of physical and psychological symptoms of uncertain cause occurring in the days preceding the onset of menstruation in some women. (The uncertainty of the cause and even of the number of women who suffer from this sometimes debilitating disorder stems from the fact that PMS is probably a cluster of syndromes with different causes, although they probably all involve hormone imbalances. Some of the specific symptoms can be alleviated by treating them rather than the [unknown] cause. For example, diuretics may help symptoms of bloating, while avoiding caffeine-containing foods, such as coffee and chocolate, may relieve headaches. Sometimes doctors prescribe mood-altering drugs for women who suffer anxiety or depression.)

61. (a) **Incorrect.** **Normal levels have identifiable effects in women** (such as stimulating growth of pubic hair and sex drive).

(b) **Correct.** (Although this fact is not explicitly stated in the text, you should have deduced it from knowing that the actions of androgens, such as the stimulation of pubic-hair growth, are first seen at the time of puberty.)

(c) **Correct.**

62. (a) **Correct.**

(b) **Correct.**

(c) **Correct.**

(d) **Incorrect.** **Females do not ejaculate during orgasm.** (A few reports in the popular press several years ago claimed that some women have remnants of a prostate-like gland that produces female ejaculation during orgasm. There is no scientific basis for these reports.)

63. **True.** (Again, although this is not explicitly stated by the text, you should be able to deduce that the parasympathetically induced vasocongestion that leads to erection in the male must be responsible for the similar changes described for the female.)

64. **False.** The "fertile period" (that period of time in a woman's menstrual cycle during which intercourse is likely to lead to pregnancy) normally lasts about ~~one week~~ **2 to 3 days** (although it may vary).

    You may be wondering how sperm can survive for two or more days in the female tract, and where they spend that time. Once out of reach of the harmfully acidic (to them and to disease-causing microorganisms) secretions of the vagina, sperm may remain in the cervix or in the uterus. Specifically, there are invaginations (inpocketings) in the lining of the cervix and uterus called *crypts* that are quite hospitable to sperm. After an ejaculation, therefore, some of the sperm remain in these crypts while others sally forth immediately in search of the ovum. There are no long-distance chemical signals from the ovum "telling" the sperm what to do, and half of the sperm that reach the uterine tubes travel down the wrong one.

65. (a) **Incorrect.** **The oocyte is extruded onto the ovarian surface and is swept into the uterine tube by movements of the fimbriae.**
    (b) **Incorrect.** **The cilia beat toward the uterus.**
    (c) **Correct.**

66. (a) **Incorrect.** **Actions of cilia in the cervix and currents produced by beating cilia elsewhere in the tract aid in sperm transport.**
    (b) **Incorrect.** **The vast majority of ejaculated sperm die before reaching the uterine tube.**
    (c) **Incorrect.** **Sperm transport is dependent upon effects of estrogen on cervical mucus.**
    (d) **Correct.**

67. (a) **Incorrect.** **It requires prior activation in the female tract.**
    (b) **Incorrect.** **Sperm must be motile to reach the ovum, where activation occurs.** (But sperm activation does include changes in motility, as described in the text.)
    (c) **Correct.**
    (d) **Correct.**

68. In general terms, **fertilization** is the fusion of a sperm with an ovum. Early events in this process include binding of proteins on the **sperm plasma membrane** to receptors on the **zona pellucida**. After moving through the **zona pellucida** with the aid of enzymes from the **acrosome**, the sperm fuses to the **ovum (secondary oocyte) plasma membrane**, triggering several events in this order: 1) activation of **secretory vesicles** in the periphery of the ovum that secrete **enzymes** into the **zona pellucida**, causing

changes in its structure and blocking **polyspermy**; 2) drawing in of the **sperm (head)** into the ovum; 3) completion of the second **meiotic division** and extrusion of the second **polar body**; and 4) formation of **pronuclei** from the sperm and ovum that migrate to the center of the fertilized cell, now called a **zygote**, and replicate.

(The importance of blocking polyspermy is highlighted by the fact that secretory-vesicle activation is the first step after the sperm binds to the ovum. Rarely, more than one sperm will bind simultaneously to the ovum's plasma membrane, and the resulting zygote will have more than two 23-chromosome sets. This condition is lethal after a few mitotic divisions. Also lethal is the equally rare case when the second polar body, with its 23 chromosomes and virtually no cytoplasm, does not get out of the way before the male and female pronuclei fuse; again the zygote has three sets of chromosomes, which is incompatible with life.)

69. **No. Fertilization must occur in the uterine tube** (or rarely, on the surface of the ovary) **because the journey of the ovum, fertilized or not, to the uterus takes three to four days. An ovum loses its ability to be fertilized after about 15 hours.**

70. **True.**

71. **False.** While passing down the uterine tube, the zygote undergoes several ~~meiotic~~ **mitotic** cell divisions. (Before each division, the 23 pairs of chromosomes replicate, resulting in "ordinary" — mitotic — cell division. Cleavage of the zygote is unusual, however, in that there is no cell *growth* preceding each division. Thus, the very large single-celled zygote becomes two cells that are half as large, then four one-quarter as large, and so on. By the time the zygote reaches the uterus, it comprises 16 to 32 cells, but it is still the same size as when it was only one cell. In fact, the zona pellucida is still intact around the zygote until just prior to implantation. The many cells derived from the single one make good use of all the nutrients stored in the ovum, so that the cleaved cells can begin to differentiate.)

72. Cleavage of the zygote yields, after a few days, a multicellular, fluid-filled structure called the **blastocyst**. This structure has two general types of differentiated cells: the **inner cell mass**, which will develop into the embryo and some of its associated membranes, and the **trophoblast**, which initiates the embedding of the structure into the endometrium, a process called **implantation**. These latter cell types also eventually provide the fetal portion of the **placenta**. (The term "trophoblast" is derived from the Greek *blastos*, meaning "germ," and *trophe*, meaning "nourishment." Thus, the trophoblast is the germ layer that nourishes.)

73. (a) **Incorrect.** **Implantation outside of the uterus is *ectopic*.**

    (b) **Correct.** (However, implantation arrests the luteal phase. In other words, implantation occurs at a time when the corpus luteum is secreting maximal amounts of hormones. Instead of declining, as in a "normal" luteal phase, these hormones continue to be secreted at ever-increasing rates.)

(c) **Correct.**

(d) **Incorrect.** **The trophoblastic enzymes digest endometrial cells.**

74. (a) **Correct.** (This is one of the placenta's critical functions.)

(b) **Incorrect.** **It is composed of trophoblast-derived tissues (chorion) and maternal tissues (endometrium).**

(c) **Correct.** (The fetus is in an aquatic environment prior to birth, and it lungs do not begin to function in the exchange of gases until after birth. The fetal kidneys do begin to filter fetal plasma and to form urine before birth, but the nitrogenous wastes from fetal blood are disposed of by the mother's own kidneys, after these wastes diffuse into the maternal blood in the placenta.)

(d) **Incorrect.** **There is no mixing of fetal and maternal blood.** (This fact is important because a fetus may have a blood type that is incompatible with that of its mother, as you will learn in the next chapter.)

(e) **Incorrect.** **Many harmful agents, including drugs and viruses, can cross the placenta.** Street drugs, such as cocaine and alcohol, have caused severe defects in babies whose mothers used them during pregnancy. Prescription drugs are also potentially dangerous. For example, the anti-acne drug Accutane is known to cause birth defects and must not be taken by pregnant women. Many physicians strongly recommend that pregnant women ingest nothing but nutritious food and water (and, generally, vitamin supplements). The period of embryonic development (from conception through the second month) is the most dangerous in terms of drugs or other agents causing damage. The fetal period is one primarily of growth and maturation of organs and systems that were first "outlined" during the embryonic stage and is somewhat less susceptible to harm.

75. **Maternal: Blood from the uterine arteries drains into pools surrounding the chorionic villi and is in turn drained by uterine veins.**

Fetal: Blood from the fetus travels to the placenta in the umbilical arteries, which branch into arterioles and capillaries in the chorionic villi. The capillaries form the (single) **umbilical vein, which carries blood from the placenta back to the fetus.**

76. **False.** Blood in the umbilical ~~arteries~~ **vein** is well-oxygenated, whereas that in the umbilical ~~vein~~ **arteries** is poorly oxygenated.

Thus, the umbilical vein and arteries are analogous to the *pulmonary* veins and arteries in postnatal humans (see text Figure 13-8). Both the pulmonary and umbilical arteries carry poorly oxygenated blood away from the heart to the source of the oxygen — which is the placenta in the case of the fetus. Recall from Chapter 13 that fully oxygenated blood is bright red and poorly oxygenated blood is bluish. If you have an early printing of the text, you may have realized that the artist for text Figure

18-24 apparently would have answered this question incorrectly! (Actually, this mistake is common to many textbooks.)

77. **False.** During pregnancy, the fetus is suspended by the umbilical cord in the ~~lumen of the uterus~~ **amniotic cavity**. The conceptus (a word that encompasses all stages of development from zygote to full-term fetus) is in the uterine lumen only for about three days before implantation. The rest of development takes place first while the conceptus is embedded in endometrium and then while it is surrounded by the amniotic sac.

78. **False.** ~~Chorionic villus sampling~~ **Amniocentesis** is the removal of a sample of amniotic fluid for diagnosis of genetic or chromosomal disorders of the fetus.

    Alternatively, the answer below is also correct:

    False. Chorionic villus sampling is the removal of a sample of ~~amniotic fluid~~ **a chorionic villus** for diagnosis of genetic or chromosomal disorders of the fetus.

79. (a) **Correct.**
    (b) **Incorrect.** **It is similar to LH.**
    (c) **Incorrect.** **It is secreted by trophoblast cells.** (After the placenta is formed, CG is secreted specifically by the cells in the chorion — hence the name. Note that CG is critically important even before a chorion differentiates, however. The CG molecules secreted by the implanting trophoblast cells diffuse into capillaries in the endometrium, to be carried eventually to the corpus luteum.)
    (d) **Correct.**
    (e) **Correct.**

80. (a) **Correct.**
    (b) **Correct.**
    (c) **Correct.**
    (d) **Correct.**
    (e) **Incorrect.** **It inhibits milk synthesis.**

81. (a) **Incorrect.** **Progesterone has no effect on myometrial growth.**
    (b) **Correct.**
    (c) **Correct.**
    (d) **Correct.**
    (e) **Incorrect.** **It inhibits milk synthesis.**

    Note that estrogen and progesterone work together in some functions, such as maintaining the endometrium, promoting breast development, and inhibiting GnRH secretion and milk synthesis. They have opposite effects on myometrial contractility, however.

82. (a) **Incorrect.** **The major estrogen of pregnancy is estriol** (after the placenta takes over as the primary source of estrogen synthesis).

(b) **Correct.**

(c) **Incorrect.** **The placenta requires precursors from the *fetal* adrenal cortex for estrogen synthesis.** (As is true for the synthesis of any estrogen, the immediate precursor of estriol is an androgen. The estriol precursor is produced in large amounts in a special zone of the fetal adrenal gland that regresses after birth. As the text mentions, measurement of maternal plasma or urinary levels of estriol provides a means for monitoring fetal health and development.)

83. **True.** Placental lactogen is a "diabetogenic" hormone. That is, its metabolic actions are anti-insulin, much like those of pituitary growth hormone. Thus, maternal tissues (except for the nervous system) cannot readily make use of glucose for energy or for storage, making more of the nutrient available for the fetus. (Obviously this effect of placental lactogen would be counterproductive if it were to act on the fetus. Peptide hormones do not cross the placenta to enter the fetal circulation in appreciable amounts.)

Note: The metabolic changes and functional hypoglycemia resulting from the actions of placental lactogen and other hormones are probably responsible for the nausea ("morning sickness") experienced by many women during pregnancy. This condition, which can be extremely debilitating, is best treated by eating small amounts of carbohydrate-rich foods (crackers, for example) at frequent intervals.

84. **False.** Symptoms of toxemia of pregnancy include ~~glucose~~ **protein** in the urine, abnormal fluid retention, and hypertension. (Protein in the urine is a sign of kidney damage; the glomerulus becomes leaky and proteins are filtered, but they cannot be reabsorbed and so they are excreted. The real danger of toxemia is that the hypertension may precipitate stroke.)

85. Recall the relationship **P = F x R. A decrease in resistance (R) results in decreased arterial blood pressure, which is sensed by the arterial, atrial, and renal baroreceptors. Arterial baroreceptor reflexes then stimulate increased sympathetic activity, leading to increased arteriolar vasoconstriction, decreased glomerular filtration rate, and increased renin secretion. Renal baroreceptors directly stimulate renin secretion as well, leading to increased angiotensin II and aldosterone production. Atrial baroreceptors trigger increased antidiuretic hormone secretion. The combination of decreased glomerular filtration rate, increased aldosterone levels, and increased antidiuretic hormone levels leads to salt and water retention, and thus increased plasma volume. Increased plasma volume accompanied by increased sympathetic discharge increases cardiac output and normalizes arterial blood pressure. A decreased hematocrit stimulates erythropoietin secretion and increases erythrocyte production.** (As a result of this gradual increase in blood volume, a woman normally has about two extra liters of blood at the end of

pregnancy. This extra blood can be lifesaving because hemorrhage during parturition is not uncommon.)

86. (a) **Correct.**

    (b) **Incorrect.** **It normally occurs around the 40th week of pregnancy.**

    (c) **Incorrect.** **Sympathetic stimulation is not necessary.**

    (d) **Incorrect.** **It is aided** (perhaps initiated) **by a *decrease* in the progesterone/estrogen ratio.** (Note the gradual decrease in this ratio, particularly during the last two months of pregnancy, as shown by the increasing space between the blue and red curves in text Figure 18-26.)

    (e) **Correct.**

87. **False.** Relaxin is a ~~tranquilizer given to pregnant women to make the birth process less stressful~~ **hormone secreted by the corpus luteum that softens the cervix, allowing it to dilate.**

88. (a) **Correct.** (Increased oxytocin secretion is part of a reflex involving a neuronal afferent pathway and a hormonal efferent pathway.)

    (b) **Incorrect.** **It is the coordinated uterine contractions that cause the cervix to dilate.**

    (c) **Incorrect.** **The reverse is true.**

    (d) **Correct.**

    (e) **Correct.** This signal is important from an evolutionary standpoint because it limits the size the fetus can attain before parturition is initiated. Prior to the development of modern surgical techniques and anesthesias, a fetus that was too large to move through the birth canal and pelvic outlet would die and probably kill its mother during the birth process. The fact that fetal size affects the timing of the birth process also explains why twins and other members of multiple-birth sets are usually born somewhat prematurely: when there is more than one fetus, the combined body mass is greater sooner, and parturition occurs early. Infants born as early as the 24th week have survived, but they require the utmost in care and technology.

89. **True.** (A woman who has never been pregnant has immature mammary glands. Breast size is not an indicator of mammary development, since most of the tissue in women with large breasts is fat.)

90. Mammary glands consist of ducts that end in clusters of **alveoli**, the site of milk synthesis. At their other end, the ducts open onto the **nipple** of the breast. The stimulus of suckling reflexly inhibits secretion of **prolactin inhibiting hormone** and/or stimulates secretion of **prolactin releasing hormone** by the hypothalamus, and thus increases secretion of **prolactin** by the anterior pituitary. This latter

hormone acts on **alveolar** cells, causing them to **synthesize milk**. The same suckling stimulus causes **oxytocin** to be released from the posterior pituitary. This hormone acts on **myoepithelial** cells, causing them to **contract** and resulting in **milk ejection**. (Note the prefix "myo" — again denoting a contractile cell. This time, however, the cell is muscle*like*, not a true muscle cell.)

91. **False.** The major constituents of milk are water, protein, fat, and ~~glucose~~ **lactose**. (Another important constituent of milk is the *antibodies* that help protect the infant from disease while its immune system is maturing. Such passive immunity is discussed in the next chapter.)

92. **False.** A woman can~~not~~ become pregnant ~~while~~ **even though** she is lactating. (Lactation and the accompanying bursts of prolactin secretion do inhibit ovulation, but only when they occur frequently. Lactation has been called "nature's contraceptive," but it does not work well in that capacity in a society in which women do not nurse their infants on demand but rather supplement breast feeding with bottled formula or other foods.)

93. **b,c,g,h** require a doctor's prescription for purchase

    h most effective reversible method

    d has highest failure rate

    e,j often irreversible

    a,b*,c,(e),f,g,h provide barrier to sperm (answer **b** is correct here, but you were not expected to know that. The IUD available in the United States contains progesterone, and its contraceptive actions are similar to those of the minipill except that the IUD has no systemic effects. Answer **e** is included here because, in a sense, vasectomy creates a barrier for sperm in the male.)

    c?,h inhibit ovulation (There is some question about whether the minipill inhibits ovulation in most women.)

    b requires yearly replacement (because the progesterone supply becomes exhausted. This is *not* true for plain plastic IUDs.)

    h should not be used by women over 35 or by those who smoke (Apparently, the estrogen component of the combination pill, which affects blood clotting, potentiates the harmful vascular effects of cigarette smoking, particularly in older women. Cigarette smoking itself is very unsafe, with or without estrogen.)

    f,i kill sperm

    a protects against sexually transmitted diseases (Actually, any method that employs spermicide also offers some protection against STDs. The most common spermicide kills the AIDS virus, for example.)

    c,h have protective effects against endometrial cancer (It is the progestational component of the combination pill that is protective against endometrial and ovarian cancer, and cysts in the ovaries and breasts. Taking either of these pills eliminates the effects of *unopposed* estrogen that normally occur during the follicular phase of the menstrual cycle. Apparently, repeated stimulation of endometrial proliferation or follicular growth can increase the chance of abnormal growth — cancer — in the uterus or ovaries, respectively. As the text mentions, the case is not completely clear with regard to associations of oral contraceptives with breast cancer, but most studies have not shown increased risk. In any case, a woman who has had breast cancer should not take the pill.)

d reliability improved by keeping track of basal body temperature

94. **True.** (There *is* such a compound, called *RU486*, that is being used in France to induce early abortion. It causes the endometrium to fail and to be sloughed away because of contractions of the myometrium. Scientists in Europe are studying the possibility of using it as a once-a-month pill to induce endometrial sloughing whether or not pregnancy has occurred. RU486 has not been introduced in this country because of the controversial nature of the abortion issue.)

95. **Yes, it would. Inhibin selectively inhibits the secretion of FSH, which is necessary for spermatogenesis but not for testosterone secretion. *But* there is a very good reason inhibin is not being used as a male contraceptive. Inhibin is a large peptide hormone that cannot bc taken as a pill (unlike the steroids in the oral contraceptives available for women) because it would be digested into its component amino acids. Thus, the hormone would have to injected or infused, a major drawback.**

96. (a) **Incorrect.** **Individuals who do not have a Y chromosome are genetic females.** (Technically, only one gene is thought to make the difference, as you will see in the next section.)

    This question points out that individuals can have different combinations of sex chromosomes and still have a clear assigned sex. For example, a person with one X and no Y chromosome is female (but sterile). On the other hand, a person with the XXY combination is male (but again, sterile). The combination XXX would be female as well, and so on.

    (b) **Correct.** (Again, such individuals may not be XY, but XXY or XYY, for example.)

    (c) **Correct.** (This imbalance in gene number is implicit in the statement that the X chromosome is larger than the Y. Now that you know this, we can finally explain why males have disorders that affect them comparatively more often than females: If the defective gene that causes the disorder is on the X chromosome, then males, who normally have only one X, have no "good" gene to balance or mask the defective gene. Genes on the X chromosome are called "sex-linked.")

    (d) **Correct.** (Cells in females with one X would not contain chromatin, while cells in males with XXY would.)

97. **True.**

98. **False.** The gene on the Y chromosome that determines whether gonads will be testes or ovaries codes for a protein called ~~Mullerian inhibiting hormone~~ **testes determining factor**.

99. (a) **Incorrect. They begin to differentiate into testes during the seventh week.**
    (b) **Incorrect. The Mullerian ducts regress. The Wolffian ducts develop into the male duct structures.**
    (c) **Incorrect. Placental chorionic gonadotropin stimulates fetal testosterone secretion.**
    (d) **Correct.**
    (e) **Incorrect. Testosterone must be converted to dihydrotestosterone in order to stimulate growth of the penis and scrotum.**

100. (a) **Incorrect. In the normal female there is no Mullerian inhibiting hormone, and the Mullerian ducts develop into the uterus and uterine tubes** (and upper two thirds of the vagina).
    (b) **Incorrect. They develop from the Mullerian ducts.**
    (c) **Correct.** (Estrogen may be required for normal female development, but it is always present in high levels in fetuses of either sex. Recall that maternal estrogen levels are high throughout pregnancy. Estrogen freely diffuses through the placenta into the fetal compartment.)
    (d) **Incorrect. The gonads differentiate into ovaries only at about the eleventh week.**

101. (a) **Correct.**
    (b) **Incorrect. They have testes.**
    (c) **Incorrect. Their Mullerian ducts regress because of the presence of Mullerian inhibiting hormone.**
    (d) **Incorrect. They lack androgen receptors.**
    (e) **Incorrect. Spermatogenesis requires testosterone. The Sertoli cells lack androgen receptors, and so spermatogenesis cannot occur.**
    (f) **Correct.** (You may be wondering whether genetic females can have a defect in androgen receptors and so be "androgen insensitive." Such a female would have no pubic hair, for example, and a diminished sex drive. However, it is almost impossible for a genetic female to have androgen insensitivity, because, by a bizarre twist of fate, the genes for the androgen receptors are on the X chromosome! Thus, in order to exhibit androgen insensitivity, a female would have to inherit defective genes from both her mother and her father — and that is nearly impossible because a man with a defective gene for the androgen receptor cannot make sperm. A mutation could occur in the sperm themselves, however, which is why the situation is only *nearly* impossible.)

102. (a) **Correct.**
   (b) **Incorrect.** **Masculinization by androgens cannot convert ovaries** (or undifferentiated gonads) **to testes.**
   (c) **Incorrect.** **Androgens do not affect the Mullerian duct.**
   (d) **Correct.** (This fact was not stated in the text, but from your knowledge of steroid biosynthetic pathways you should be able to deduce that a defective enzyme in the pathway to cortisol synthesis, for example, could cause the fetal adrenals to secrete abnormally high levels of androgens. See text Figure 10-4.)
   (e) **Incorrect.** **In human females, there is little evidence that masculinization during fetal life has any effect on adult behavior.**

103. (a) **Incorrect.** **Answer (c) is correct.**
   (b) **Incorrect.** **Answer (c) is correct.**
   (c) **Correct.** (The pituitary and gonads are capable of functioning during childhood if GnRH is present. This fact is demonstrated by the children who undergo precocious puberty as a result of brain lesions. Precocious puberty is a very troubling experience for both the children and their parents, and is often difficult to treat surgically. A new therapy is available that takes advantage of down-regulation of GnRH receptors: a long-acting synthetic analogue of the hypothalamic hormone is injected into the children daily, much as diabetic children must have insulin injections daily. In the case of the GnRH analogue, its constant presence abolishes the effect of the pulsatile secretion of hypothalamic GnRH, and the too-early sexual maturation is halted.)
   (d) **Incorrect.** **The menarche is the last pubertal event. The first is usually growth of the breasts.**
   (e) **Correct.** (Actually, the first event of puberty in females is also growth of the gonads, but this is not an easily demonstrable sign.)

104. **True.**

105. (a) **Correct.**
   (b) **Incorrect.** **It occurs as a result of ovarian failure.**
   (c) **Incorrect.** **The climacteric is the phase of life beginning with menstrual irregularity and ending with the menopause.** (It is also known as the "change of life." As the text mentions, the male counterpart to the climacteric is less well characterized, and it does not have a definitive beginning and ending.)

(d) **Correct.**

Note: When oral contraceptives first became available, there was speculation that the women who took them would have a delayed climacteric. The reasoning ran that these women would have had fewer ovulations and thus should have viable follicles remaining in their ovaries for a longer period of time. However, this delay in ovarian aging did not turn out to be the case. The ovaries seem to have their own timetable, and the number of ovulations experienced has no effect on that schedule.

106. **True.**

PRACTICE

True or false (correct the false statements):

P1. One function of semen is to buffer the alkaline secretions of the vagina.

P2. The primary storage site for sperm is the seminal vesicles.

P3. Vasectomy results in sterility because it inhibits sperm production.

P4. Erection of the penis is a result of vasodilation brought about by a sympathetic reflex.

P5. Luteinizing hormone stimulates testosterone secretion by the cells of the seminiferous tubules.

P6. Spermatogenesis and testosterone production are interdependent events, meaning that neither process can occur without the other.

P7. Sertoli cells in the testis are analogous to theca cells in the ovary in that both types control the microenvironment in which germ cells develop.

P8. Inhibin secreted by granulosa cells selectively inhibits secretion of FSH by the pituitary.

P9. The secretory phase of the uterus coincides with the follicular phase of the ovary.

P10. Progesterone increases the thickness of the myometrium in preparation for implantation.

P11. Implantation of the trophoblast occurs about seven days after ovulation.

P12. Detection of LH in the urine or blood of a woman is an indicator that she is pregnant.

P13. Oxytocin stimulates myometrial contractions most strongly when plasma progesterone levels are high relative to estrogen.

P14. Milk ejection is a neural reflex arc involving afferent input to the hypothalamus from receptors in the nipples and a neural efferent output to the myoepithelial cells in the ducts of the mammary glands.

P15. Oral contraceptives act by inhibiting ovulation, interfering with endometrial proliferation, and inducing changes in the composition of the cervical mucus.

P16. The presence of sex chromatin in a smear of epithelial cells indicates that the donor of the cells is a genetic male.

P17. An individual with androgen insensitivity syndrome is a genetic male with female internal reproductive organs.

P18. One consequence of menopause in many women is osteoporosis, which is a result of decreased androgen secretion by the ovaries.

Multiple choice (correct each incorrect choice):

P19-21. Ovulation

P19. of the dominant follicle occurs after the second meiotic division of the ovum.

P20. requires a surge of FSH about 18 hours earlier.

P21. requires synthesis of prostaglandins by the follicle.

P22-24. Fertilization

P22. normally occurs in the uterus.

P23. follows proteolysis of the corpus luteum.

P24. of two ova results in monozygotic twins.

P25-27. After fertilization, the successful development of a mature, full-term fetus depends on

P25. the maintenance of the corpus luteum for nine months.

P26. secretion of chorionic gonadotropin by the trophoblast cells.

P27. the direct transfer of blood cells between mother and fetus by way of the placenta.

P28-30. Normal physiological consequences of pregnancy to the mother include

P28. increased appetite.

P29. increased blood volume.

P30. increased blood pressure.

P31-33. Lactation

P31. requires growth and maturation of the mammary glands under the influence of estrogen, progesterone, prolactin, and placental lactogen.

P32. does not take place during pregnancy because there is insufficient prolactin available.

P33. can continue indefinitely as long as the mother continues to nurse her infant.

P34-40. Indicate whether the following are functions of androgens (A), estrogens (E), or both (B).

P34. ____ stimulation of skeletal muscle growth

P35. ____ stimulation of long bone growth

P36. ____ cessation of long bone growth

P37. ____ stimulation of hair growth

P38. ____ stimulation of hair loss

P39. ____ stimulation of LH release

P40. ____ inhibition of LH release

19: DEFENSE MECHANISMS OF THE BODY

Human beings and other animals could never have evolved to the extent that we have were it not for the elaborate defense mechanisms that protect us from disease-causing microorganisms and keep our own cells from running amok. Among the body's most important defenders are the leukocytes, which constitute the major effectors of the *immune system*.

Imagine the effects of a microorganism that attacks immune-system cells — the very cells that protect us against such organisms! You do not need to imagine these effects, unfortunately; you can simply observe the symptoms of the most frightening disease of our time: AIDS. The AIDS virus kills the cells (*helper T cells*) that activate two kinds of cells in the immune system: the *B cells*, which produce microorganism-inactivating proteins (antibodies), and the *cytotoxic (killer) T cells*, which destroy other virus-infected and cancerous cells. Thus, the AIDS virus kills its victims indirectly, by immobilizing the immune system and allowing other, normally easily defended-against infections to wreak havoc on the victim's body.

One of the triumphs of modern medicine has been the development of *vaccines* that protect humans and animals from viral diseases, such as polio and feline leukemia. (How vaccines work is one subject covered in this chapter.) As of this writing, there is not yet a vaccine against AIDS, although there are drugs available that slow the progress of the disease. Over the past century, particularly the past few decades, our knowledge of immune-system functions has been greatly expanded, as has our understanding of the type of virus that causes AIDS. Thanks in particular to long years of study of chickens, mice, cats, rats, and monkeys, we have knowledge today that allows hope for a cure, and soon, for this killer disease.

In this chapter you will learn the fundamentals of the bewilderingly complex system we have developed to defend ourselves against our worst enemies — viruses, bacteria, and other parasites.

Instructions for answering questions in this Study Guide

1. True or false: Correct each false statement. Whenever you must correct a false statement, there will almost always be more than one way to do so. In the Guide, the answer requiring the least correction will generally be given, with the understanding that other correct answers are possible. It is usually not sufficient, in terms of demonstrating understanding, to simply insert a "no" or "not," however.

2. Multiple choice: Any, all, or none of the responses may be correct. Explain why each incorrect response is incorrect.

3. Fill in the blanks: Choose the best word or words to complete each statement.

4. Directions will be given for other types of questions as they appear.

SECTION A. IMMUNOLOGY: DEFENSES AGAINST FOREIGN MATTER

❐ Introduction (text page 654)

1. The immune system
   (a) protects against infection by microbes.
   (b) must recognize the identity of a foreign cell or substance in order to inactivate it.
   (c) must recognize a foreign cell or substance as foreign in order to inactivate it.
   (d) can recognize and destroy "self" cells that have been changed or damaged.

2. True or false: Immune surveillance refers to the search for and destruction of bacteria in the blood.

3. True or false: Bacteria and viruses are unicellular organisms.

4. Viruses
   (a) consist of a nucleic acid surrounded by a carbohydrate shell.
   (b) require a host cell in order to reproduce themselves.
   (c) may reside in a host cell for years without killing it.
   (d) may cause a host cell to become cancerous.

❐ Cells Mediating Immune Responses (text pages 654-655)

5. The cells that mediate immune responses
   (a) constitute the immune system.
   (b) include polymorphonuclear granulocytes.
   (c) include erythrocytes.
   (d) are all leukocytes or derived from leukocytes.
   (e) function primarily in the blood.

6. Regarding the immune-system cells,
   (a) there are three subfamilies of lymphocytes.
   (b) plasma cells are derived from T cells.
   (c) macrophages are derived from basophils.
   (d) monocytes have functions similar to the functions of macrophages but are found in different compartments of the body.
   (e) phagocytosis, cell killing, and the secretion of chemicals are the primary mechanisms whereby immune-system cells exert their protective effects.

❐ **Nonspecific Immune Responses** (text pages 655-661)

7. True or false: The body's first line of defense against invasion by microbes is a physical and chemical barrier.

8. True or false: An important part of the barrier against invaders from the external environment is "friendly" microbes.

9. True or false: The local response to injury in the body is called infection.

10. The process whereby a cell engulfs material and then destroys what it engulfs is called ______________. The cell that does this is called, in general terms, a ______________. The primary immune-system cells that perform this function are ______________, ______________, and ______________.

11. The chemical mediators of inflammatory responses
    (a) may be secreted by cells residing in the area of injury.
    (b) may be secreted by cells that enter the affected area.
    (c) may be generated by enzymatic cleavage of proteins in plasma.
    (d) are all peptides.

12. True or false: Histamine is a mediator of inflammatory responses that is generated by enzymatic cleavage of a plasma precursor.

13. Monokines
    (a) are secreted by lymphocytes.
    (b) include interleukin 1.
    (c) play a role in specific immune responses only.

14. The first event of an inflammatory response is
    (a) vasodilation.
    (b) increased capillary permeability to plasma proteins.
    (c) secretion or generation of chemical mediators.

15. True or false: A consequence of inflammation is increased capillary filtration and edema, which causes swelling and pain.

16. Chemotaxis
    (a) refers to the chemical attraction of phagocytes to an inflammation site.
    (b) brings neutrophils and monocytes to the scene of microbial invasion or injury.
    (c) is dependent upon chemotaxins.
    (d) begins with neutrophil exudation.
    (e) activates macrophages.

17. During phagocytosis, the foreign body engulfed by the phagocyte is referred to as a ___________________. After fusion with a _________________ (which organelle?), the entire structure is called a ________________________. Inside this latter structure, the macromolecules of the engulfed particle are destroyed by __________________________ and _____________________. Phagocytosis is described as a positive feedback phenomenon because chemicals secreted by phagocytes ____.

18. True or false: A chemical that causes binding of a phagocyte to a microbe is called a chemotaxin.

19. True or false: A phagocyte can kill a microbe only after phagocytosis.

20. Complement is
    (a) a specific protein.
    (b) a general name for a group of proteins that all have the same function.
    (c) the name of a specific family of proteins that interact in a defined, complex sequence.

21. MAC
    (a) refers to "membrane attack complex."
    (b) is an attack by microbes on mucous membranes of the body.
    (c) is a complex of five activated complement proteins.
    (d) kills microbes by destroying the integrity of their plasma membranes.

22. Functions of complement include
    (a) stimulation of vasodilation and capillary permeability.
    (b) chemotaxis.
    (c) stimulation of histamine release.
    (d) opsonization.
    (e) direct cell killing.

23. Kinins are
    (a) part of the complement family.
    (b) vasodilators.
    (c) chemotaxins.

24. Activation of complement
    (a) does not occur during nonspecific immune responses.
    (b) occurs at a later stage of the cascade in a nonspecific immune response than in a specific one.

25. Tissue repair following an inflammatory response
    (a) is always complete and restores the injured part of the body to normal.
    (b) may result in scar tissue formation.
    (c) includes fibroblast proliferation and collagen secretion.

26. True or false: If an infection results in the formation of a granuloma, the wisest course is to have the granuloma surgically drained.

❐ **Specific Immune Responses** (text pages 661-677)

27. True or false: Specific immune responses differ from nonspecific ones in that the former are mediated by leukocytes.

28. The lymphoid organs are
    (a) the same as the lymphatic system.
    (b) the site of synthesis, maturation, and activation of lymphocytes.
    (c) subdivided into central and peripheral lymphoid organs.

29. True or false: Bone marrow and thymus are the primary sites of lymphocyte action.

30. Regarding the lymphoid organs,
    (a) the thymus is the largest single lymphoid organ.
    (b) the thymus secretes hormones.
    (c) large numbers of macrophages and lymphocytes are found in the lymph nodes and spleen.
    (d) the tonsils act as a first line of defense against microbes in food.

(e) once a lymphocyte reaches a peripheral lymphoid organ, it remains there for its lifetime.

31. There are three subpopulation of lymphocytes: those that mature in the thymus, called ______________________; those that mature in the bone marrow, called ______________________; and those whose origin is unclear, one subset of which are the ______________________. Cells in the first subpopulation can be further classified as ______________________, ______________________, and ______________________ cells.

32. An antigen
(a) is any foreign molecule.
(b) may be a single molecule.
(c) may be part of a larger structure.
(d) must usually be processed by cells such as macrophages before it can be recognized by specific lymphocytes.

33. True or false: B and T cells are preprogrammed to recognize specific antigens.

34. Following the encounter with and recognition of antigens, lymphocytes undergo ______________________ and ______________________, a process termed lymphocyte ______________________. The next step in the immune response is ______________________.

35. True or false: The two broad categories of specific immune responses are antibody-mediated and humoral.

36. Antibodies
(a) are secreted by helper T cells.
(b) are proteins.
(c) are immunoglobulins.
(d) are composed of two polypeptide chains, a heavy chain and a light chain.
(e) all share the same amino acid sequence in the heavy chain.

37. True or false: In cell-mediated immune responses, immune cells directly kill antigen-bearing cells.

38. B cells
    (a) are essential to the humoral immune response.
    (b) are the most numerous lymphocytes in blood.
    (c) have a wider range of targets than do T cells.

39. True or false: Helper T cells are important only for cell-mediated immune responses.

40. Receptors on T cells
    (a) recognize specific antigens.
    (b) are antibodies.
    (c) require the presence of certain plasma membrane proteins in order to bind to antigen.

41. True or false: Unlike cytotoxic T cells, natural killer cells do not require antigen to be complexed with Class I major histocompatibility complex proteins in order to bind to the antigen.

42. MHC proteins
    (a) are found in the membranes of all cells.
    (b) determine the likelihood that tissue from one individual will be compatible with that of another.
    (c) are like fingerprints; only identical twins have identical MHC proteins.

43. Regarding antigen processing by macrophages or macrophage-like cells,
    (a) antigen is processed in phagolysosomes before being presented on the cell's surface.
    (b) the antigen can be presented in association with MHC Class I proteins.
    (c) antibodies on B cells bind to the presented antigen.
    (d) cytotoxic T cells bind to the antigen-MHC complex.

44. B cells
    (a) can act as antigen-presenting cells.
    (b) can secrete interleukin 1.
    (c) can secrete interleukin 2.
    (d) are activated by lymphokines.
    (e) require secretion of IL-1 for activation.

45. Compared to undifferentiated B cells,
    (a) plasma cells have more cytoplasm.
    (b) plasma cells have increased amounts of smooth endoplasmic reticulum.
    (c) plasma cells travel more freely in the blood plasma.
    (d) plasma cells are shorter lived.
    (e) memory cells respond faster to antigen.

46. True or false: The primary effect of antibody attack is to enhance the inflammatory response.

47. Match the immune globulins on the right with the correct descriptions on the left.

____	mediate allergic responses	(a) IgA
____	are secreted into milk	(b) IgD
____	most abundant	(c) IgE
____	do not circulate	(d) IgG
____	gamma globulins	(e) IgM
____	confer immunity against multicellular parasites	
____	confer most humoral immunity against bacteria and viruses	

48. Binding of IgG or IgM to an antigen
    (a) activates the complement system by the alternate complement pathway.
    (b) facilitates production of MAC.
    (c) stimulates phagocytosis by enabling phagocytes to be bound to the antigen.
    (d) can neutralize effects of harmful proteins and viruses.
    (e) is an example of allosteric protein modulation.

49. True or false: A person who receives a vaccine made from killed virus particles will, if the vaccination is successful, acquire passive immunity against that virus.

50. True or false: The primary difference between active and passive immunity is accounted for by the presence of memory cells in the former.

51. Cytotoxic T cells
    (a) kill viruses directly.
    (b) bind to antigen complexed to MHC Class I protein.
    (c) kill the body's own cells.
    (d) are activated by IL-1.
    (e) secrete pore-forming protein into the circulation.

52. Natural killer cells
    (a) do not respond to lymphokines.
    (b) have no MHC restriction.
    (c) secrete IL-2.
    (d) secrete gamma interferon.
    (e) secrete pore-forming protein.

53. True or false: Pore-forming protein and the MAC are similar in structure and function.

54. Effector macrophages
    (a) are activated by gamma interferon.
    (b) are more effective phagocytes than are ordinary macrophages.
    (c) preferentially kill abnormal body cells.
    (d) are activated by tumor necrosis factor.

55. Interleukin-2
    (a) is a lymphokine.
    (b) is secreted by helper T cells.
    (c) acts as an autocrine.
    (d) acts as a paracrine.

56. Interferon
    (a) is a family of protein mediators.
    (b) interferes with viral replication in cells.
    (c) shows specificity in that it binds only to cells infected by a specific virus.
    (d) may be secreted by any cell that is infected with a virus.
    (e) enters cells and directly affects their protein-assembly functions.

57. Graft rejection
    (a) occurs primarily because donor macrophages recognize the host tissue as foreign.
    (b) can be combatted by drugs that inhibit lymphocyte proliferation.
    (c) can be combatted by drugs that inhibit lymphokine secretion.
    (d) is a great concern for anyone who has had an organ transplant.

❐ Systemic Manifestations of Inflammation and Infection (text page 677)

58. Fever
    (a) is induced by monokines.
    (b) is induced by hormones.
    (c) has no adaptive value.

59. Which of the following are systemic effects of IL-1 and other monokines in response to infection or tissue injury?
    (a) increased plasma levels of iron.
    (b) increased plasma levels of amino acids.
    (c) increased secretion of colony-stimulating factors.
    (d) increased appetite.
    (e) secretion of acute-phase proteins by the liver.
    (f) stimulation of helper T cells.

❐ Factors That Alter the Body's Resistance to Infection (text pages 677-680)

60. Explain how each of the following tends to decrease resistance to infection:
    (a) too little protein in diet

    ___

    ___

    ___

    (b) liver damage

    ___

    ___

    ___

    (c) leukemia

    ___

    ___

    ___

    (d) cyclosporin

    ___

    ___

    ___

(e) HIV

61. AIDS is

(a) caused by a retrovirus.

(b) caused by a virus that contains RNA.

(c) transmitted by intimate contact with infected blood, sexual intercourse with an infected partner, or by an infected mother to her fetus or breast-fed infant.

(d) a highly infectious disease.

62. True or false: Use of condoms reduces the risk of AIDS transmission during sexual intercourse.

63. Which of the following drugs might be useful in treating AIDS? Explain your answer.

(a) a drug that inhibits the reverse transcription of RNA to DNA

(b) interferon

(c) cyclosporin

64. True or false: Antibiotics are drugs that kill or interfere with replication of bacteria and viruses.

❒ Allergy (Hypersensitivity Reactions) (text pages 680-681)

65. True or false: The immune system can be activated only by harmful antigens.

66. Allergic responses
   (a) are inappropriate responses by the immune system to stimuli that are not antigens.
   (b) may cause damage to the body.
   (c) are mediated only by antibodies.
   (d) may be fatal.

67. The allergic reactions mediated by antibodies are referred to as ______________________. These reactions are a result of production of ____________ (what class?) antibodies, which bind to ______________. Subsequent binding of the antibodies to ____________________ causes immediate release of chemical mediators such as ______________ from the cells, and synthesis of other mediators such as ______________. A more delayed part of the response is the migration of ______________, which are attracted to the affected area by ________________ released by __________ cells. A severe form of allergic reaction that results in systemic responses including severe ______________ (vasodilation/vasoconstriction, ________________ (hypotension/hypertension), and bronchiolar ____________ (dilation/constriction) is called ________________.

68. Describe the rationale for treating allergies with the following:
   (a) antihistamines

   ____

   ____

   (b) corticosteroids

   ____

   ____

   (c) cyclooxygenase blockers

   ____

   ____

   (d) repeated injections of antigen

   ____

   ____

❐ Autoimmune Disease (text pages 681-683)

69. Some of the following are autoimmune diseases. For those that are, state the defect (that is, which tissue is attacked by the immune system):
    (a) type I diabetes Autoimmune Pancreatic Cells
    (b) type II diabetes NO
    (c) multiple sclerosis Autoimmune CNS myelin
    (d) muscular dystrophy ____________
    (e) myasthenia gravis Autoimmune ACh receptors
    (f) rheumatoid arthritis yes cartilage synovial Joint tissue

70. True or false: Cyclosporin may be useful for treating autoimmune disorders.
T

❐ Transfusion Reactions and Blood Types (text pages 683-686)

71. A, B, and O proteins are
    (a) MHC proteins.
    (b) found on leukocyte plasma membranes.
    (c) antigens.

72. Regarding blood types and blood-type interactions, people with
    (a) type O blood have natural antibodies to A and B antigens.
    (b) type B blood have natural antibodies to A and O antigens. NOT O
    (c) type AB blood can safely (if other factors are compatible) receive a blood transfusion from someone with type O blood.
    (d) type B blood can safely (if other factors are compatible) receive a blood transfusion from someone with type AB blood.

73. True or false: Rh factor is a protein found in the plasma membrane of erythrocytes in rhesus monkeys. T

74. True or false: People lacking Rh factor have natural antibodies to it.
F but it will be seen as foreign + antibodies will develop.

75. Regarding the Rh-factor compatibility of a mother with her fetus,
    (a) an Rh-negative fetus is at risk if its mother is Rh positive.
    (b) an Rh-positive fetus is at risk if its mother is Rh negative.
    (c) an Rh-incompatible pregnancy is at little risk if it is the first one a woman has had.

76. Explain the rationale for passive immunization of an Rh-negative woman with anti-Rh factor antibodies within a few hours of giving birth to an Rh-positive baby.

Anti-Rh antibodies will destroy any Rh+ cells in her circulation before her immune system launches an attack.

SECTION B. METABOLISM OF FOREIGN CHEMICALS

❒ Introduction (text page 686)

77. A foreign chemical may
    (a) be absorbed into the blood from the skin.
    (b) be stored in tissue depots.
    (c) have biological effects on tissues remote from the site of absorption.
    (d) be changed into different chemicals by mechanisms in the body.

❒ Absorption (text pages 687-688)

78. True or false: The skin constitutes a good barrier for foreign chemicals that are nonpolar.

79. True or false: Some foreign chemicals are actively transported across the lining of the gastrointestinal tract into the blood.

80. True or false: Airborne chemicals can gain entrance to the body via the lungs.

❒ Storage Sites (text page 688)

81. True or false: Most foreign chemicals are stored bound to blood cells or dissolved in plasma.

❒ Excretion (text page 688)

82. Excretion of foreign chemicals
    (a) in the urine is impossible if the molecules are too large to be filtered.
    (b) in the urine is inefficient if the molecules are very lipid-soluble.
    (c) may be accomplished by the lungs.

❒ Biotransformation (text pages 688-689)

83. The biotransformation of foreign chemicals
    (a) may make a toxic compound less toxic.
    (b) may make an innocuous compound toxic.
    (c) occurs predominantly in the liver.
    (d) generally results in chemicals that are more lipid-soluble than the original chemicals.

84. The MES
    (a) refers to a group of hepatic enzymes.
    (b) stands for "mitochondrial enzyme system."
    (c) is located in the granular endoplasmic reticulum of hepatic cells.
    (d) is important for metabolism of endogenous steroids.
    (e) may be induced by the chemicals it transforms.
    (f) may be inhibited by the chemicals it transforms.

SECTION C: HEMOSTASIS: THE PREVENTION OF BLOOD LOSS

❒ Introduction (text page 689)

85. Hemostasis is
    (a) the maintenance of relatively stable conditions in the internal environment.
    (b) important for homeostasis.
    (c) dependent upon blood platelets.

86. True or false: Hemostatic mechanisms can generally stop blood loss from a small vein that is cut.

87. True or false: The first reaction of a blood vessel to damage is to dilate. This reaction has adaptive value because it brings leukocytes to the area to fight off harmful microorganisms.

❒ **Formation of a Platelet Plug** (text pages 689-690)

88. Platelet aggregation is
    (a) the first step in the formation of a platelet plug.
    (b) stimulated by chemical mediators released by platelet granules.
    (c) stimulated by ATP.
    (d) stimulated by von Willebrand factor.
    (e) stimulated by the conversion of arachidonic acid in platelet plasma membranes to prostacyclin.

89. True or false: The function of the platelet plug is to block the blood vessel so that blood cannot flow through it.

90. True or false: Chemical mediators released from platelet granules and plasma membranes stimulate contraction of vascular smooth muscle.

❒ **Blood Coagulation: Clot Formation** (text pages 690-694)

91. Blood clotting
    (a) is the formation of a thrombus.
    (b) occurs after formation of a platelet plug in a damaged vessel.
    (c) is necessary for hemostasis.
    (d) requires the presence of erythrocytes.
    (e) is a result of a cascade of enzyme activation.
    (f) is an example of positive feedback.

92. True or false: A blood clot is essentially an intricate network of interlacing strands of thrombin.

93. In the intrinsic clotting pathway, exposure of inactive ______________ to ______________ fibers underlying damaged endothelium activates the enzyme. The activated ______________ then activates ______________, which in turn activates ______________. This last enzyme requires the presence of two cofactors, ______________ and ______________ (from platelet membranes), to convert ______________ to active form. The last two enzymes can also be activated by a combination of ______________ from the blood with ______________

from interstitial fluid. This latter route for activating enzyme ____________________ is known as the ____________________ pathway. In the presence of cofactors ____________________ and ____________________, the enzyme converts ____________________ to ____________________, which in turn catalyzes the formation of ____________________ from ____________________ molecules and also the activation of ____________________, which stabilizes the clot.

94. Bleeding disorders can result from
   (a) genetic defects.
   (b) liver disease.
   (c) vitamin deficiency.

95. True or false: In the clotting cascade, an enzyme activated toward the end of the cascade stimulates its own activation.

❒ **Anticlotting Systems** (text pages 694-695)

96. True or false: In the clotting cascade, an enzyme activated toward the end inhibits its own activation.

97. True or false: Regarding platelet aggregation, PGI_2 and thrombin have the same effect.

98. Protein C
   (a) is present in active form at all times in the blood plasma.
   (b) inhibits clot formation.
   (c) inactivates factors V and VIII.
   (d) inactivates thrombin.

99. True or false: Thrombin is inactivated when it binds to heparin.

100. After a blood clot is formed, it is eventually dissolved by the action of the enzyme ____________________, which digests ____________________. This enzyme is activated by a number of proteins called ____________________. One of these proteins, an enzyme called ____________________, is synthesized by endothelial cells and has very weak activity unless it is bound to ____________________.

101. True or false: One of the actions of thrombin is the inhibition of t-PA action.

❐ Anticlotting Drugs (text pages 695-696)

102. Match each description on the left with the correct term from the list on the right:

____	inhibits cyclooxygenase	(a) heparin
____	activates plasminogen	(b) thromboxane
____	drug that dissolves blood clots	(c) coumarin
____	inhibits synthesis of clotting factors	(d) t-PA
____	daily use shown to reduce incidence of heart attacks	(e) aspirin
		(f) vitamin K
____	cofactor for inactivation of thrombin	(g) fibrin

SECTION D. RESISTANCE TO STRESS

❐ Introduction (text pages 696-697)

103. Stress
   (a) refers to a wide variety of noxious stimuli.
   (b) includes pain, fear, physical trauma, infection, and shock.
   (c) invariably results in aldosterone secretion by the adrenal cortex.
   (d) leads to release of hypothalamic CRH.
   (e) usually activates the parasympathetic nervous system.

104. True or false: Stress results in increased vasopressin secretion and decreased urine flow.

❐ Functions of Cortisol in Stress (text pages 697-698)

105. During stress, cortisol
   (a) stimulates protein synthesis.
   (b) antagonizes the actions of insulin.
   (c) increases glucose utilization.
   (d) increases glucose availability to the brain.
   (e) dampens the inflammatory response to tissue injury or infection.
   (f) has as yet unknown protective effects to help the body survive stress.

106. Severe stress during childhood retards growth. Explain.

____

____

107. True or false: ACTH is thought to play a direct role in learning during stressful situations.

108. True or false: Stress should be avoided to every extent possible by people with immune system dysfunction.

109. True or false: A patient with AIDS would be likely to react to an infection with a greater elevation of plasma cortisol than a person without the disease.

❐ **Functions of the Sympathetic Nervous System in Stress** (text pages 698-699)

110. Activation of the sympathetic nervous system during stress
    (a) increases one's ability to respond to situations where physical activity is required.
    (b) may be maladaptive for today's psychosocial stresses.
    (c) increases heart and breathing rate.
    (d) inhibits blood flow to the skeletal muscles and viscera.
    (e) inhibits blood clotting systems.

❐ **Other Hormones Released During Stress** (text page 699)

111. During stress,
    (a) insulin secretion is stimulated and glucagon secretion is inhibited.
    (b) hormones that decrease salt and water excretion are secreted.
    (c) prolactin and growth hormone are co-secreted from the anterior pituitary with ACTH.

ANSWERS

Boldface type indicates the answers you should have given. Words in medium-face (ordinary) type explain or expand upon the answer.

1. (a) **Correct.** ("Microbes" is a general term for life forms that are so small they can be viewed only through a microscope. Look back at text Figure 3-2 and note that we can see bacteria with a light microscope but need an electron microscope to see viruses.)
   (b) **Incorrect.** **Nonspecific immune responses require only recognition that the cell or substance is foreign.** (Specific responses do require recognition of the foreign cell's identity, however.)
   (c) **Correct.**
   (d) **Correct.**

2. **False.** Immune surveillance refers to the search for and destruction of ~~bacteria in the blood~~ **cells that have become cancerous**.

3. **False.** Bacteria ~~and viruses~~ are unicellular organisms, **but viruses are not cells**. (Viruses are completely parasitic; they cannot perform any of a cell's functions on their own. Review Chapter 1 for the properties of cells if you have forgotten.)

4. (a) **Incorrect.** **They consist of a nucleic acid surrounded by a *protein* shell.** (As you will see later on in this chapter, viruses have either a strand of DNA or a strand of RNA to direct protein synthesis in the host cell. RNA-containing viruses are called *retroviruses*. The AIDS-causing virus is a retrovirus.)
   (b) **Correct.**
   (c) **Correct.**
   (d) **Correct.** One virus recently linked to cancer is the *feline leukemia virus* (FLV), which causes leukemia, a cancer of leukocytes, in cats. FLV is also a retrovirus and has other similarities to the AIDS virus. As was mentioned on page 468 of this Study Guide, there is now a vaccine available to protect cats from FLV. The research that brought us this vaccine is very important in the war against AIDS.

5. (a) **Correct.**
   (b) **Correct.** (If you didn't review text pages 356-357, you may have missed this.)
   (c) **Incorrect.** **Erythrocytes are not part of the immune system.**
   (d) **Correct.**
   (e) **Incorrect.** **They function primarily in the tissues.**

6. (a) **Correct.**

   (b) **Incorrect.** **Plasma cells differentiate from B cells.**

   (c) **Incorrect.** **Macrophages are derived from monocytes.** (This is easy to remember because the first letters are the same. However, mast cells also start with "m," and these cells are derived from basophils. Sorry, no helpful hints here.)

   (d) **Correct.** (Monocytes exert their effects as monocytes — as opposed to being precursors for macrophages — in the blood. Macrophages act in the tissues.)

   (e) **Correct.** (This is the most important fact to keep straight at this point. You will learn more about "who does what" as we go along — but whatever a particular immune-system cell does, it accomplishes the task by one or a combination of these three mechanisms.)

7. **True.**

8. **True.**

9. **False.** The local response to injury in the body is called ~~infection~~ **inflammation**. (As the text points out, inflammation is a generalized response to something injurious to the body, including infection. "Inflammation" and "infection" are *not* synonymous.)

10. The process whereby a cell engulfs material and then destroys what it engulfs is called **phagocytosis**. The cell that does this is called, in general terms, a **phagocyte**. The primary immune-system cells that perform this function are **neutrophils**, **monocytes**, and **macrophages**. (As you might guess from the name, macrophages are very important phagocytes. They are the largest and "hungriest" of the three cell types. But macrophages act locally in tissues; most do not travel freely. The major importance of neutrophils and monocytes is that, because they circulate in the blood, they can be called upon, like an emergency rescue crew, to go to the scene of an invasion or injury. Recall, too, that neutrophils are the most numerous of the leukocytes).

11. (a) **Correct.**

    (b) **Correct.**

    (c) **Correct.**

    (d) **Incorrect.** **Eicosanoids and histamine are nonpeptide chemical mediators of inflammatory responses.** (The mediators derived from plasma proteins are peptides, however.)

12. **False.** Histamine is a mediator of inflammatory responses that is ~~generated by enzymatic cleavage of a plasma precursor~~ **secreted by mast cells residing in the affected area**.
   Alternatively, the answer below is also correct:
   False. ~~Histamine is a~~ **Kinins are** mediator**s** of inflammatory responses that ~~is~~ **are** generated by enzymatic cleavage of a plasma precursor.

13. (a) **Incorrect.** **Monokines are secreted by monocytes and macrophages.** (Note from text Table 19-4 that lymphocytes secrete molecules called *lymphokines*. The general term for lymphokines and monokines is *cytokines* — hormone-like chemicals secreted by immune-system cells. This designation loses the specificity conferred by the separate names but is useful nevertheless.)
    (b) **Correct.** (As you will see later, interleukin 2 is a lymphokine.)
    (c) **Incorrect.** **They have roles in both specific and nonspecific responses.**

14. (a) **Incorrect.** (see answer c)
    (b) **Incorrect.** (see answer c)
    (c) **Correct.** (The other two events are consequences of this first one, but all three continue to occur during the course of the inflammatory response.)

15. **True.** (Pain is also a consequence of chemicals such as histamine stimulating pain receptors.)

16. (a) **Correct.**
    (b) **Correct.**
    (c) **Correct.**
    (d) **Incorrect.** **It begins with binding of neutrophils to capillary endothelium.**
    (e) **Correct.**

17. During phagocytosis, the foreign body engulfed by the phagocyte is referred to as a **phagosome**. After fusion with a **lysosome**, the entire structure is called a **phagolysosome**. Inside this latter structure, the macromolecules of the engulfed particle are destroyed by **lysosomal enzymes** and **hydrogen peroxide**. Phagocytosis is described as a positive feedback phenomenon because chemicals secreted by phagocytes **act as chemotaxins, vasodilators, and stimulators of capillary permeability and thus draw other phagocytes to the area.**

18. **False.** A chemical that causes binding of a phagocyte to a microbe is called ~~a chemotaxin~~ **an opsonin.**

19. **False.** A phagocyte can kill a microbe ~~only~~ after phagocytosis **or by secretion of hydrolytic enzymes into the extracellular fluid**.

20. (a) **Incorrect.** (See answer c)
    (b) **Incorrect.** (See answer c)
    (c) **Correct.** (Note the difference between this question and question b. Complement proteins do not all have the same function, but they act *in complementary fashion* with other players in the immune system to destroy microbes.)

21. (a) **Correct.**
    (b) **Incorrect.** (Answers a, c, and d in combination describe MAC.)
    (c) **Correct.**
    (d) **Correct.**

22. (a) **Correct.**
    (b) **Correct.**
    (c) **Correct.**
    (d) **Correct.**
    (e) **Correct.**

23. (a) **Incorrect.** **They are activated by the activation of complement proteins.**
    (b) **Correct.**
    (c) **Correct.**

24. (a) **Incorrect.** **The alternate complement pathway is activated in nonspecific responses to microbes that have certain carbohydrates in their cell membranes.** (Recall here a difference between specific and nonspecific immune responses: In the former the body has encountered a particular invader before, whereas in the latter it has not. The nonspecific response has elements [such as the requirement for the presence of certain carbohydrates mentioned here] that are not completely nonspecific, however.)
    (b) **Correct.**

25. (a) **Incorrect.** **Tissue repair may restore the injured part to normal or it may leave a scar, a granuloma, or an abscess, none of which is "normal."** (Granulomas and abscesses are also examples of incomplete tissue repair.)
    (b) **Correct.**
    (c) **Correct.**

26. **False.** If an infection results in the formation of ~~a granuloma~~ **an abscess**, the wisest course is to have the ~~granuloma~~ **abscess** surgically drained.

    Alternatively, the answer below is also correct:

    False. If an infection results in the formation of a granuloma, ~~the wisest course~~ **one of the least wise things to do** is to have the granuloma surgically drained.

27. **False.** Specific immune responses differ from nonspecific ones in that the former are mediated by ~~leukocytes~~ **lymphocytes**. (Both responses are mediated by leukocytes, one category of which are lymphocytes.)

28. (a) **Incorrect.** **The lymph nodes are both lymphoid organs and part of the lymphatic system, but otherwise there is no overlap.**
    (b) **Correct.**
    (c) **Incorrect.** **They are subdivided into *primary* and peripheral lymphoid organs.**

29. **False.** Bone marrow and thymus are the primary sites of lymphocyte ~~action~~ **maturation**.
    Alternatively, the answer below is also correct:
    False. ~~Bone marrow and thymus~~ **Lymph nodes, spleen, tonsils, and linings of the intestinal, respiratory, genital, and urinary tracts** are the primary sites of lymphocyte action.

30. (a) **Incorrect.** **The spleen is the largest.**
    (b) **Correct.**
    (c) **Correct.**
    (d) **Correct.** (At least, for the people who still have them. Tonsils easily become swollen and painful during bouts of infection, and are often removed if the condition becomes chronic. You may be interested to know that George Washington died in 1799 of asphyxiation from a chronic condition called "peritonsillar abscess" or *quinsy*. In his day, tonsillectomies were never done and abscesses in that location could not be lanced. Instead, Washington's physicians tried to cure his problem by putting leeches on his neck. It didn't work.)
    (e) **Incorrect.** **Lymphocytes constantly recirculate in the blood and lymph to all the body tissues.**

31. There are three subpopulation of lymphocytes: those that mature in the thymus, called **T lymphocytes**; those that mature in the bone marrow, called **B lymphocytes**; and those whose origin is unclear, one subset of which are the **natural killer cells**. Cells in the first subpopulation can be further classified as **helper**, **suppressor**, and **cytotoxic T** cells. (As you might expect, the "T" designation refers to the site of maturation of the T cells. The "B" refers not to bone marrow, however, but to the bursa of Fabricius, a sac-like organ in chickens, where these cells were first demonstrated. Much of our knowledge of these cells goes back to work done on chickens a few decades ago.)
    Note: This "T cell" and "B cell" nomenclature sometimes causes confusion because people associate B cells with insulin-secreting cells in the pancreas. The context should make clear what type of cells are being described.

32. (a) **Incorrect.** **An antigen is any foreign molecule that can trigger a specific immune response.** (As the text points out, most antigens are proteins.)
    (b) **Correct.**
    (c) **Correct.**
    (d) **Correct.**

33. **True.** (This concept — that we have cells with receptors ready to bind to foreign molecules never encountered before — is difficult to comprehend at first. It is as if cells in the primary lymphoid organs have the capacity to "know" what foreign molecules exist that may invade the body. There are at least a million different lymphocytes programmed to recognize at least a million different antigens.)

34. Following the encounter with and recognition of antigens, lymphocytes undergo **mitosis** and **differentiation**, a process termed lymphocyte **activation**. The next step in the immune response is **to attack all antigens of the kind that initiated the response**.

35. **False.** The two broad categories of specific immune responses are antibody-mediated and ~~humoral~~ **cell-mediated**. (Antibodies are by definition "humoral" mediators.)

36. (a) **Incorrect.** **They are secreted by plasma cells.**
    (b) **Correct.**
    (c) **Correct.**
    (d) **Incorrect.** **They are composed of four chains: two heavy and two light.**
    (e) **Incorrect.** **All antibodies in a particular class share the same amino acid sequence in the F_c portion of the heavy chain.** (The other end of each heavy chain forms part of the antigen-binding site, and so its amino acid sequence varies from antibody to antibody.)

37. **True.** ("Cytotoxic" means "cell poisoner.")

38. (a) **Correct.**
    (b) **Incorrect.** **T cells are the most numerous.**
    (c) **Correct.**

39. **False.** Helper T cells are important ~~only~~ for **both** cell-mediated **and antibody-mediated** immune responses.

40. (a) **Correct.**
    (b) **Incorrect.** **They bind to antigens, but they are not antibodies.**
    (c) **Correct.** (They require MHC proteins to be complexed with the antigen.)

41. **True.**

42. (a) **Incorrect. Erythrocytes do not have them.**
    (b) **Correct.**
    (c) **Correct.** (This fact is very important. The MHC proteins identify one's own cells as being "self.")
43. (a) **Correct.**
    (b) **Incorrect. Antigen-presenting cells have Class II HMC proteins.**
    (c) **Correct.**
    (d) **Incorrect. Helper T cells bind to the complex.**

44. (a) **Correct.**
    (b) **Correct.**
    (c) **Incorrect. IL-2 is secreted only by helper T cells in response to stimulation by IL-1.**
    (d) **Correct.**
    (e) **Correct.** (They require IL-1 indirectly; unless this cytokine is secreted by macrophages or by the B lymphocytes themselves, helper T cells will not be stimulated to secrete IL-2 and thus will not become activated to proliferate and secrete the lymphokines necessary for B cell activation.)

45. (a) **Correct.**
    (b) **Incorrect. They have increased amounts of granular endoplasmic reticulum.**
    (c) **Incorrect. Plasma cells do not circulate as freely.**
    (d) **Correct.**
    (e) **Correct.** (When exposed to the antigen that triggered the original B-cell activation, memory cells begin an immediate clonal proliferation of plasma cells and more memory cells.)

46. **True.**

47. **c** mediate allergic responses
    a are secreted into milk
    d most abundant
    a do not circulate
    d gamma globulins
    c confer immunity against multicellular parasites
    d,e confer most humoral immunity against bacteria and viruses

48. (a) **Incorrect. It activates the classical complement pathway.**
    (b) **Correct.**
    (c) **Correct.**
    (d) **Correct.**

(e) **Correct.** (Binding of antigen to antibody changes the configuration of the C_1 receptor and that of the phagocyte-binding site so that they can interact with their respective proteins. This fact is very important. If, for example, C_1 could bind to any free antibody, the complement system would be constantly and inappropriately activated. The signal for complement activation is binding of antigen to the antibody, not simply antibody production.)

49. **False.** A person who receives a vaccine made from killed virus particles will, if the vaccination is successful, acquire ~~passive~~ **active** immunity against that virus.

50. **True.**

51. (a) **Incorrect.** **They can kill only the cells infected by viruses.** (Antibody molecules then complex with the freed viruses and allow them to be killed by phagocytes.)
    (b) **Correct.**
    (c) **Correct.**
    (d) **Incorrect.** **They are activated by IL-2 and gamma interferon.**
    (e) **Incorrect.** **They secrete pore-forming protein only into the space separating them from the bound target cell.**

52. (a) **Incorrect.** **They are activated by IL-2 and gamma interferon.**
    (b) **Correct.**
    (c) **Incorrect.** **Only helper T cells secrete IL-2.**
    (d) **Correct.**
    (e) **Correct.**

53. **True.**

54. (a) **Correct.**
    (b) **Correct.**
    (c) **Correct.**
    (d) **Incorrect.** **They secrete tumor necrosis factor.**

55. (a) **Correct.**
    (b) **Correct.**
    (c) **Correct.** (It stimulates helper T cell proliferation.)
    (d) **Correct.**

56. (a) **Correct.**
    (b) **Correct.** (This function explains its name.)
    (c) **Incorrect.** **Interferon is nonspecific and will bind to cells whether they are infected with a virus or not.**

(d) **Incorrect.** **It is secreted by a number of cell types but not by all that can be infected.**

(e) **Incorrect.** **It binds to receptors in cell membranes and activates enzymes via a second messenger.**

57. (a) **Incorrect.** **It occurs primarily because the host T cells recognize the donor tissue as being foreign.**

(b) **Correct.** (This is how steroid hormones inhibit graft rejection.)

(c) **Correct.** (This is how cyclosporin inhibits graft rejection.)

(d) **Correct.** (Until the discovery of cyclosporin, the outlook for most organ recipients was grim. But now many people with heart, lung, kidney, liver, and bone marrow transplants can look forward to full lives. The greatest problem now facing people who need organ transplants is the shortage of available organs.)

58. (a) **Correct.**

(b) **Correct.** (Monokines acting at a distance are hormones by definition.)

(c) **Incorrect.** **A modest elevation in temperature is helpful for all inflammatory responses and is probably harmful for many microbes.**

59. (a) **Incorrect.** **Plasma iron concentrations are decreased because the liver releases smaller amounts into the blood.** (This reduction inhibits bacterial proliferation.)

(b) **Correct.**

(c) **Correct.** (This increase stimulates synthesis of leukocytes.)

(d) **Incorrect.** **Appetite is decreased.**

(e) **Correct.**

(f) **Incorrect.** **This is a *local* response to IL-1.**

60. (a) **The immune system is very dependent upon protein synthesis to function properly, and so too little protein in the diet would be detrimental to, for example, proliferation of lymphocytes and other cells, production of antibodies and other mediators, and tissue repair.**

(b) **Liver damage would reduce the ability of the liver to respond to monokines such as IL-1 by producing acute phase proteins, which enhance the inflammatory response and immune cell function.**

(c) **In leukemia, the total number of leukocytes is increased but these cells are immature or otherwise nonfunctional.**

(d) **Cyclosporin inhibits production of IL-2 and other lymphokines by helper T cells. Thus, it inhibits proliferation of lymphocytes.** (The type of cyclosporin used most often for organ transplants has a more pronounced effect on T cell proliferation than on B cells, leaving these patients with reduced but still functional humoral immune responses.)

(e) **HIV (human immunodeficiency virus) causes AIDS. It selectively destroys helper T cells and thus interferes with both cell-mediated and antibody-mediated specific immune responses.**

61. (a) **Correct.**
    (b) **Correct.** (This is the meaning of "retrovirus.")
    (c) **Correct.** ("Intimate contact with blood" means that infected blood has to breach the skin or other physical barriers of the body.)
    (d) **Incorrect.** **AIDS is difficult to transmit.** (It is the least communicable of the sexually transmitted diseases and is much less infectious than, say, flu virus or polio virus. Even intimate contact with an AIDS patient that does not involve activities listed in Question c is not dangerous.

62. **True.**

63. (a) **Probably useful.** **This reverse transcription is necessary for viral RNA to become part of the host cell's DNA. A drug that inhibits that process would hamper replication of the virus.** (There is such a drug, called *azidothymidine* — AZT — and it is currently the most widely used drug for treating AIDS. The drug inhibits the enzyme *reverse transcriptase.*)
    (b) **Probably useful.** **Interferon interferes with viral replication. Therefore it should be useful for treating AIDS.**
    (c) **Not useful.** **Cyclosporin inhibits helper T cell proliferation and would make AIDS worse.**

64. **False.** Antibiotics are drugs that kill or interfere with replication of bacteria and ~~viruses~~. (We have very few drugs that specifically attack viruses, and those that do are referred to as antiviral drugs, not antibiotics.)

65. **False.** The immune system can be activated ~~only~~ by ~~harmful~~ **harmless** antigens **as well as harmful ones**. (As well as by the host's own tissues, as we shall see in the next section.)

66. (a) **Incorrect.** **They are inappropriate responses by the immune system to harmless antigens.**
    (b) **Correct.**
    (c) **Incorrect.** **They are mediated by T cells as well as by antibodies.**
    (d) **Correct.**

67. The allergic reactions mediated by antibodies are referred to as **immediate hypersensitivity**. These reactions are a result of production of **IgE** antibodies, which

bind to **mast cells**. Subsequent binding of the antibodies to **antigen** causes immediate release of chemical mediators such as **histamine** from the cells, and synthesis of other mediators such as **prostaglandins** (or leukotrienes or platelet-activating factor). A more delayed part of the response is the migration of **eosinophils**, which are attracted to the affected area by **chemotaxins** released by **mast** cells. A severe form of allergic reaction that results in systemic responses including severe **vasodilation**, **hypotension**, and bronchiolar **constriction** is called **anaphylactic shock**.

68. (a) **Histamine is a very common mediator of the inflammatory response and is always secreted by mast cells activated by antigen-bound IgE. Therefore blockers of histamine's actions can relieve allergy symptoms.**

(b) **Corticosteroids in high doses inhibit the enzyme phospholipase A_2, which converts membrane phospholipid to arachidonic acid, the precursor of prostaglandins and leukotrienes. These eicosanoids are active mediators of allergy-induced inflammatory responses.**

(c) **Cyclooxygenase catalyzes the conversion of arachidonic acid to prostaglandins. Blocking their formation dampens the inflammatory response** (more selectively than do inhibitors of phospholipase A_2. Scientists are also trying to develop specific lipoxygenase inhibitors, which would selectively suppress conversion of arachidonic acid to leukotrienes. These latter molecules are the most potent of the eicosanoids in stimulating inflammatory responses.)

(d) **Minute, repeated injections of antigen can sometimes desensitize an allergy sufferer, either by stimulating suppressor T cells that inhibit IgE secretion or by stimulating production of IgG antibodies to the offending antigen, which bind up the antigen and get it out of the way of symptom-causing IgE's.**

69. (a) **Autoimmune.** **Pancreatic B cells.** (This may be an example of a viral attack that damages so many cells that the organ is destroyed.)

(b) **Not autoimmune.**

(c) **Autoimmune.** **Central nervous system myelin.**

(d) **Not autoimmune.** (Recall from the Study Guide introduction to Chapter 11 that muscular dystrophy is caused by lack of the protein *dystrophin*.)

(e) **Autoimmune.** **Acetylcholine receptors on motor end plates.**

(f) **Autoimmune.** **Joint tissue.** (Approximately 6.5 million people in the United States are afflicted with this painful, crippling disease.)

70. **True.** (In fact, it has been very helpful in many cases of multiple sclerosis and rheumatoid arthritis, for example.)

71. (a) **Incorrect. They are not the same as MHC proteins.**
    (b) **Incorrect. They are found on erythrocytes.**
    (c) **Correct.**

72. (a) **Correct.**
    (b) **Incorrect. People with type B blood have natural antibodies to A antigens but not to O antigens.**
    (c) **Correct.**
    (d) **Incorrect. The anti-A natural antibodies would attack and destroy the transfused AB blood cells.**

73. **True.** (It is also found in other primates, including humans.)

74. **False.** People lacking Rh factor have ~~natural~~ antibodies to it **only if they are exposed to Rh-positive blood.**

75. (a) **Incorrect. The danger lies in the mother lacking Rh factor and making antibodies to it that would attack and destroy her fetus's blood cells, not the reverse.**
    (b) **Correct.** (But see answer to Question c.)
    (c) **Correct.**

76. **The anti-Rh factor antibodies destroy any fetal blood cells with Rh antigen that escaped into the maternal circulatory system during parturition, preventing her own immune system from becoming sensitized to the antigen and making anti-Rh antibodies that would threaten the life of any subsequent Rh-positive fetuses.**

    You may be wondering why the subject of maternal-fetal blood-type incompatibility was not brought up during the discussion of the A,B,O blood types. After all, a woman with type O blood, for example, has natural antibodies to A and B antigens. What if her fetus is type B? Fortunately, two facts make this type of incompatibility normally of little consequence: first, the anti-A and anti-B antibodies are of the class IgM, which, unlike IgG's, do not readily cross the placenta; second, the A and B antigens are not strongly expressed in fetal erythrocytes.

77. (a) **Correct.**
    (b) **Correct.**
    (c) **Correct.**
    (d) **Correct.**

78. **False.** The skin constitutes a good barrier for foreign chemicals that are ~~non~~**polar**.

79. **True.** (This is true of chemicals that are similar in structure to the nutrients that are actively transported.)

80. **True.**

81. **False.** Most foreign chemicals are stored bound to ~~blood~~ **protein in** cells **or bone** or dissolved in ~~plasma~~ **fat**.

82. (a) **Incorrect. Even large molecules may be secreted into the tubular fluid and then excreted.**
 (b) **Correct.**
 (c) **Correct.**

83. (a) **Correct.**
 (b) **Correct.**
 (c) **Correct.**
 (d) **Incorrect. Biotransformation makes the chemicals less lipid-soluble.**

84. (a) **Correct.**
 (b) **Incorrect. It stands for "*microsomal* enzyme system."** (You may be wondering about the term "microsome," which is not the name of a cellular organelle. Rather, it refers to proteins associated with plasma and organellar membranes. When cells are gently disrupted by grinding and then subjected to centrifugation, cell nuclei settle out of solution after being spun at relatively low speeds, followed by mitochondria at higher speeds and finally by the "microsomal fraction" — the membranes and their associated proteins.)
 (c) **Incorrect. They are located in the smooth endoplasmic reticulum of hepatic cells.**
 (d) **Correct.**
 (e) **Correct.**
 (f) **Incorrect. It may be inhibited by other chemicals in the environment.**

85. (a) **Incorrect. It is the stoppage of blood loss.** (*Hemo* means "blood.")
 (b) **Correct.**
 (c) **Correct.**

86. **True.**

87. **False.** The first reaction of a blood vessel to damage is to ~~dilate~~ **constrict**. This reaction has adaptive value because it ~~brings leukocytes to the area to fight off harmful microorganisms~~ **slows blood loss.**

88. (a) **Incorrect. The first step in plug formation is attachment of platelets to collagen.**
 (b) **Correct.**
 (c) **Incorrect. It is stimulated by ADP.**

(d) **Correct.** (This factor helps in the first step of platelet activation.)

(e) **Incorrect.** **It is stimulated by the conversion of arachidonic acid in platelet plasma membranes to thromboxane A_2.** (Prostacyclin inhibits platelet aggregation.)

89. **False.** The function of the platelet plug is to ~~block the blood vessel so that blood cannot flow through it~~ **seal the break in the vessel wall.**

90. **True.**

91. (a) **Correct.**

(b) **Correct.**

(c) **Correct.**

(d) **Incorrect.** **Erythrocytes are usually trapped in the clot, but only fibrin and platelets are necessary for clot formation.**

(e) **Correct.**

(f) **Correct.**

92. **False.** A blood clot is essentially an intricate network of interlacing strands of ~~thrombin~~ **fibrin**.

93. In the intrinsic clotting pathway, exposure of inactive **factor XII** to **collagen** fibers underlying damaged endothelium activates the enzyme. The activated **factor XII** then activates **factor XI**, which in turn activates **factor IX**. This last enzyme requires the presence of two cofactors, **factor VIII** and **PF_3** (from platelet membranes), to convert **factor X** to active form. The last two enzymes can also be activated by a combination of **factor VII** from the blood with **tissue factor** from interstitial fluid. This latter route for activating enzyme **factor X** is known as the **extrinsic** pathway. In the presence of cofactors **factor V** and **PF_3**, the enzyme converts **prothrombin** to **thrombin**, which in turn catalyzes the formation of **fibrin** from **fibrinogen** molecules and also the activation of **factor XIII**, which stabilizes the clot.

The numbering sequence for the factors involved in blood clotting is not completely straightforward, as you can see. However, it is easier for many people to remember these numbers than to remember the names they have been given in the course of research on how blood clots, such as "Hageman factor," "Christmas factor," and the like. Factor VIII is also known as *hemophilic* factor because this cofactor is the protein missing in the most common form of the bleeding disorder known as hemophilia. Incidentally, hemophilia is much more common in men than in women because the gene that codes for factor VIII is on the X chromosome.

Note: PF_3 may be easier to remember as "platelet factor 3." The "3" designation means that there are other factors secreted by platelets, including growth factors that act as paracrines in stimulating wound healing.

94. (a) **Correct.** (As in hemophilia.)
 (b) **Correct.**
 (c) **Correct.** (Vitamin K)

95. **True.** (This is true of thrombin.)

96. **True.** (This is also true of thrombin.)

97. **False.** Regarding platelet aggregation, PGI_2 ~~and~~ **inhibits it while** thrombin ~~have the same effect~~ **stimulates it**.

98. (a) **Incorrect. It is activated by thrombin.**
 (b) **Correct.**
 (c) **Correct.**
 (d) **Incorrect. It has no effect on thrombin.**

99. **False.** Thrombin is inactivated when ~~it~~ **antithrombin III** binds to heparin.

100. After a blood clot is formed, it is eventually dissolved by the action of the enzyme **plasmin**, which digests **fibrin**. This enzyme is activated by a number of proteins called **plasminogen activators**. One of these proteins, an enzyme called **tissue plasminogen activator**, is synthesized by endothelial cells and has very weak activity unless it is bound to **fibrin**.

101. **False.** One of the actions of thrombin is the inhibition of **an inhibitor of** t-PA action. (As text Table 19-15 indicates, this action of thrombin, like its other inhibitory actions, is mediated by protein C.)

102. **e** inhibits cyclooxygenase
 d activates plasminogen
 d drug that dissolves blood clots
 c inhibits synthesis of clotting factors
 e daily use shown to reduce incidence of heart attacks
 a cofactor for inactivation of thrombin

103. (a) **Correct.**
 (b) **Correct.** (Physical trauma need not be perceived as painful to be a stress in this sense. For example, an anesthetized patient feels no pain during surgery, but his or her cortisol levels are nevertheless elevated. This elevation has clinical metabolic consequences, as the text explains in the next section.)
 (c) **Incorrect. It invariably results in cortisol secretion.**
 (d) **Correct.**
 (e) **Incorrect. It usually activates the sympathetic nervous system.**

104. **True.** (Recall that "vasopressin" is another name for antidiuretic hormone.)

105. (a) **Incorrect. It stimulates protein breakdown.**
(b) **Correct.**
(c) **Incorrect. It inhibits glucose utilization by non-brain cells.**
(d) **Correct.**
(e) **Correct.** (probably)
(f) **Correct.**

106. **Severe stress causes elevation of plasma cortisol levels, inhibiting the protein anabolism needed for growth.**

107. **True.**

108. **True.**

109. **False.** A patient with AIDS would be likely to react to an infection with a ~~greater~~ **lesser** elevation of plasma cortisol than a person without the disease (because the person with AIDS would secrete lower amounts of cytokines to stimulate ACTH release).

110. (a) **Correct.**
(b) **Correct.**
(c) **Correct.**
(d) **Incorrect. It increases blood flow to the skeletal muscles while at the same time inhibiting flow to the viscera.**
(e) **Incorrect. It stimulates blood coagulability by enhancing platelet aggregation.**

111. (a) **Incorrect. The reverse is true.**
(b) **Correct.**
(c) **Incorrect. Endorphin and ß-lipotropin are co-secreted with ACTH.** (However, the secretion of growth hormone and prolactin is increased during stress, as the text mentions. As it also mentions, the role of prolactin during stress in unclear, but recent evidence points to an inhibitory effect on the immune system.)

PRACTICE

True or false (correct the false statements):

P1. Cytotoxic T cells kill body cells by phagocytosis.

P2. Helper T cells bind to antigen complexed with MHC class II protein, while cytotoxic T cells bind to antigen complexed with MHC class I protein.

P3. HIV preferentially attacks cytotoxic T cells.

P4. A person with type B blood can generally accept a transfusion of type AB blood.

P5. Biotransformation results in the metabolism of foreign chemicals into forms that are more lipid-soluble.

P6. Phagocytosis and blood clotting are both positive feedback mechanisms.

P7. The value of having a complex cascade system for blood clotting is that a defect of one participant in the cascade can be corrected by another down the chain.

P8. During stress, cortisol stimulates glycolysis.

Multiple choice (correct each incorrect choice):

P9-11. Antibodies

P9. are secreted by B lymphocytes.

P10. are opsonins.

P11. activate complement.

Fill in the blanks:

P12-16. The three stages of a typical specific immune response to a microbial invasion are ________________, ________________, and ________________. If the body has been previously exposed to the specific microbe, the second response is ________________ (faster/slower) than the first. This change in response time is dependent upon the presence of ________________ cells.

P17. Multiple sclerosis, myesthenia gravis, and type I diabetes have very different effects on the body but they share a common cause: they are all ________________ diseases.

Matching:

P18-23. The events below describe the local inflammatory response to a bacterial invasion. Place them in the correct order:

___	P18.	(a) Chemotaxis
___	P19.	(b) Increased protein permeability in capillaries and veins
___	P20.	(c) Phagocytosis or extracellular killing
___	P21.	(d) Secretion or enzymatic generation of chemical mediators
___	P22.	(e) Tissue repair
___	P23.	(f) Increased blood flow to infected area

P24-33. For each description on the left, indicate which one or more of the chemical mediators on the right fit the description. (The number in parentheses indicates the number of mediators that fit each description.)

P24.	___(2)	monokine	(a)	complement
P25.	___(2)	lymphokine	(b)	gamma interferon
P26.	___(2)	activates helper T cells	(c)	histamine
P27.	___(2)	secreted by activated helper T cells	(d)	interleukin 1
P28.	___(2)	secreted by macrophages	(e)	interleukin 2
P29.	___(3)	secreted by mast cells	(f)	kinins
P30.	___(1)	inhibits viral replication	(g)	leukotriene
P31.	___(2)	chemotaxin	(h)	tumor necrosis factor
P32.	___(3)	involved in allergic responses	(i)	prostaglandins
P33.	___(2)	mediates systemic responses to infection		

P34-40. Match the correct term or terms on the right with the description on the left. (The number in parentheses indicates the number of mediators that fit each description.)

P34.	___(1)	catalyzes formation of fibrin	(a)	thromboxane A2
P35.	___(3)	stimulates platelet aggregation	(b)	factor VII
P36.	___(1)	inhibits platelet aggregation	(c)	plasmin
P37.	___(1)	activates factors VIII and V	(d)	heparin
P38.	___(2)	deactivates factors VIII and V	(e)	protein C
P39.	___(1)	digests fibrin	(f)	antithrombin III
P40.	___(1)	activated by exposure to collagen	(g)	prostacyclin
			(h)	von Willebrand factor
			(i)	factor XII
			(j)	thrombin

20: CONSCIOUSNESS AND BEHAVIOR

Throughout the text and this Study Guide, reference has been made implicitly and explicitly to knowledge gained by experimentation: the experimental science pioneered by Harvey and Bernard and continued to this day by researchers seeking to explain the mysteries of life. The most complex aspect of life is the workings of the human brain, and in particular the brain functions that give rise to conscious experience. In the areas of consciousness and behavior, more than in any others we have covered, much has been learned through accidents and observations of disease: the so-called experiments of nature.

One of the most striking (literally) of these is the case of a young man named Phineas P. Gage, who worked as a railroad construction foreman in New England. On September 13, 1848, Gage was tamping gunpowder into a hole when the powder exploded, sending the three-and-a-half-foot long tamping rod through his skull. The rod entered just beneath his left eye and exited through a hole in the top of his head on the right side, destroying his anterior frontal lobes. To the amazement of everyone, Gage survived and eventually made a complete physical recovery. Furthermore, his memory, sensory, and motor functions seemed to be unimpaired, and he appeared to have retained his intellectual faculties. But the man who survived the terrible accident was not the same polite, dependable, likeable person he had been. His physician, John Harlow, wrote:

> *The equilibrium or balance, so to speak, between his intellectual faculties and animal propensities seems to have been destroyed. He is fitful, irreverent, indulging at times in the grossest profanity (which was not previously his custom), manifesting but little deference for his fellows, impatient of restraint or advice when it conflicts with his desires, at times pertinaciously obstinate, yet capricious and vacillating, devising many plans of future operation, which are no sooner arranged than they are abandoned ... In this regard his mind was radically changed, so radically that his friends and acquaintances said that he was 'no longer Gage.'*

Thus, Gage suffered a complete change in personality. He could no longer control himself enough to work, and he spent his last years drifting around, exhibiting himself and the tamping rod at circus sideshows. He died in 1860. His skull (and the tamping iron) are on display at the Harvard Medical School.

The function of the frontal lobes is not as clearly understood as is that of, say, the visual cortex, but we know that Gage's physician was correct in noting that Gage had lost balance between his intellect and his "animal propensities" (in other words, his emotions). The frontal lobes link the "thinking brain" with the "emotional brain," investing thought with emotion and emotion with control and context. The frontal lobes are crucial for making us human.

Instructions for answering questions in this Study Guide

1. True or false: Correct each false statement. Whenever you must correct a false statement, there will almost always be more than one way to do so. In the Guide, the answer requiring the least correction will generally be given, with the understanding that other correct answers are possible. It is usually not sufficient, in terms of demonstrating understanding, to simply insert a "no" or "not," however.

2. Multiple choice: Any, all, or none of the responses may be correct. Explain why each incorrect response is incorrect.

3. Fill in the blanks: Choose the best word or words to complete each statement.

4. Directions will be given for other types of questions as they appear.

❒ Introduction (text page 705)

1. True or false: An observer can determine the conscious experiences of another person by noting the person's behavior and EEG tracings.

❒ States of Consciousness (text pages 705-710)

2. An electroencephalogram
   (a) is a record of action potentials in the brain.
   (b) records the potential difference between two points on the scalp's surface.
   (c) is a pattern of complex waves with amplitudes similar to those of action potentials.
   (d) reflects neural activity going on beneath the recording electrode.
   (e) is a useful clinical tool.

3. With regard to EEG patterns, the one associated with a state of being relaxed and awake but with eyes closed is the ____________________ rhythm. When the person being tested opens her/his eyes or begins to think hard about something, the pattern changes to one of ____________________ (higher/lower) frequency and ______________ (higher/lower) amplitude called the __________________ rhythm. A high-amplitude "spike and wave" pattern is associated with the neurological disease called ____________________.

4. True or false: EEG arousal is the recording of brain activity of someone who has just been awakened from sleep.

5. True or false: Sleep can be divided into two distinct stages based on behavior and EEG recordings.

6. Indicate whether the following are descriptive of or occur during slow-wave sleep (S), paradoxical sleep (P), both (B), or neither (N):
   (a) ___ delta EEG rhythm
   (b) ___ rapid eye movements
   (c) ___ alpha EEG rhythm
   (d) ___ absence of postural muscle tone
   (e) ___ theta EEG rhythm
   (f) ___ dreaming
   (g) ___ similar to beta EEG

(h) ____ pulsatile secretion of growth hormone

(i) ____ sleepwalking

7. Regarding sleep,
   (a) it progresses from stage 1 to stage 4 of slow-wave sleep and then into REM sleep.
   (b) a typical adult has four or five cycles of alternating slow-wave and paradoxical sleep patterns per night.
   (c) onset of sleep is a combination of activation of one brainstem region and simultaneous inhibition of another.
   (d) one region of the brainstem is responsible for inducing both slow-wave and paradoxical sleep.
   (e) serotonin secretion is important for REM sleep.

8. A person in a coma
   (a) usually has an alpha-rhythm EEG.
   (b) may recover if the cause is a drug overdose.
   (c) may be considered brain dead even though he or she can breathe without artificial respiration, if other criteria are met.
   (d) may be considered brain dead even though spinal reflexes are present.

❐ **Conscious Experiences** (text pages 710-712)

9. True or false: A progressive decrease in an orienting response to a repeated stimulus is called adaptation.

10. Habituation to a stimulus
    (a) is due to receptor fatigue.
    (b) is a result of decreased neurotransmitter release secondary to decreased calcium influx at synaptic terminals.
    (c) can be overcome by a stronger stimulus of the same type.

11. True or false: The brainstem nucleus most strongly implicated in the mechanism for directed attention is the nucleus accumbens.

12. Regarding its role as a neurotransmitter in the CNS, norepinephrine
    (a) is secreted by brainstem neurons in response to sensory stimulation.
    (b) amplifies weak sensory signals and dampens strong ones so that more information can reach conscious levels.
    (c) is important for maintaining directed attention.

13. Which of the following statements regarding the neural mechanisms for conscious experiences is/are correct?
    (a) Every neural response in the brain is experienced consciously.
    (b) "Mind" is synonymous with "brain."
    (c) Conscious experiences probably require complex interactions of large networks of neurons.
    (d) Unconscious neural activity has no effect upon conscious experiences.
    (e) The minimal time neural activity must be present in order to reach consciousness is about one-half second.
    (f) In the absence of sensory stimulation, the brain may "make up" stimulation on its own.

❒ **Motivation and Emotion** (text pages 712-715)

14. True or false: Primary motivated behavior is behavior that is based on changes or anticipated changes in the internal environment.

15. True or false: Most behaviors in human beings are directed by primary motivation.

16. True or false: As a rule, rewards positively reinforce the behaviors that produced them, whereas punishments reduce the behaviors that produced them.

17. True or false: Appetitive motivation can lead to primary motivated behaviors, whereas aversive motivation cannot.

18. Regarding brain self-stimulation experiments in laboratory animals,
    (a) the animals are anesthetized throughout the experiment.
    (b) an increase of bar-pressing behavior when current flows to an electrode in the brain means that the electrode is in an area associated with aversive motivation.
    (c) the area of the brain that provides the greatest reward for bar pressing is the lateral hypothalamus.

(d) positive self-stimulation may prompt the animals to engage in appetitive behaviors.

19. The fiber tract that influences the hypothalamus and is implicated in the brain's reward system is the ______________________. Neurotransmitters that mediate information flow in this system are the two catecholamines ______________________ and ______________________. Another neurotransmitter important for motivation is the opiate ______________________.

20. True or false: Administration of chlorpromazine to a rat that has an electrode implanted in the lateral hypothalamus will cause an increase in the rat's self-stimulation.

21. True or false: Many emotional states have behavioral correlates.

22. Regarding the brain areas that direct emotion,
    (a) the hypothalamus is the site of the conscious feeling of emotion.
    (b) the hypothalamus integrates emotional behaviors.
    (c) the limbic system delivers information about emotion from the cerebral cortex to the hypothalamus.
    (d) the cerebral cortex is responsible for control over emotions.

23. Damage to the septum of the limbic system causes a tame animal to become vicious, whereas destruction of the amygdala will make the same animal docile again. Which of the following statements may explain this result or correctly follow from it?
    (a) The septum is required for the expression of rage.
    (b) In a normal animal, the septum may inhibit the amygdala.
    (c) Stimulation of the septum in a normal animal would be likely to provoke rage.

❐ **Altered States of Consciousness** (text pages 715-718)

24. Regarding schizophrenia,
    (a) it is a family of mental disorders that involves disturbances of thinking, perceiving, control of motor activity, and mood.
    (b) its symptoms can include paranoid delusions and hallucinations.
    (c) it probably has a hereditary component.
    (d) it has no physical cause.

25. Schizophrenia is treated by drugs such as chlorpromazine, which block dopamine receptors in the brain. One troubling side-effect of this treatment is symptoms of Parkinson's disease. Explain.

____

____

____

____

26. True or false: The affective disorders are primarily disturbances of thought processes.

27. Bipolar affective disorder
    (a) involves both mania and depression.
    (b) is most effectively treated with drugs that increase availability of norepinephrine.
    (c) is an exaggeration of normal changes in mood.

28. True or false: Hyperactivity of neurons in the locus ceruleus has been implicated as a cause of some forms of depression.

29. Psychoactive drugs
    (a) can be taken to relieve altered states of consciousness.
    (b) can be taken to experience altered states of consciousness.
    (c) always interact with neurotransmitter receptors.
    (d) may stimulate neuronal activity in the reward systems of the brain.

30. Tolerance for psychoactive drugs
    (a) refers to the fact that different people require different doses of a drug to experience an effect.
    (b) may be caused by stimulation of the MES.
    (c) may result from negative feedback of the drug on neurotransmitter release.

31. Symptoms of withdrawal when drug use is stopped
    (a) may result from lower-than-normal secretion of neurotransmitter.
    (b) are psychological, not physical.
    (c) may be alleviated by taking a drug that interacts with the same receptor as the original drug.

32. True or false: A brain structure thought to be involved in dependence upon certain euphorigenic drugs is the locus ceruleus.

33. True or false: All of the psychogenic drugs that induce drug dependence act on the same neurotransmitter/receptor system.

34. True or false: Alcohol is a stimulant and antidepressant.

❐ **Learning and Memory** (text pages 718-721)

"Each cubic inch of the cerebral cortex probably contains more than ten thousand miles of nerve fibers, connecting the cells together. If the cells and fibres of one human brain were all stretched out end to end they would certainly reach to the moon and back. Yet the fact that they are not arranged end to end enabled man to go there himself."

Colin Blakemore, "Mechanics of the Mind"

35. Learning
    (a) involves electrical activity in the brain.
    (b) involves a change in neuronal circuitry.
    (c) requires formation of a memory trace.
    (d) is generally facilitated if the material to be learned is emotionally neutral.

36. Unlike long-term memory, working memory
    (a) has unlimited capacity.
    (b) exists in the form of either graded or action potentials.
    (c) can be disrupted by drugs, electroconvulsive shock, or a blow to the head.

37. True or false: Retrograde amnesia is the loss of memory for all events that happened before a serious brain trauma such as a blow to the head.

38. The transfer of working memory into long-term memory is
    (a) called memory retrieval.
    (b) inhibited by such hormones as ACTH, epinephrine, and vasopressin.
    (c) thought to be caused by a relatively nonspecific "fix" signal.

39. Long-term memories are thought to be of two general kinds: ____________________, which is a memory of how to do something, and ________________________, which is the ability to recall what it is that one knows how to do. Among the brain areas thought to be involved in the memory trace are the ________________________ and ____________________ in the limbic system and the ________________________ and ________________________. The ____________________________ is known to be especially important for memory consolidation.

40. True or false: Memory consolidation probably occurs via an increase in the effectiveness of existing synapses and also via formation of new ones.

41. True or false: During learning, DNA levels in the brain have been shown to increase.

42. In the experiment described on text page 720 comparing rats raised in enriched environments with similar (age, sex, and so on) rats raised in impoverished environments,
    (a) the animals raised in isolation had bigger brains.
    (b) the animals in the enriched environment could learn new tests faster.
    (c) environment affected the learning ability of young rats but not that of older rats.
    (d) there was a positive correlation between the amount of brain changes and learning ability.

❒ **Cerebral Dominance and Language** (text pages 721-722)

Because true language is thought to be unique to human beings (bird song, the songs of the humpbacked whales, and chimpanzee sign language notwithstanding), determining the location of language centers in the brain has depended almost completely upon those "experiments of nature" mentioned at the beginning of this chapter.

The first inkling of the existence of such centers was a result of observations by a French physician named Pierre-Paul Broca, who noted that several stroke victims in his hospital had pronounced impairment of speech (although they seemed to understand spoken and written speech) accompanied by right-side paralysis of the body. The first of Broca's patients — a man known as "Tan" because that was the only word he could say following his stroke — died in 1861, and Broca found at autopsy that the man had severe damage to his left posterior frontal lobe, near what is now known to be the motor cortex. Broca performed several other autopsies on patients with this *motor aphasia* and found similar lesions — always in the frontal lobe on the left side.

A few years later, a German neurologist named Karl Wernicke reported on a different sort of *conceptual aphasia* associated with left temporal lobe damage. Wernicke's patients could speak but what they said was fluent nonsense. The two brain areas came to be known by the names of their discoverers. Although the localization of language function is not so discrete as was originally thought (see text Figure 20-14), these descriptions are still clinically useful.

What about the right side of the brain? As the text mentions, it specializes in recognizing material seen as a whole, not in a sequence. People with lesions in the right hemisphere may lose the ability to read maps or to recognize faces, or even recognize that faces *are* faces (as was recounted by physician Oliver Sacks in "The Man Who Mistook His Wife for a Hat"). Such disabilities are called "propagnosias." In the most severe form of propagnosias, patients feel that the left side of their own body does not belong to them. This results in "left-side neglect" — men don't shave the left side of their face, some patients don't clothe the left side of their body, and some even awaken at night trying to throw the strange arm or leg out of bed.

In normal people, it is important to note, the two hemispheres cooperate and coordinate their efforts through fiber tracts such as the corpus callosum.

43. True or false: Damage to the language centers of the brain invariably leads to defects in speech.

❐ **Conclusion** (text page 722)

44. True or false: A generalization of brain function is that simple mental tasks are performed by primitive areas of the brain and complex ones by the cerebral cortex.

ANSWERS

Boldface type indicates the answers you should have given. Words in medium-face (ordinary) type explain or expand upon the answer.

1. **False.** An observer can determine the ~~conscious experiences~~ **state of consciousness** of another person by noting the person's behavior and EEG tracings. (Remember the quote at the beginning of Study Guide Chapter 9: "Each one of us can be certain that our own thoughts exist, but that is the only thing we can be absolutely certain about." An EEG does not tell us someone's thoughts.)

2. (a) **Incorrect. Action potentials contribute little to the EEG.**
   (b) **Correct.**
   (c) **Incorrect. The amplitudes are generally a thousand-fold less than those of action potentials.**
   (d) **Correct.**
   (e) **Correct.**

3. With regard to EEG patterns, the one associated with a state of being relaxed and awake but with eyes closed is the **alpha** rhythm. When the person being tested opens her/his eyes or begins to think hard about something, the pattern changes to one of **higher** frequency and **lower** amplitude called the **beta** rhythm. A high-amplitude "spike and wave" pattern is associated with the neurological disease called **epilepsy**.

4. **False.** EEG arousal is the recording of brain activity of someone who has ~~just been awakened from sleep~~ **been relaxed and becomes alert** (shifts from alpha to beta rhythms).

5. **True.**

6. (a) **S**; (b) **P**; (c) **S** (in stage 1); (d) **P**; (e) **S**; (f) **P**; (g) **P**; (h) **S**; (i) **S**

   (Sleepwalking was not mentioned in the text. But you should be able to reason that sleepwalking cannot occur during paradoxical sleep because in that sleep stage there is almost complete loss of skeletal-muscle tone. Therefore, sleepwalking must occur — in those rare individuals who do it — during slow-wave sleep. Perhaps the inhibition of skeletal muscle activity during REM sleep is protective: It keeps us from "acting out" our dreams.)

   Sleep research has determined that the visual cortex is very active during REM sleep and dreaming. Thus, it is as if the eyes are "seeing" dreams played out on a screen behind closed eyelids. Interestingly, people who have been blind since birth dream during paradoxical sleep, but their dreams are not visual. Such subjects do not demonstrate rapid eye movements, and the increased cortical activity associated with paradoxical sleep occurs in the somatosensory or auditory parts of the cortex.

   Although we do not know the precise functions served by the two different kinds of sleep, we do know that when a person is deprived of sleep for one or more nights and then allowed to sleep undisturbed, he or she will spend an unusually large part of that sleeping episode in REM sleep, as if to catch up on his or her dreams. Furthermore, if an individual in a sleep research lab is awakened at the start of every REM episode during the night, that person will be as tired (and irritable) as a person deprived of all sleep and will also catch up on REM at the next opportunity. Even though we lack concrete evidence about why we dream, we know that dreams are somehow necessary for normal mental functioning.

7. (a) **Incorrect.** **It progresses from stage 1 to stage 4 and then back to stage 1 before entering REM.**

   (b) **Correct.**

   (c) **Correct.**

   (d) **Incorrect.** **A separate nucleus imposes REM sleep upon slow-wave sleep.**

   (e) **Incorrect.** **Serotonin secretion ceases during dreaming.** (See text page 210. Serotonin secretion is apparently important for the onset of slow-wave sleep, however.)

8. (a) **Incorrect.** **The EEG is flat.**
   (b) **Correct.** (Overdose is not the only reversible cause of coma, but it is relatively common.)
   (c) **Incorrect.** **If a person can breathe, then the brainstem is still functional, at least to the level of the respiratory control centers.**
   (d) **Correct.**

9. **False.** A progressive decrease in an orienting response to a repeated stimulus is called ~~adaptation~~ **habituation**.

10. (a) **Incorrect.** See answer (b).
    (b) **Correct.**
    (c) **Correct.** (Or by another type of stimulus.)

11. **False.** The brainstem nucleus most strongly implicated in the mechanism for directed attention is the ~~nucleus accumbens~~ **locus ceruleus**.

12. (a) **Correct.**
    (b) **Incorrect.** **It suppresses weak signals and amplifies strong ones so that the strong signals are clearer.**
    (c) **Correct.**

13. (a) **Incorrect.** **Many neural responses do not reach consciousness.**
    (b) **Incorrect.** **"Mind" encompasses processes or functions of the brain.** (As the text mentions, we have no really good definition of "mind" as yet. Webster's Third New International Dictionary gives us "the complex of man's faculties involved in perceiving, remembering, considering, evaluating, and deciding." But this definition does not explain "mind" satisfactorily. The subject has been the source of philosophical argument for centuries, and many philosophers have argued that it is impossible for the mind to understand itself. However, neuroscientists are making progress, and we are gradually learning more and more about the operation of the mind, although we may never understand it completely. By the end of this chapter, you will have a better idea of what all this means.)
    (c) **Correct.**
    (d) **Incorrect.** **Unconscious processing can either repress or stimulate mental processing that reaches consciousness.**
    (e) **Correct.** (500 ms is 0.5 second.)
    (f) **Correct.**

14. **True.** (Putting on a coat and hat before leaving a warm house to go out into a blizzard is also primary motivated behavior.)

15. **False.** Most behaviors in human beings are directed by ~~primary~~ **secondary** motivation.

16. **True.**

17. **False. Both** ~~A~~appetitive motivation **and aversive motivation** can lead to primary motivated behaviors ~~whereas aversive motivation cannot~~. (Obviously the aversive motivation that leads to running out of a burning building, for example, has homeostatic implications.)

18. (a) **Incorrect.** **The animals are awake and alert throughout the experiment** (but of course they are fully anesthetized during the surgery that is done to implant the electrodes).

    (b) **Incorrect.** **The electrode is in an area associated with rewards, or appetitive motivation.**

    (c) **Correct.**

    (d) **Correct.** But in some cases, positive self-stimulation appears to take the place of appetitive behavior. In other words, the animal will press the bar and self-stimulate its brain rather than eat if it is hungry, drink if it is thirsty, or mate, depending on where the electrode is placed. It is as if the electrical stimulation delivers the reward ordinarily obtained from appetitive behavior without going through the motions. Unlike the case with normal appetitive behaviors, however, there is no satiation with positive self-stimulation.

    The investigators who did the first electrical stimulation experiments referred to areas of the brain that led to repetitive self-stimulation as "pleasure centers" and those that produced aversive behaviors as "pain centers." But the scientists could not be sure the animals were experiencing pleasure or pain because the animals could not directly express their feelings. However, people undergoing brain surgery are sometimes given only a local anesthetic at the edges of the incision (the brain has no pain receptors) and are alert. Areas of the brain can then be electrically stimulated — in order to identify lesions that lead to epileptic seizures, for example. Patients who experience electrical stimulation of areas that lead to bar-pressing in animals report that they do in fact experience pleasure — elation, euphoria, intense feelings of well-being. Stimulation of other areas brings about feelings of anxiety and other sorts of "mental pain." Indeed, the electrical activity of the brain can influence mood. This knowledge is helping scientists who are seeking to understand the physiological basis for drug addiction and ways to cure it, as you will see in the next section.

19. The fiber tract that influences the hypothalamus and is implicated in the brain's reward system is the **medial forebrain bundle**. Neurotransmitters that mediate information flow in this tract are the two catecholamines **dopamine** and

norepinephrine. Another neurotransmitter important for motivation is the opiate **enkephalin**.

20. **False.** Administration of chlorpromazine to a rat that has an electrode implanted in the lateral hypothalamus will cause ~~an increase~~ **a decrease** in the rat's self-stimulation.

    Alternatively, the answer below is also correct:

    False. Administration of ~~chlorpromazine~~ **amphetamine** to a rat that has an electrode implanted in the lateral hypothalamus will cause an increase in the rat's self-stimulation.

21. **True.** (This is what the text calls "emotional behavior." It is this aspect of emotion that can be studied in animals.)

22. (a) **Incorrect.** **The cerebral cortex is the site of the conscious feeling of emotion** (and indeed all conscious feelings).
    (b) **Correct.** (along with some help from the cerebral cortex and brainstem.)
    (c) **Correct.** (although this statement is not a complete description of the role of the limbic system.)
    (d) **Correct.** (Especially the frontal lobes — remember Phineas Gage.) Again, the ability to master or govern our emotions is one of the areas that sets human beings apart from other animals.

23. (a) **Incorrect.** **Obviously, if it were required for the expression of rage, destruction of it would not result in rage.**
    (b) **Correct.** (This may explain the result.)
    (c) **Incorrect.** **From the result with damage to the septum, it follows that electrical stimulation of it inhibits rage.** (You may be wondering if human beings would react with inappropriate rage if they suffered septal damage from, say, a stroke or tumor. They do, and such behavior can sometimes be modified by surgically destroying parts of the amygdala. Other humans have intermittent epileptic seizures involving the amygdala and experience episodes of violence they cannot control. They "awaken" from these episodes very puzzled, contrite, and frightened. Destruction of the abnormal brain tissue can frequently help.)

24. (a) **Correct.** (In its most severe form, schizophrenia is a global impairment of the highest brain functions. It should not be confused with "split personality," or "multiple personality," another and much rarer form of mental illness. The term "fragmented personality" describes schizophrenia because the patient's thinking is so disorganized that he or she does not seem to be a "complete" person.)

(b) **Correct.** (In the descriptive words of Jon Franklin, "full-blown schizophrenics are cursed with the certain knowledge of things palpably not so.")

(c) **Correct.**

(d) **Incorrect.** **It is likely to be caused by excessive dopamine or other neurotransmitter activity.**

Our recent understanding of the importance of neurotransmitters, neuromodulators, and their receptors in various aspects of brain functioning has had the salutary effect of removing much of the stigma from mental illness. Sufferers from mental disorders and their families used to feel shame about being or causing someone to be "crazy," and treatment of these disorders in "insane asylums" was often simply physical restraint in straight jackets and padded rooms for the violent patients, and really nothing for catatonics. Although the antipsychotic drugs available today are not perfect and have side effects (see Question 25), they enable many of the more than 2.5 million schizophrenics in this country to lead relatively normal lives. Research continues to search for more specific drugs and methods of delivery, as well as clues as to the nature of an underlying genetic defect.

25. **Parkinson's disease is caused by too little dopamine in the basal ganglia** (from destruction of cells in the substantia nigra). **Large doses of chlorpromazine affect cells in the basal ganglia much the same way** (but the effect is reversible).

26. **False.** The affective disorders are primarily disturbances of ~~thought processes~~ **mood or emotion**.

27. (a) **Correct.**

(b) **Incorrect.** **It is most effectively treated with lithium carbonate, which decreases norepinephrine release from synaptic terminals and thus decreases its availability.**

Depression *without* mania is usually treated by drugs that enhance norepinephrine activity, although, as the text mentions, in some cases lithium is useful for this "unipolar" disorder. In these cases, the tricyclic antidepressants and monoamine oxidase inhibitors may not alleviate the depression. The differences in response to the various drugs for treatment of the affective disorders illustrate that there are several clinically distinct causes for the diseases. Indeed, in some cases no drug therapy is helpful, and electroconvulsive (electroshock) therapy or even a modified frontal lobotomy may be necessary.

Researchers are striving to find answers for the causes of depression and better drug therapies and eventual cures because the disease makes life so miserable for its victims. It is also quite common. At least 1 percent of our population suffers from bipolar disorder and more than 20 percent suffers from depression alone.

The true incidence is probably much higher than this because many victims do not realize that they have a treatable disease and try to tough it out on their own. This strategy can have tragic consequences, because at least 16 percent of untreated known victims of depression commit suicide. An estimated 60 to 80 percent of all suicides involve people with depressive illness, and the incidence among young people (ages fifteen to twenty-four) has increased by a shocking 150 percent in the last twenty years.

(c) **Correct.** (This is one reason the bipolar affective disorders are often not diagnosed. Everyone has changes of mood, from the occasional blues or genuine grief at the loss of a loved one, for example, to elation and feelings of being on top of the world — perhaps as a result of a good grade in a physiology exam. But these mood shifts are far different from true mania [which is often harder on the victim's family than on the victim] and the total, hopeless despair of depression.)

28. **True.**

29. (a) **Correct.**

(b) **Correct.**

(c) **Incorrect.** **Many do, but others affect neurotransmitter metabolism.**

(d) **Correct.** (This is surely one of the causes of psychological dependence on drugs such as morphine or cocaine — they produce the same sort of euphoria that occurs naturally when something good happens to us or we perform rewardable behaviors, but the drugs bypass the going-through-the-motions just as electrical self-stimulation in rats does. Rats also demonstrate drug dependence: when allowed free access to cocaine, for example, they will self-administer the drug and ignore food and water. Reward centers exist in virtually all animals with brains because the centers provide positive reinforcement for adaptive behaviors that ensure the survival of the individual and of the species. But drugs can make rewards the end in themselves, which is obviously maladaptive.)

30. (a) **Incorrect.** **It refers to the fact that, with drug use, increasing doses of the drug become necessary to achieve the same effect lower doses had initially.**

(b) **Correct.**

(c) **Correct.**

31. (a) **Correct.**

(b) **Incorrect.** **Withdrawal refers to physical symptoms associated with cessation of use of a drug on which one has become dependent.**

(c) **Correct.**

32. **False.** A brain structure thought to be involved in dependence upon certain euphorigenic drugs is the ~~locus ceruleus~~ **nucleus accumbens** (in the septal area. Apparently, cocaine and amphetamines act there to inhibit dopamine reuptake. Morphine and related opiates act nearby, also on dopaminergic neurons that connect to the nucleus accumbens. If this nucleus is destroyed, cocaine-dependent rats no longer show any interest in the drug.)

33. **False.** ~~All of the~~ Psychogenic drugs that induce drug dependence act on ~~the same~~ **a variety of** neurotransmitter/ receptor systems. (As you can see with the example of alcohol interacting with GABA receptors and cocaine acting on the dopamine system. Other addictive drugs, such as nicotine, act at acetylcholine receptors; LSD acts at serotonin receptors; mescaline acts at norepinephrine receptors. These drugs all have different effects, and people take them for different reasons. But they all have the power to cause tolerance and dependence.)

34. **False.** Alcohol is a ~~stimulant and anti-~~ **central nervous system** depressant. (The text mentions that alcohol is used medically as an anesthetic. It is rarely used that way any more, except for emergencies, but before the discovery of the modern anesthetics, alcohol was the best choice available for the poor unfortunate who had to undergo surgery. The benzodiazepines are routinely used as sedatives in modern surgical procedures not requiring complete general anesthesia.)

35. (a) **Correct.**
    (b) **Correct.**
    (c) **Correct.**
    (d) **Incorrect.** **Emotionally charged events or pieces of information are more easily learned than neutral ones.**

36. (a) **Incorrect.** **Working memory has limited capacity** (unlike long-term memory).
    (b) **Correct.**
    (c) **Correct.**

37. **False.** Retrograde amnesia is the loss of memory for all events that happened **for about 30 minutes** before a serious brain trauma such as a blow to the head. (Perhaps you have experienced such a "where am I?" sensation. It is a frightening experience. There are cases of retrograde amnesia for events over longer periods of time — months or even years — particularly following damage to the hippocampus [see discussion following Answer 39], but their cause is unclear.)

38. (a) **Incorrect.** **It is called memory consolidation.**
    (b) **Incorrect.** **It is enhanced by those hormones.**
    (c) **Correct.**

39. Long-term memories are thought to be of two general kinds: **procedural memory**, which is a memory of how to do something, and **declarative memory**, which is the

39. Long-term memories are thought to be of two general kinds: **procedural memory**, which is a memory of how to do something, and **declarative memory**, which is the ability to recall what it is that one knows how to do. Among the brain areas thought to be involved in the memory trace are the **hippocampus** and **amygdala** in the limbic system and the **cerebellum** and **cerebral cortex**. The **hippocampus** is known to be especially important for memory consolidation.

    The importance of the hippocampus can be illustrated by a famous case in neurological clinical literature — the case of Henry M. Mr. M. suffered from recurrent, serious bouts of epileptic seizures, resulting in loss of consciousness, that could not be treated with drugs available in the 1950s, when he became ill. The abnormal cells, or epileptic foci, were located in the hippocampal region on both sides of the brain. When Henry was 27 and unable to work because of his illness, a well-intentioned surgeon destroyed both hippocampal regions, and Henry M. lost forever the ability to store new memories. He could remember clearly his life up to about three years before his surgery (why he lost those three years is a puzzle) but could recall nothing of the events that occurred afterward for more than a few minutes. His psychotherapist, who worked with him regularly over a period of twenty years, remained a stranger to him — each time they met, she had to introduce herself all over again. As British neurophysiologist Colin Blakemore expressed Henry's situation, "Henry is a 'child of the moment,' ... trapped, interminably in the naïveté of infancy."

    Henry was able to learn to *do* new things, however. He learned to work at assembling packages, and he learned to play tennis, but he had no recollection of knowing how to do those things. Thus, this sad case provides evidence not only for the importance of the hippocampus for the consolidation of declarative memory but also for the fact that procedural memory does not require an intact hippocampus. (Animal studies have localized at least one kind of procedural memory trace to the cerebellum.)

40. **True.**

41. **False.** During learning, ~~DNA~~ **messenger RNA and protein** levels in the brain have been shown to increase. (When this finding was first described, it caused great concern because some scientists thought it might mean that somehow specific memories *reside* in protein — that is, anything we might learn is already coded for in our genes! This thesis was not very popular because it seemed inconceivable, and it quickly gave way to our present understanding that the protein synthesis is necessary for the growth of dendritic spines and axon terminals in the new and strengthened synapses. The synaptic circuits and neural pathways that are formed during learning are themselves the repositories of memory and are unique in every human brain, just as our individual experiences are unique.)

42. (a) **Incorrect. The opposite is true.**

    (b) **Correct.**

    (c) **Incorrect. Older rats in the enriched environment also learned faster** (or, you *can* teach an old rat new tricks!).

    (d) **Correct.**

    This sort of experiment illustrates a very important point: The more we use our brains to learn, the better we *can* learn. Just as the performance of our muscles is improved with use (exercise), so too is the performance of our brains. "Use it or lose it" does not necessarily apply to old memories, but it certainly does apply to the ability to learn new things. The more you learn, the more you *can* learn — the best kind of positive feedback!

43. **False.** Damage to the language centers of the brain invariably leads to defects in speech **unless the victim is very young**. (Plasticity remains possible even in old age, as we saw in the last section. But the enormously complex language functions apparently cannot be reformed or relearned to any great extent in adults.)

44. **False.** A generalization of brain function is that **even** simple mental tasks are performed by ~~primitive areas of the brain and complex ones by the cerebral cortex~~ **the interaction of many basic units widely distributed throughout the brain.**

PRACTICE

True or false (correct the false statements):

P1. A high-amplitude, spike-wave EEG pattern is characteristic of someone in a coma.

P2. The EEG tracing of a relaxed individual who has closed eyes and is not concentrating on anything in particular is mainly a beta rhythm.

P3. Dreaming occurs when one is in slow-wave sleep.

P4. Sleep-wake cycles are produced by interactions of nuclei in the thalamus.

P5. Behaviors that satisfy homeostatic needs are called secondary motivated behaviors.

P6. Appetitive motivation leads one to avoid repeating a behavior.

P7. Neurotransmitters involved in appetitive motivation include enkephalin, dopamine, and epinephrine.

P8. The integrator of inner emotions and emotional behaviors is the thalamus.

P9. One cause of schizophrenia is thought to be too little dopamine activity in parts of the brain.

P10. The tricyclic antidepressants are usually helpful for controlling bipolar affective disorder.

P11. Tolerance to drugs that are neurotransmitter agonists is thought to involve increased release of the neurotransmitter in response to the drug.

P12. Working memory is labile and can be lost in response to any condition that interrupts electrical activity of the brain.

P13. Stressful situations are likely to be remembered in striking detail in part because of the hormones released as a result of the stress.

P14. Brain size and the complexity of neuronal circuits are fixed and independent of environmental stimulation.

P15. The right side of the brain is specialized for identifying visual patterns and three-dimensional objects.

Multiple choice (correct each incorrect choice):

P16-18. During paradoxical sleep,

- P16. the eyes move back and forth beneath closed eyelids.
- P17. postural muscle tone is easily demonstrated.
- P18. growth hormone is secreted in pulses.

P19-25. Match the brain area from the right with the correct function/description on the left.

P19. _ important for directed attention	(a) cerebellum
P20. _ important for consolidation of declarative memory	(b) amygdala
P21. _ associated with strong emotional behavior such as rage	(c) locus ceruleus
	(d) Wernicke's area
P22. _ damage causes conceptual aphasia	(e) corpus callosum
P23. _ damage causes expressive aphasia	(f) Broca's area
P24. _ important for learning discrete movements	(g) hippocampus
P25. _ part of brain reward system associated with drug dependence	(h) substantia nigra
	(i) nucleus accumbens

ANSWERS TO PRACTICE

Page numbers refer to pages in the text in which explanations for answers can be found.

Chapter 1:

P1. **False.** The statement...is an example of ~~teleology~~ **causality**. (*page 1*)

P2. **False.** ~~Differentiation is necessary before a cell~~ **Every living cell, no matter how undifferentiated,** can... (*page 2*)

P3. **False.** Organs...composed of ~~only one kind~~ **all four kinds**... (*page 3*)

P4. **False.** The composition...is ~~essentially the same as~~ **very different from** that... (*pages 5-6*)

P5. **extracellular** (*page 5*)

P6-9. **muscle**, **nerve**, **epithelial**, **connective tissue** (in any order; *pages 2-3*)

P10. **homeostasis** (*page 5*)

Grading: *9 -10 correct = excellent*
7 - 8 correct = good
5 - 6 correct = fair
fewer than 5 correct = not passing

Chapter 2:

P1. **True.** (*page 12*)

P2. **False.** The mass ...protons and ~~electrons~~ **neutrons**. (*page 12*)

P3. **True.** (*page 12*)

P4. **False.** ...mineral elements...include Na, Ca, and ~~O~~ **P** (or K, S, Cl, Mg. O is a major element, not a mineral; *Table 2-1*)

P5. **True.** (*page 14*)

P6. **False.** ...oxygen atom forms a ~~double~~ **single** bond... (*page 14*)

P7. **True.** (*page 15*)

P8. **False.** NaCl... ~~covalent~~ **ionic** bonding of a sodium ~~atom~~ **ion** to a ~~chlorine atom~~ **chloride ion**. (*page 17 and Figure 2-5*)

P9. **False.** ~~All c~~Covalent bonds **in which electrons are not shared equally between two atoms** are polar. (But all polar bonds are, by definition, covalent; *page 16*)

P10. **False.** Water molecules... ~~ionic~~ **hydrogen** bonds... (*pages 16-17 and Figure 2-4*)

P11. **True.** (*page 17*)

P12. **False.** Solutes...called ~~hydrophilic~~ **hydrophobic**. (*page 18*)

P13. **False.** Molecules... ~~ambidextrous~~ **amphipathic**. (*page 18*)

P14. **False.** The molarity...concentration of the ~~solvent~~ **solute**. (*pages 19-20*)

P15. **False.** A solution...more ~~acidic~~ **alkaline** ... (*page 20*)

P16. **False.** Organic... ~~oxygen~~ **carbon**-containing compounds. (*page 21*)

P17. **True.** (*page 22 and Table 2-5*)

P18. **True.** *(page 23 and Table 2-5)*

P19. **False.** Saturated...~~double~~ **single** bonds. *(page 23)*

P20. **False.** ...base cytosine binds with ~~uracil~~ **guanine.** *(page 32)*

P21. **Correct.** *(page 24)*

P22. **Incorrect.** **Proteins are composed of amino acids.** *(page 24 and Table 2-5)*

P23. **Incorrect.** See answer to P22. *(page 24 and Table 2-5)*

P24. **Incorrect.** **Protein subunits are linked by peptide bonds.** *(page 27 and Figure 2-14)*

P25. **Incorrect.** **Protein conformation is dependent upon the sequence of subunits forming the protein.** *(pages 29-31)*

P26. **Correct.** *(pages 29-30 and Figure 2-17)*

P27. **Correct.** *(page 30 and Figure 2-17)*

P28. **Correct.** *(page 21)*

P29. **Correct.** *(page 31 and Table 2-5)*

P30. **Correct.** *(page 32 and Figure 2-20)*

Grading: *26-30 correct = excellent*
21-25 correct = good
15-20 correct = fair
fewer than 15 correct = not passing

Chapter 3:

P1. **False.** In general, the ~~larger an animal is, the larger are its individual cells~~ **size of an animal's cells is not related to the size of the animal.** *(page 37)*

P2. **True.** *(page 37)*

P3. **False.** ...fluid in the cytoplasm **and nucleus.** *(pages 38-39)*

P4. **True.** *(page 39)*

P5. **False.** The special functions...composition of the ~~phospholipids~~ **integral proteins** ... *(page 43)*

P6. **False.** ~~Desmosomes~~ **Gap junctions** are... *(page 45 and Figure 3-9)*

P7. **True.** *(page 45)*

P8. **False.** Free ribosomes...proteins for ~~export (secretion) from the cell~~ **the cell's own use.** *(pages 46-47)*

Alternatively, the answer below is also correct:

False. ~~Free r~~Ribosomes **bound to endoplasmic reticulum**...*(pages 46-47)*

P9. **True.** *(page 48)*

P10-13. Rough-surfaced endoplasmic reticulum, which is also called **granular** endoplasmic reticulum, is important for the synthesis and packaging of **protein** molecules. The rough apagesearance stems from the association of **ribosomes** with the endoplasmic reticulum membrane. Smooth endoplasmic reticulum is important for the synthesis of **lipid** molecules. *(page 46)*

P14. **Incorrect.** **Cytosol refers to the fluid bathing the organelles.** *(pages 38-39)*

P15. **Incorrect. Membranes regulate the passage of molecules into and out of cells and organelles.** *(page 39)*

P16. **Incorrect. Membranes consist primarily of protein and (phospho)lipids.** *(page 39)*

P17. **Incorrect. Cytoskeleton refers to a network of filaments within cells that helps maintain cell shape and produce cell movements.** *(pages 48-50)*

P18. **Correct.** *(pages 48-50)*

P19. **Correct.** *(page 50)*

P20. **Correct.** *(page 50)*

Grading: 18-20 correct = excellent
14-17 correct = good
10-13 correct = fair
fewer than 10 correct = not passing

Chapter 4:

P1. **False.** ...the larger the number of different ligands...the ~~greater~~ **lesser** the specificity...*(pages 54-55)*

Alternatively, the answer below is also correct:

False. ...the ~~larger~~ **smaller** the number...the greater the specificity...*(pages 54-55)*

P2. **True.** *(page 56)*

P3. **False.** The greater the ligand concentration...the ~~higher~~ **lower** the affinity of the binding site...*(page 56 and Figure 4-6)*

P4. **False.** The role of modulator molecules is **either** to enhance **or to reduce** the binding affinity of the functional site...*(pages 57-58)*

P5. **False.** Proteins...are called ~~substrates~~ **enzymes**. *(page 59)*

P6. **False.** The site...is **ribosomes** (which may or may not be associated with the endoplasmic reticulum). *(pages 63-65 and Figures 4-15 and 4-16)*

P7. **False.** A ribosome is composed of ~~one molecule of~~ **two subunits, each containing** RNA and several proteins. *(page 63)*

P8. **False.** The most likely time for a genetic mutation to occur is **during DNA replication, prior to** ~~during~~ mitosis. *(pages 69-70)*

P9. **False.** A tumor is a result of ~~malignant~~ unregulated growth in a tissue **that may be either malignant or benign**. *(pages 71-72)*

Alternatively, the answer below is also correct:

False. A ~~tumor~~ **cancer** is... *(pages 71-72)*

P10. **True.** *(page 71)*

P11. **True.** *(pages 72-73 and page 32)*

P12. **Correct.** *(page 59)*

P13. **Correct.** (*page 59*)

P14. **Incorrect. They mediate phosphorylation, the attachment of phosphate groups to proteins.** (*page 59*)

P15. **Correct.** (*page 59*)

P16. **Incorrect. There is more than one code for most amino acids.** (*page 60*)

P17. **Correct.** (*page 60*)

P18. **Correct.** (*page 60*)

P19. **Correct.** (*page 60*)

P20. **Incorrect. It is called transcription.** (*page 61*)

P21. **Incorrect. It requires RNA polymerase.** (*pages 61-62*)

P22. **Correct.** (*page 67*)

P23. **Incorrect. It occurs in the nucleus.** (Ribosomal RNA synthesis occurs in the nucleolus.) (*pages 61 and 63*)

P24-25. The DNA sequence that codes for the codon AUG is **TAC**. (*page 62 and Figure 4-11*) The anticodon on tRNA that binds to codon AUG is **UAC**. (*page 64*)

P26-28. During RNA processing, nucleotide sequences corresponding to "nonsense" sequences of DNA, called **introns**, are split from the coding regions, known as **exons**. The latter are then spliced together to form **(mature or processed) mRNA**. (*pages 62-63 and Figure 4-12*)

P29-30. Mutated genes that allow cell division to escape normal regulatory mechanisms are called **oncogenes**. Invasion of normal tissues by these unregulated cells is called **metastasis**. (*page 72*)

Grading: 26-30 correct = excellent
21-25 correct = good
15-20 correct = fair
fewer than 15 correct = not passing

Chapter 5:

P1. **False.** The activation energy...is ~~often~~ **not** given... (*pages 78-79*)

Alternatively, the answer below is also correct:

False. The ~~activation~~ **energy differential between reactants and products** is ~~often~~ **sometimes** given... (*page 79*)

P2. **True.** (*page 81*)

P3. **True.** (*page 82*)

P4. **False.** During oxidative...~~hydrogen atoms~~ **pairs of electrons**... (*page 86 and Figure 5-11*)

P5. **False.** In the absence of oxygen, ...~~fatty acids~~ **glucose**. (*page 87*)

P6. **True.** (*page 93 and Figure 5-18*)

P7. **True.** (*pages 97-98 and Figure 5-19*)

P8. **False.** ~~Although~~ **Many** amino acids can be metabolized to form glucose, **and** glucose ~~cannot~~ **can also** be metabolized to form **several** amino acids. (*pages 99-100 and Figure 5-25*)

P9. **False.** An essential nutrient...~~may or may not be~~ **is not** synthesized... (*page 101*)

P10. **Correct.** (*page 82 and Figure 5-3*)

P11. **Correct.** (*page 82*)

P12. **Correct** (*page 82 and Figure 5-5*)

P13. **Correct.** (*page 84 and Figure 5-8*)

P14. **Incorrect. It occurs in mitochondria.** (*page 86*)

P15. **Correct.** (*pages 86-87*)

P16. **Incorrect. It requires inorganic phosphate.** (*pages 86-87*)

P17. **Correct.** (*pages 86-87*)

P18. Increasing the concentration of a substrate for a reaction will increase the rate of that reaction until the point of **enzyme saturation** is reached. (*page 82 and Figure 5-3*)

P19. The slowest reaction in a metabolic pathway is called the **rate-limiting reaction.** (*pages 83-84*)

P20-21. ATP is formed from the **phosphorylation** of ADP. This reaction requires **7** kcal/mol of energy. (*page 86*)

P22. The series of reactions that result in ATP formation in the absence of oxygen is called **anaerobic glycolysis**. (*page 87*)

P23-26. In the Krebs cycle, a molecule of **acetyl coenzyme A** reacts with a molecule of oxaloacetic acid to form **citric acid**. At later stages in the cycle, **(pairs of) hydrogen** atoms are transported to coenzymes to react ultimately with oxygen, while molecules of **carbon dioxide** are generated as waste products. (*page 90 and Figure 5-15*)

P27-28. Most of the body's energy is stored in the form of **triacylglycerol** in **adipose tissue** cells. (*pages 94-95*)

P29-30. The nitrogen from metabolized amino acids is excreted from the body as **urea** in **the urine**. (*pages 99-100*)

Grading: *26-30 correct = excellent*
21-25 correct = good
15-20 correct = fair
fewer than 15 correct = not passing

Chapter 6:

P1. **True.** (*page 110*)

P2. **False**. Although permeability to ~~mineral ions~~ **nonpolar molecules** does not vary much from one cell to another, different cells vary considerably in their permeability to ~~nonpolar molecules~~ **mineral ions.** (*pages 111-112*)

P3. **False**. Movement of lipid-soluble...is ~~mediated by~~ **independent of** specific proteins... (*page 111*)

Alternatively, the answer below is also correct:

False. Movement of ~~lipid-soluble~~ **polar** molecules...(*pages 112-113*)

P4. **True.** (*page 112*)

P5. **True.** (*page 116, Figure 6-11, and Table 6-2*)

P6. **True.** (*pages 118-119 and Figure 6-14*)

P7. **False**. ...electrical difference such that the inside of cells is ~~positive~~ **negative** with respect to the outside. (*page 113*)

P8. **True.** (*page 121 and Figure 6-18*)

P9. **True**. (*page 126*)

P10. **False**. Active transport~~, facilitated diffusion, and osmosis all~~ require**s** the expenditure of metabolic energy**, whereas facilitated diffusion and osmosis do not.** (*pages 117 and 122 and Table 6-2*)

P11. **False.** ...it would **neither swell nor shrink**. (*pages 123 and 126*)

P12. **False**. Pinocytosis...by which molecules can ~~leave~~ **enter** cells... (*page 128*)

Alternatively, the answer below is also correct:

False. ~~Pinocytosis~~ **Exocytosis** is a method...(*page 128*)

P13. **True**. (*page 129*)

P14. **True**. (*page 132*)

P15. **Correct** . (*page 111*)

P16. **Incorrect. It enters by active transport**. (*pages 116-117 and Figure 6-11*)

P17. **Correct**. (*implied on page 114*)

P18. **Incorrect. This is a property of mediated transport systems.** (*page 114*)

P19. **Correct**. (*page 111*)

P20. **Incorrect. It is not directly dependent upon ATP.** (*Table 6-2*)

P21. **Incorrect. With simple diffusion, C_i cannot be greater than C_o.** (*pages 109-111 and Figure 6-5*)

P22. **Correct**. (*implied, especially page 129*)

P23. **Incorrect. As with simple diffusion, C_i cannot be greater than C_o.** (*page 116, text Figure 6-11 and Table 6-2*)

P24. **Correct.** (*page 117, Figure 6-11, and Table 6-2*)

P25. **Correct.** (*page 116*)

P26. **Incorrect. The reverse is true.** (*page 122*)

P27. **Incorrect. It would take years.** (*page 110*)

P28. **Incorrect. It will shrink.** (*page 126*)

P29. **Correct.** (*page 126*)

P30. **Incorrect. It will increase.** (*implied on page 126*)

Grading: 26-30 correct = excellent
21-25 correct = good
15-20 correct = fair
fewer than 15 correct = not passing

Chapter 7:

P1. **False.** A control system...~~negative~~ **positive** feedback. (*page 143*)

P2. **True.** (*page 144*)

P3. **True.** (*page 145*)

P4. **False. Some** body changes...are ~~independent of~~ **dependent upon** changes in lifestyle... (*page 146*)

P5. **False.** When loss...~~positive~~ **negative** balance... (*page 147 and Table 7-2*)

P6. **True.** (*page 151*)

P7. **Incorrect. They help to minimize changes, but cannot prevent them.** (*pages 141-143*)

P8. **Incorrect. They operate by negative feedback.** (*page 143*)

P9. **Correct.** (*pages 148-150 and Table 7-3*)

P10. **Correct.** (*page 144*)

P11. **Correct.** (*page 144*)

P12. **Incorrect. They lead to stability.** (*page 144*)

P13. **Incorrect. They help to minimize fluctuations.** (*page 144*)

P14. **Incorrect. It carries information from the integrating center.** (*page 149*)

P15. **Correct.** (*page 149*)

P16. **Correct.** (*page 149*)

P17. **Correct.** (*page 153*)

P18. **Correct.** (*page 155*)

P19. **Correct.** (*page 155*)

P20. **Correct.** (*pages 155-156*)

P21. **c** (*page 156*)

P22. **b** (*page 156*)

P23. **c** (*page 160*)

P24. **b** (*page 158*)

P25. **a** (*page 157 and Figure 7-17*)

P26. **d** (*page 156 and Figure 7-13*)

P27. **c** (*page 151*)

P28. **b** (*page 160*)

P29. **a** (*page 151*)

P30. **e** (*page 160 and Figure 7-20*)

Grading: 26-30 correct = excellent
21-25 correct = good
15-20 correct = fair
fewer than 15 correct = not passing

Chapter 8:

P1. **True.** (*page 180, Table 8-6, and Figure 8-16*)

P2. **True.** (*page 184 and Table 8-7*)

P3. **True.** (*page 186*)

P4. **False.** ...are referred to as ~~action~~ **graded** potentials. (*pages 191-194, Table 8-9, and Figures 8-27 and 8-28*)

P5. **True.** (*page 197*)

P6. **True.** (*page 198 and Figure 8-34*)

P7. **False.** ...in ~~either~~ **only one** direction...(*page 201*)

P8. **True.** (*pages 200, 203-204*)

P9. **False.** The transmission...~~occurs at about the same rate~~ **is slower than** the conduction...(*pages 201-202*)

P10. **False.** In general, **summation of more than** one action potential arriving at an excitatory presynaptic terminal ~~will~~ **is necessary for** the generation of... (*page 203*)

P11. **False.** ...is called ~~spatial~~ **temporal** summation. (*page 203 and Figure 8-43*)

Alternatively, the answer below is also correct:

False. ...due to the successive **or simultaneous** stimulation of ~~the same~~ **different** presynaptic ~~fiber~~ **fibers**...(*page 203 and Figure 8-43*)

P12. **False.** ...~~each excitatory synapse plays an equal role with every other synapse~~... **synapses close to the initial segment of the axon are more important than synapses farther from it** in determining... (*pages 204-205*)

P13. **True.** (*page 207 and Table 8-12*)

P14. **False.** A drug...would be likely to ~~inhibit~~ **enhance** the activity... (*can be inferred from Figure 8-46 and page 209*)

P15. **False**. Alzheimer's... ~~dopamine~~ **acetylcholine**. (*page 208*)

P16. **False**. Receptors...are found in the ~~CNS~~ **peripheral nervous system**. (*page 211*)

P17. **True**. (*page 212*)

P18. **True**. (*page 212*)

P19. **Incorrect. It is gray matter**. (*page 176*)

P20. **Incorrect. It is part of the forebrain**. (*page 175*)

P21. **Correct**. (*pages 174 and 176*)

P22. **Incorrect**. (*See answer 24.*)

P23. **Incorrect**. (*See answer 24.*)

P24. **Correct**. (*page 186*)

P25. **Correct**. (*page 189*)

P26. **Correct**. (*page 189*)

P27. **Incorrect. This is true for chloride ion in most neural cells, but not other ions.** (*pages 189-191*)

P28. **Correct**. (*pages 190-191 and Figure 8-25*)

P29. **Correct**. (*pages 195 and 202*)

P30. **Incorrect. It is closer to the equilibrium of K ion**. (*page 190*)

P31. **Correct**. (*page 197*)

P32. **Correct**. (*page 195*)

P33. **Correct**. (*page 196*)

P34. **Correct**. (*page 198*)

P35. **Correct**. (*page 199*)

P36. **Incorrect**. (*can be inferred from discussion on pages 198-199*)

P37. The most abundant cells in the CNS are **glial** cells. (*page 171*)

P38-41. Groups of cell bodies in the CNS are called **nuclei**, while similar cell groups in the peripheral NS are referred to as **ganglia**. Similarly, axon fibers travelling together in the two divisions of the nervous system are called **pathways or tracts** in the CNS and **nerves** in the peripheral NS. (*page 172*)

P42. A neuron will not generate an action potential unless excited beyond its **threshold** potential. (*pages 196-197*)

P43-48. During an action potential in an axon, the membrane potential changes from **-70** mV to **+30 to 40** mV (*page 194 and Figures 8-30 and 8-32*). During this time, **Na** ions rush into the cell as a result of a change in permeability. The membrane then becomes more permeable to **K** ions, which then

leave the cell, returning the potential to its resting level (*page 195*). The absolute refractory period corresponds to the period of **Na** ion permeability changes (*page 198*).

P49. Transmission at a chemical synapse requires activation of voltage-regulated **Ca^{2+}** channels. (*page 201*)

P50. The decrease or cessation of action potentials in a sensory neuron despite continued stimulation of the receptor is called **adaptation**. (*page 213*)

Grading: 43-50 correct = excellent
34-42 correct = good
25-33 correct = fair
fewer than 25 correct = not passing

Chapter 9:

P1. **True.** (*page 222*)

P2. **False.** In general, the larger the receptive field...and the lesser degree of overlap... the ~~greater~~ **lesser** the precision... (*page 223*)

P3. **False.** The ability to discriminate... is greatest in the ~~back, thigh and forearm~~ **lips, thumbs, and fingers.** (*page 223*)

P4. **False.** A somatosensory map of the left side...is present in the ~~frontal~~ **parietal** lobe of the right side...(*page 221 and Figure 9-3*)

P5. **True.** (*Figure 9-12*)

P6. **True.** (*page 230*)

P7. **False.** The majority of the fibers...~~suprachiasmatic nucleus~~ **lateral geniculate nucleus of the thalamus**...(*page 238*)

P8. **False.** The two parts... are the lens and the ~~iris~~ **cornea.** (*page 231*)

P9. **False.** An important function of the bones of the ~~inner~~ **middle** ear... (*page 243*)

P10. **True.** (*page 246*)

P11. **True.** (*pages 243-244 and 246*)

P12. **False.** The primary sensory area...for the sense of ~~taste~~ **smell** is found...(*page 250*)

Alternatively, the answer below is also correct:

False. ..for the sense of taste is found in the ~~limbic system~~ **parietal cortex.** (*page 249*)

P13. **False.** Processing...one primary sensory area...to ~~another~~ **its association areas.** (*page 251*)

P14. **M** (*Figure 9-11*)

P15. **M** (*page 244*)

P16. **N** (*page 234 and Figure 9-26*)

P17. **C** (*page 249*)

P18. **B** (*page 229*)

P19. **M** (*page 247*)

P20. **N** (*page 229*)

P21. **C** (*page 249*)

P22. **f** (*page 233*)

P23. **b** (*page 233*)

P24. **d** (*page 233*)

P25. **e** (*page 234*)

P26. **f** (*Figure 9-23*)

P27-30. The light-sensitive chemicals in the photoreceptors are molecules called **photopigments** (*page 234*). There are **four** (*page 234*) different kinds of these molecules in the human eye. Each of these molecules is composed of an identical light-sensitive **chromophore** molecule (*pages 234-235*) and a surrounding protein called **opsin** (*page 234*), which varies from one kind of photoreceptor to another.

Grading: 26-30 correct = excellent
21-25 correct = good
15-20 correct = fair
fewer than 15 correct = not passing

Chapter 10:

P1. **True.** (*page 262*)

P2. **False.** In general, ~~steroid~~ **peptide** hormones bind to receptors on cell membranes whereas ~~peptide~~ **steroid** hormones bind to receptors inside cells. (*page 264 and Figure 10-8*)

P3. **True.** (*page 264*)

P4. **True.** (*page 264*)

P5. **False.** Most molecules of thyroxine ~~and thyrotropin~~ in blood are bound to proteins, **but molecules of thyrotropin are not**. (*page 262*)

P6. **True.** (*pages 265-266*)

P7. **False.** The name...is ~~"amines"~~ **"steroids."** (*pages 258-259*)

P8. **True.** (*page 259*)

P9. **False.** The adrenal ~~cortex~~ **medulla** is...(*page 258*)

P10. **False.** Receptors for estradiol **and** T_3 ~~and~~ **but not** vasopressin...(*page 264*)

P11. **True.** (*page 267*)

P12. **False.** The pituitary portal...anterior pituitary with the ~~target glands of anterior pituitary hormones~~ **hypothalamus.** (*page 270 and Figure 10-18*)

P13. **True.** (*pages 269-270 and Figure 10-17*)

P14. **Correct.** (*page 267*)

P15. **Incorrect. Epinephrine is secreted by the adrenal medulla.** (*page 258*)

P16. **Correct.** (*page 267*)

P17. **Incorrect. Sleeping (early stages) is a stimulus.** (*page 265 and Figure 10-9*)

P18. **Correct.** (*page 276 and Figure 10-24*)

P19. **Incorrect. This inhibits growth hormone secretion.** (*page 277 and Figure 10-24*)

P20. **Increase of TSH and prolactin secretion.** (*Figure 10-19*)

P21. **Decrease of growth hormone and TSH secretion.** (*Figure 10-19*)

P22. **Increase of prolactin secretion.** (*page 273 and Figure 10-19*)

P23. **Decrease of ACTH secretion.** (*page 274 and Figure 10-21*)

P24. **Increase of LH and FSH secretion.** (*Figure 10-19*)

P25-26. Elevated TH levels in blood **decrease** (*page 275*) the sensitivity of pituitary cells to TRH. This is an example of **long**-loop negative feedback (*Figure 10-23*).

P27-29. Elevated prolactin levels in blood will cause an **increase** (*page 276 and Figures 10-19 and 10-23*) in PIH secretion from the **hypothalamus** (*Figure 10-19*). This is an example of **short**-loop negative feedback (*page 276*).

P30. Depressed levels of TH in blood resulting from a defect in the cells that secrete TSH is an example of **secondary** (*page 278*) hyposecretion of TH.

Grading: 26-30 correct = excellent
21-25 correct = good
15-20 correct = fair
fewer than 25 correct = not passing

Chapter 11:

P1. **False.** During skeletal-muscle contraction, ...decreases because of ~~shortening~~ **sliding** of the filaments. (*pages 288-289*)

P2. **True.** (*Figure 11-8*)

P3. **False.** Curare blocks neuromuscular transmission by preventing the ~~release~~ **binding** of neurotransmitter ~~from the motor neuron~~ **to its receptor.** (*pages 297-298*)

P4. **True.** (*page 299*)

P5. **False.** The latent period of an isotonic twitch is ~~shorter~~ **longer** than... isometric twitch. (*page 300*)

P6. **False.** Muscles in the back... have a ~~higher~~ **lower** proportion of fast-glycolytic fibers... (*page 308*)

P7. **False.** Muscles in the hands have ~~larger~~ **smaller**... (*page 308*)

P8. **False.** The biceps muscle is ~~an extensor~~ **flexor**, and... triceps is an ~~flexor~~ **extensor.** (*pages 311-312*)

P9. **False.** A skeletal muscle generates...it is ~~stretched to twice~~ **at** its resting length. (*page 302*)

P10. **False.** ~~Endurance exercise such as long-distance swimming~~ **Short-duration, high-intensity exercise, such as weight lifting,** causes a preferential increase in glycolytic enzymes... *(page 310)*

P11. **True.** *(page 307)*

P12. **False.** Myesthenia gravis is caused by a ~~genetic defect in the acetylcholinesterase pathway~~ **decrease in the number of ACh receptors**. *(page 313)*

P13. **True.** *(pages 297 and 318 and Table 11-6)*

P14. **True.** *(page 317)*

P15. **False.** ~~Multiunit~~ **Single-unit**... *(page 319 and Figure 11-42)*

P16. **Correct.** *(page 309 and Figure 11-32)*

P17. **Correct.** *(pages 301-302 and Figure 11-26)*

P18. **Incorrect. The diameter of muscle fibers increases, not the number.** *(page 310)*

P19. __j__ neurotransmitter *(page 296 and Figure 11-20)*

P20. __d__ generate(s) action potential in muscle *(page 296)*

P21. __f__ store(s) neurotransmitter *(page 296 and Figure 11-20)*

P22. __h__ terminate(s) action of neurotransmitter *(page 297 and Figure 11-20)*

P23. __a__ bind(s) to troponin *(page 293 and Figure 11-17)*

P24. __b__ needed for cross-bridge movement *(pages 290-291 and Table 11-1)*

P25. __g__ binding with neurotransmitter causes end-plate depolarization *(page 296)*

P26. __k__ develop(s) muscle tension *(pages 290-291 and 301)*

P27. __i__ store(s) calcium ions *(page 293 and Figure 11-16)*

P28. __c__ inhibit(s) binding of actin to myosin *(page 292 and Figure 11-14)*

P29. __e__ conduct(s) action potentials into muscle-fiber interior *(page 293)*

P30. __b__ required for release of actin from myosin *(pages 290-291)*

Grading: 26-30 correct = excellent
21-25 correct = good
15-20 correct = fair
fewer than 25 correct = not passing

Chapter 12:

P1. **False.** ...all the motor neurons innervating **a** muscle~~s in a limb~~. *(page 325)*

P2. **True.** *(pages 332-333 and Figure 12-8)*

P3. **False.** ...left foot, flexor muscles on the ~~right~~ **left** leg and extensor muscles on the ~~left~~ **right** leg will be stimulated... *(page 333)*

P4. **True.** *(page 335 and Figure 12-12)*

P5. **False.** Electrodes placed on the scalp over the ~~primary~~ **supageslementary** motor **area** ~~cortex~~ can record... *(page 336)*

P6. **False.** The thalamus... information ~~to~~ **from** the basal ganglia ~~from~~ **to** the motor cortex. (*page 337 and Figure 12-14*)

P7. **True.** (*page 339 and Figure 12-12*)

P8. **False.** ...the ~~multineuronal~~ **corticospinal** pathways have greater influence over motor neurons controlling muscles involved in fine movements, and the ~~corticospinal~~ **multineuronal** pathways are more involved in coordination of large muscle groups. (*page 340*)

P9. **Incorrect. It is a stretch reflex.** (*page 330*)

P10. **Correct.** (*pages 341-342*)

P11. **Correct.** (*page 332*)

P12. **Correct.** (*page 339*)

P13. **Incorrect. Its influence is indirect.** (*page 338*)

P14. **Incorrect. Its input is relayed by the thalamus.** (*page 338*)

P15. **Correct.** (*pages 338-339*)

P16. **Correct.** (*page 341*)

P17. **Incorrect. The efferent pathways are the alpha motor neurons.** (*page 341*)

P18. **Incorrect. The integrating centers are in the brainstem and spinal cord.** (*page 341*)

P19. **Incorrect. They are present in blind people.** (*page 342*)

P20. **P** (*page 337*)

P21. **C** (*page 338*)

P22. **P** (*page 338*)

P23. **P** (*page 338*)

P24. **C** (*pages 338-339*)

P25. **N** (*implied from discussion on pages 337-339*)

Grading: 21-25 correct = excellent
17-20 correct = good
13-16 correct = fair
fewer than 12 correct = not passing

Chapter 13:

P1. **False.** A person...is probably suffering from ~~iron~~ **folic acid or vitamin B_{12}** deficiency. (*page 355*)

P2. **True.** (*page 356*)

P3. **True.** (*pages 365-366*)

P4. **False.** The QRS wave...depolarization of the ~~atria~~ **ventricles**. (*page 369*)

P5. **False.** An electrocardiogram would be useful...~~valves~~ **conducting system** of the hear.t (*pages 369 and 375*)

P6. **True.** (*page 370*)

P7. **False.** The left ventricle ~~has to pump more~~ **pumps the same amount of** blood ~~than~~ **as** the right ventricle ~~because~~ **even though** the left has to... (*page 375*)

P8. **True.** (*page 375*)

P9. **True.** (*page 376*)

P10. **False.** Starling's law...higher ~~heart rate~~ **stroke volume**. (*page 377*)

P11. **False.** The stroke volume...increased by ~~recruiting more cardiac muscle fibers into activity~~ **increasing end-diastolic volume and/or contractility.** (*pages 377 -378*)

P12. **False.** Mean arterial pressure...~~determining the average of the systolic and diastolic pressures~~ **adding 1/3 the pulse pressure to the diastolic pressure**. (*page 383*)

P13. **True.** (*pages 362-363 and page 397*)

P14. **False.** Total peripheral resistance...degree of vasoconstriction in the ~~veins~~ **arterioles.** (*page 401*)

P15. **False.** An athlete...~~both~~ an increased mean arterial blood pressure and ~~increased~~ **decreased** total peripheral resistance. (*Figure 13-68*)

P16. **False.** Hypertension...drugs that ~~increase~~ **decrease** total peripheral resistance. (*page 417*)

P17. **True.** (*page 418*)

P18. **False.** ~~Increased~~ **Decreased** concentrations of plasma proteins... (*Table 13-9*)

P19. **Incorrect.** **Cardiac output is greater in B.** (*page 375*)

P20. **Correct.** (*page 401*)

P21. **Correct.** (*Table 13-6*)

P22. **Incorrect.** ***Decreased* frequency** (*page 403 and Figure 13-62*)

P23. **Incorrect.** ***Increased* frequency** (*page 403 and Figure 13-62*)

P24. **Correct.** (*pages 403 and 406 and Figure 13-62*)

P25. **Incorrect.** **Increased contractility due to increased stimulation of sympathetic nerves** (*page 406 and Figure 13-62*)

P26. **Correct.** (*page 406 and Figure 13-62*)

P27-29. In tissue capillaries, most net movement of nutrients from the blood occurs by the process of **diffusion** (*pages 392-393*). Fluid movement occurs in response to differences in **hydrostatic** pressure and **osmotic** pressure between the capillaries and interstitial fluid. (*pages 394-396*)

P30. **NC** brain (*Figure 13-67*)

P31. **I** heart (*Figure 13-67*)

P32. **I** skeletal muscles (*Figure 13-67*)

P33. **D** kidneys (*Figure 13-67*)

P34. **D** stomach (*Figure 13-67*)

P35. **I** skin (*Figure 13-67*)

Grading: 30-35 correct = excellent
24-29 correct = good
18-23 correct = fair
fewer than 17 correct = not passing

Chapter 14:

P1. **False.** ...ciliated cells...are ~~stimulated to have increased activity~~ **inhibited.** (*page 429*)

P2. **True.** (*pages 435-436 and Figures 14-11 and 14-12*)

P3. **False.** ...the part of the respiratory cycle...at the end of ~~inspiration~~ **expiration**, prior to ~~expiration~~ **inspiration.** (*pages 435-436*)

P4. **True.** (*Figure 14-11*)

P5. **False.** Emphysema is ... characterized by ~~low~~ **high** lung compliance.. .(*page 439 and Table 14-10*)

P6. **False.** Doubling the rate of breathing will cause a ~~greater~~ **lesser** increase in alveolar ventilation... (*page 442 and Table 14-5*)

P7. **True.** (*pages 444 and 457*)

P8. **True.** (*pages 456-457 and Figure 14-29*)

P9. **True.** (*page 464*)

P10. **False.** Acclimatization to high altitudes includes **increasing ventilation**. (*page 465*)

P11. **Correct.** (*page 438 and Table 14-4*)

P12. **Correct.** (*page 438 and Table 14-4*)

P13. **Correct.** (*page 441*)

P14. **Incorrect. It falls.** (*page 459 and Figure 14-31*)

P15. **Correct.** (*page 459 and Figure 14-31*)

P16. **Incorrect. It shifts to the right.** (*page 451 and Figure 14-24*)

P17. **Incorrect. They are increased by hypoventilation.** (*page 463*)

P18. **Correct.** (*page 454*)

P19. **Incorrect. It can still stimulate the central chemoreceptors.** (*page 456*)

P20-22. Increasing concentrations of **oxygen** in the alveoli cause pulmonary arterioles to dilate, and increasing concentrations of **carbon dioxide** in the alveoli cause bronchioles to dilate. These changes help ensure efficient **blood-gas matching** (ventilation-perfusion equality). (*page 448 and Figure 14-20*)

P23-26. During exercise, the O_2-hemoglobin curve shifts **down** and to the **right**. Two causes for this shift are **increased temperature** and **increased hydrogen ion concentration.** (*page 451 and Figure 14-24*)

P27-30. The efficient transport of CO_2 by the blood is possible because of the enzyme **carbonic anhydrase**, which is found in the **erythrocytes** and catalyzes the reaction $CO_2 + H_2O \leftrightarrow H_2CO_3$. This reaction is responsible for the transport of **60** percent of the CO_2 in venous blood. (*pages 452 and 454*)

Grading: 26-30 correct = excellent
21-25 correct = good
15-20 correct = fair
fewer than 15 correct = not passing

Chapter 15:

P1. **True.** (*page 473*)

P2. **False.** The glomerulus is involved in filtration~~, secretion and reabsorption~~; the peritubular capillaries ~~serve only to exchange O_2 and CO_2 with the cells~~ **are important for secretion and reabsorption.** (*pages 475-476 and Figure 15-5*)

P3. **True.** (*page 479*)

P4. **False.** Only substances that are filtered **and/or secreted** by the kidneys... (*page 480*)

P5. **True.** (*page 480*)

P6. **False.** ~~Unl~~**Like** water, sodium is filtered and reabsorbed **but not** secreted. (*page 482*)

P7. **False.** ...the concentration of glucose in the distal tubular fluid is ~~about one-third that in Bowman's capsule~~ **zero.** (*page 484*)

P8. **True.** (*pages 486-488*)

P9. **False.** The walls of the ~~ascending~~ **descending** limb... (*page 487*)

P10. **True.** (*page 487*)

P11. **True.** (*page 487 and Figure 15-15*)

P12. **False.** The main force... is the ~~low hydrostatic pressure in the surrounding interstitial space~~ **osmotic gradient between collecting ducts and interstitium**. (*page 487*)

P13. **True.** (*page 491*)

P14. **False.** ~~High~~ **Low** levels of sodium chloride... (*page 491*)

P15. **False.** A ~~fall~~ **rise** in the osmolarity... (*page 494*)

P16. **False.** A substance that interferes...potassium ~~reabsorption~~ **secretion**. (*page 497*)

P17. **False.** The primary stimulus...is ~~increased plasma potassium~~ **decreased plasma calcium** levels. (*page 499*)

P18. **False.** One response to increased hydrogen ion production...is ~~decreased~~ **increased** reabsorption of bicarbonate ion... (*page 502*)

P19. **False.** ~~Bicarbonate~~ **hydrogen** ions... (*page 502 and Figures 15-29 and 15-30*)

P20. **True.** (*page 505*)

P21. **Incorrect. It is passive but depends on both solute** (not just sodium) **reabsorption and the presence of ADH.** (*pages 483-485*)

P22. **Correct.** (*pages 484 and 505*)

P23. **Incorrect. It is equal to the amount filtered minus the amount reabsorbed.** (*page 482*)

P24. **Incorrect. Decreased sympathetic activity.** (*page 489 and Figure 15-16*)

P25. **Correct.** (*page 489*)

P26. **Correct.** (*pages 491-492*)

P27. **Incorrect. It would decrease.** (*Figure 15-22*)

P28. **Correct.** (*page 505*)

P29. **Correct.** (*page 494*)

P30. **Correct.** (*page 494 and Figure 15-22*)

P31. **Incorrect. It would be increased.** (*page 494 and Figure 15-22*)

P32. **Correct.** (*page 494 and Figure 15-22*)

P33. **Incorrect. It is independent of the renin-angiotensin system.** (*pages 498-501*)

P34. **Incorrect. It is regulated by the actions of hormones on the gastrointestinal tract, bone, and kidney.** (*page 498*)

P35. **Incorrect. It requires parathyroid hormone and vitamin D.** (*pages 498-501*)

Grading: 30-35 correct = excellent
24-29 correct = good
18-23 correct = fair
fewer than 17 correct = not passing

Chapter 16:

P1. **True.** (*page 517*)

P2. **True.** (*page 518*)

P3. **True.** (*page 522*)

P4. **False.** Polysaccharides must be broken down to ~~disaccharides~~ **monosaccharides**... (*page 522*)

P5. **False.** ...fat digestion is ~~normal provided~~ **impaired even though** bile is still produced. (*page 522*)

P6. **True.** (*pages 523-524 and Figure 16-11*)

P7. **True.** (*page 524*)

P8. **False.** "Heartburn"...due to ~~pressure of the stomach against the heart~~ **regurgitation of gastric juice into the esophagus**. (*page 530*)

P9. **False.** Gastric chief cells secrete ~~pepsin~~ **pepsinogen**. (*page 531*)

P10. **False.** The secretion of gastrin... ~~stimulated~~ **inhibited**... (*pages 531-532 and Table 16-2*)

P11. **True.** (*page 537*)

P12. **False.** Secretion of ~~secretin~~ **CCK** is stimulated by the presence of fat in the duodenum, whereas ~~CCK~~ **secretin** release... (*page 538 and Table 16-2*)

P13. **False.** Bile...contains ~~the major enzymes for digesting~~ **bile salts for emulsifying** fats.

Alternatively, the answer below is also correct:

False. ~~Bile secreted by the liver~~ **Secretions of the pancreas** contains ... (*pages 538-540*)

P14. **False.** Bile ~~pigments~~ **salts** are essential... (*page 540*)

P15. **False.** The breakdown products...bile ~~salts~~ **pigments**...(*page 540*)

P16. **True.** (*pages 541-542*)

P17. **False.** During a meal, ~~peristalsis~~ **segmentation** is ... (*page 543*)

P18. **False.** Lactose intolerance... ~~amylase~~ **lactase** deficiency. (*page 546*)

P19. **Correct.** (*page 517*)

P20. **Incorrect. It dissolves proteins but does not break them down.** (*page 517*)

P21. **Correct.** (*pages 532-533*)

P22. **Correct.** (*page 517*)

P23. **Incorrect. It secretes enzymes in response to CCK.** (*page 538*)

P24. **Correct.** (*page 517*)

P25. **Correct.** (*Table 16-3*)

P26. **Incorrect. It is decreased by increased osmolarity of the intestinal contents.** (*Table 16-13*)

P27. **Correct.** (*page 531 and Figure 16-17*)

P28. **Incorrect. It helps inhibit HCl secretion.** (*Table 16-3*)

P29. **Correct.** (*page 538*)

P30. **Correct.** (*page 542*)

Grading: 26-30 correct = excellent
21-25 correct = good
15-20 correct = fair
fewer than 15 correct = not passing

Chapter 17:

P1. **False.** Plasma glucose... ~~higher~~ **lower** in the postabsorptive period...~~because of~~ **even though** gluconeogenesis **occurs**... (*page 562 and Figure 17-5*)

P2. **False.** Insulin secretion...whereas glucagon secretion is inhibited by ~~these stimuli~~ **elevated glucose levels and stimulated by elevated amino acid levels.** (*pages 562-564*)

P3. **False.** ~~Epinephrine~~ **Glucagon is...** (*page 563*)

P4. **True.** (*page 563*)

P5. **False.** Cortisol is permissive for...~~insulin~~ **glucagon and epinephrine.** (*page 567*)

P6. **False.** During exercise, glucose uptake...is increased because of ~~increased insulin secretion~~ **local chemical changes in muscle.** (*page 569*)

P7. **False.** Type ~~1~~ **2** diabetes... (*page 570*)

P8. **False.** A baby with untreated ~~growth hormone~~ **thyroid hormone** deficiency... (*page 578*)

Alternatively, the answer below is also correct:

False. A baby with untreated growth hormone...impaired growth (seen after the first few months of life). ~~and mental development.~~ (*page 577*)

P9. **False.** The resting energy requirement...is ~~independent of~~ **dependent upon** body size. (*page 582*)

P10. **False.** Basal metabolic rate is increased by **both** epinephrine and ~~decreased by~~ thyroid hormones. (*page 583*)

P11. **True.** (*page 584 and Table 17-10*)

P12. **False.** Anorexia nervosa is excessive thinness ~~usually caused by hyperthyroidism~~ **associated with pathological fear of gaining weight**. (*page 588*)

P13. **False.** Heat loss from evaporation occurs ~~only when one is sweating~~ **constantly**. (*page 591*)

P14. **False.** The integrating centers...located in the ~~brainstem~~ **hypothalamus**. (*page 592*)

P15. **False.** The cause of ~~hyperthermia~~ **fever** ... (*page 593*)

P16. **Incorrect. Insulin facilitates the diffusion of glucose into all cells except those of the brain.** (*page 560*)

P17. **Correct.** (*page 556*)

P18. **Correct.** (*page 556 and Figure 17-1*)

P19. **Incorrect. Glucose is stored as triacylglycerol in adipose tissue.** (*page 556*)

P20. **Incorrect. Glucose is stored as glycogen in skeletal muscle.** (*page 555*)

P21. **Correct.** (*page 557*)

P22. **Correct.** (*page 557*)

P23. **Correct.** (*page 565*)

P24. **Incorrect. Glucagon stimulates gluconeogenesis in liver cells.** (*page 564*)

P25. **Correct.** (*page 562*)

P26. **Correct.** (*Figure 17-11*)

P27. **Correct.** (*page 570*)

P28. **Incorrect. Blood pH would be decreased.** (*page 570*)

P29. **Correct.** (*page 569*)

P30. **Incorrect. Triacylglycerol stores would be decreased.** (*Figure 17-11*)

P31. **Correct.** (*page 577*)

P32. **Correct.** (*page 577*)

P33. **Correct.** (*pages 567-568*)

P34. **Incorrect. Growth hormone decreases glucose utilization.** (*page 568*)

P35. **Correct.** (*page 577*)

P36. **Incorrect. Growth hormone is the most important hormone for postnatal growth but is not important for fetal growth.** (Insulin is important for fetal growth.) (*pages 577-578*)

P37. **Correct.** (*page 573*)

P38. **Incorrect. Cholesterol is carried to cells by LDL.** (*page 573*)

P39. **Incorrect. Its synthesis is inversely related to the level of dietary cholesterol.** (*page 574*)

P40. **Correct.** (*pages 573-574 and Figure 17-14*)

Grading: 34-40 correct = excellent
27-33 correct = good
20-26 correct = fair
fewer than 20 correct = not passing

Chapter 18:

P1. **False.** One function... buffer the ~~alkaline~~ **acidic** secretions... (*pages 605-606*)

P2. **False.** The primary storage site... is the ~~seminal vesicles~~ **vas deferens and epididymis.** (*page 608*)

P3. **False.** Vasectomy results...inhibits sperm ~~production~~ **transport out of the penis.** (*page 610*)

P4. **False.** Erection...~~sympathetic~~ **parasympathetic** reflex. (*page 611 and Figure 18-9*)

P5. **False.** Luteinizing hormone... **Leydig** cells ~~of the seminiferous tubules~~. (*page 612 and Figure 18-10*)

P6. **False.** Spermatogenesis ~~and~~ **depends upon** testosterone production ~~are interdependent events, meaning that neither process can occur without the other~~ **but testosterone production is independent of spermatogenesis**. (*page 613*)

P7. **False.** Sertoli cells... analogous to ~~theca~~ **granulosa** cells... (*page 621*)

P8. **True.** (*page 622*)

P9. **False.** The secretory phase... the ~~follicular~~ **luteal** phase... (*page 624*)

P10. **False.** Progesterone ~~increases the thickness of the myometrium~~ **stimulates the secretion of the endometrium**... (*page 625*)

P11. **False.** Implantation of the ~~trophoblast~~ **blastocyst**... (*pages 630-631*)

P12. **False.** Detection of ~~LH~~ **CG**...(*page 634*)

P13. **False.** Oxytocin...plasma progesterone levels are ~~high~~ **low**...(*page 638*)

P14. **False.** Milk ejection is a ~~neural~~ **neurohormonal** reflex arc... and a ~~neural~~ **hormonal** efferent output... (*page 640 and Figure 18-31*)

P15. **True.** (*page 641*)

P16. **False.** The presence of sex chromatin... genetic ~~male~~ **female**. (*page 643*)

P17. **False.** An individual with androgen insensitivity syndrome...with ~~female~~ **no** internal reproductive organs **except testes**. (*page 644*)

P18. **False.** One consequence of menopause...decreased ~~androgen~~ **estrogen** secretion... (*page 647*)

P19. **Incorrect. It occurs shortly after the first meiotic division.** (*page 616*)

P20. **Incorrect. It requires a surge of LH.** (*page 622*)

P21. **Correct.** (*Table 18-5*)

P22. **Incorrect. It normally occurs in the uterine tube**. (*page 628*)

P23. **Incorrect. It follows proteolysis of the zona pellucida.** (*page 629*)

P24. **Incorrect. It results in dizygotic twins.** (*pages 629-630*)

P25. **Incorrect. The corpus luteum is necessary only for the first two to three months.** (*page 634*)

P26. **Correct.** (*page 634*)

P27. **Incorrect. There is no direct transfer of blood.** (*page 633*)

P28. **Correct.** (*Table 18-9*)

P29. **Correct.** (*Table 18-9*)

P30. **Incorrect. Blood pressure should remain normal.** (*Table 18-9*)

P31. **Correct.** (*pages 638-640*)

P32. **Incorrect. It does not take place during pregnancy because estrogen and progester-one inhibit the milk-synthetic actions of prolactin.** (*page 640*)

P33. **Correct.** (*page 640*)

P34. **A** (*Table 18-3*)

P35. **B** (*Tables 18-3 and 18-8*)

P36. **B** (*Tables 18-3 and 18-8*)

P37. **A** (*Tables 18-3 and 18-8*)

P38. **A** (*page 613*)

P39. **E** (*Table 18-4*)

P40. **B** (*Tables 18-3 and 18-4*)

Grading: 34-40 correct = excellent
27-33 correct = good
20-26 correct = fair
fewer than 20 correct = not passing

Chapter 19:

P1. **False.** Cytotoxic T cells kill body cells by ~~phagocytosis~~ **secreting pore-forming protein, which disrupts their membranes.** (*Figure 19-15*)

P2. **True.** (*Figure 19-15*)

P3. **False.** HIV...~~cytotoxic~~ **helper** T cells. (*page 679*)

P4. **False.** A person with type B blood ~~can~~ **cannot**...(but the reverse is true.) (*Table 19-12*)

P5. **False.** Biotransformation...more lipid-**in**soluble. (*page 688*)

P6. **True.** (*pages 659 and 692*)

P7. **False.** The value...is that ~~a defect of one participant in the cascade can be corrected by another down the chain~~ **amplification is gained at many steps.** (*page 692*)

Alternatively, the answer below is also correct:

False. ~~The value~~ **A disadvantage** of having...a defect of one participant in the cascade can ~~be corrected by another down the chain~~ **block the entire cascade.** (*page 693*)

P8. **False.** ... cortisol stimulates ~~glycolysis~~ **gluconeogenesis.** (*Table 19-16*)

P9. **Incorrect. They are secreted by plasma cells.** (*Figure 19-9*)

P10. **Correct.** (*page 670*)

P11. **Correct.** (*page 670*)

P12-16. The three stages of a typical specific immune response to a microbial invasion are **antigen encounter and recognition** (*page 663*), **lymphocyte activation** (*page 663*), and **attack** (*page 663*). If the body has been previously exposed to the specific microbe, the second response is **faster** (*page 671*) than the first. This change in response time is dependent upon the presence of **memory** cells (*page 671*) cells.

P17. Multiple sclerosis, myesthenia gravis and type I diabetes have very different effects on the body, but they share a common cause: they are all **autoimmune** (*page 682*) diseases.

P18. **d** (*page 657*)

P19. **f** (*Table 19-3*)

P20. **b** (*Table 19-3*)

P21. **a** (*page 658*)

P22. **c** (*Table 19-3*)

P23. **e** (*Table 19-3*)

P24. **d** (*Table 19-13*); **h** (*Table 19-13*)

P25. **b** (*page 674*); **e** (*page 674*)

P26. **d** (*Table 19-13*); **e** (*page 668*)

P27. **b** (*page 674*); **e** (*page 674*)

P28. **d** (*Table 19-13*); **h** (*Table 19-4*)

P29. **c** (*Table 19-13*); **g** (*page 681*); **i** (*page 681*)

P30. **b** (*page 675*)

P31. **a** (*page 660*); **f** (*Figure 19-3*)

P32. **c** (*page 681*); **g** (*page 681*); **i** (*page 681*)

P33. **d** (*page 677*); **h** (*page 677*)

P34. **j** (*Figure 19-24*)

P35. **a** (*page 690*); **j** (*page 692*); **h** (*page 689*)

P36. **g** (*page 690*)

P37. **j** (*Figure 19-29*)

P38. **e** (*Figure 19-29*); **j** (*Figure 19-29*)

P39. **c** (*page 695*)

P40. **i** (*Figure 19-26*)

Grading: 34-40 correct = excellent
27-33 correct = good
20-26 correct = fair
fewer than 20 correct = not passing

Chapter 20:

P1. **False.** A high-amplitude, spike-wave EEG pattern ... someone ~~in a coma~~ **undergoing an epileptic seizure**. (*page 706*)

P2. **False.** The EEG tracing...is mainly a**n** ~~beta~~ **alpha** rhythm. (*page 707*)

P3. **False.** Dreaming...~~slow-wave~~ **paradoxical (REM)** sleep. (*Table 20-2*)

P4. **False.** Sleep-wake cycles ... ~~thalamus~~ **brainstem**. (*page 709*)

P5. **False.** Behaviors ... ~~secondary~~ **primary**...(*page 713*)

P6. **False.** ~~Appetitive~~ **Aversive** motivation...(*page 713*)

Alternatively, the answer below is also correct:

False. Appetitive motivation ... ~~avoid~~ **repeat**~~ing~~...(*page 713*)

P7. **False.** Neurotransmitters...and ~~epinephrine~~ **norepinephrine**. (*page 714*)

P8. **False.** The integrator ... the ~~thalamus~~ **limbic system**. (*page 714*)

P9. **False.** One cause of schizophrenia...too ~~little~~ **much** dopamine activity...(*page 715*)

P10. **False.** ~~The tricyclic antidepressants are~~ **Lithium carbonate is**... (*page 716*)

Alternatively, the answer below is also correct:

False. The tricyclic antidepressants ...~~bipolar affective disorder~~ **depressions** . (*page 716*)

P11. **False.** Tolerance to drugs...~~increased~~ **decreased** release...(*page 717*)

P12. **True.** (*page 719*)

P13. **True.** (*page 719*)

P14. **False.** Brain size ... are **not** fixed and ~~independent of~~ **depend on**...(*page 720*)

P15. **True.** (*page 722*)

P16. **Correct.** (*page 707*)

P17. **Incorrect. Postural muscle tone is strongly inhibited.** (*page 707*)

P18. **Incorrect. GH secretion occurs during slow-wave sleep.** (*page 708*)

P19. **c** (*page 711*)

P20. **g** (*page 720*)

P21. **b** (*page 714*)

P22. **d** (*page 721*)

P23. **f** (*pages 721-722*)

P24. **a** (*page 720*)

P25. **i** (*page 718*)

Grading: 22-25 correct = excellent
18-21 correct = good
14-17 correct = fair
fewer than 13 correct = not passing

CREDITS